Critical Issues in Mathematics Education

Critical Issues in Mathematics Education

edited by

Paul Ernest
University of Exeter

Brian Greer
Portland State University

Bharath Sriraman
The University of Montana

INFORMATION AGE PUBLISHING, INC.
Charlotte, NC • www.infoagepub.com

Library of Congress Cataloging-in-Publication Data

Critical issues in mathematics education / edited by Paul Ernest, Brian
Greer, Bharath Sriraman.
   p. cm. – (The Montana mathematics enthusiast: monograph series in
mathematics education)
 Includes bibliographical references.
 ISBN 978-1-60752-039-9 (pbk.) – ISBN 978-1-60752-040-5 (hardcover)
1. Mathematics–Study and teaching. 2. Mathematics–Philosophy. 3.
Mathematics teachers–Attitudes. I. Ernest, Paul. II. Greer, Brian. III.
Sriraman, Bharath.
 QA11.2.C75 2009
 510.71–dc22

                                                        2009014763

Permission to photocopy, microform, and distribute print or electronic copies may be
obtained from:
  Bharath Sriraman, Ph.D.
  Editor, *The Montana Mathematics Enthusiast*
  The University of Montana
  Missoula, MT 59812
  Email: sriramanb@mso.umt.edu
  (406) 243-6714

Printed in the United States of America

CONTENTS

SECTION 1

MATHEMATICS EDUCATION: FOR WHAT AND WHY?

v

SECTION 2

GLOBALIZATION AND CULTURAL DIVERSITY

SECTION 3

MATHEMATICS, EDUCATION, AND SOCIETY

SECTION 4

SOCIAL JUSTICE IN, AND THROUGH, MATHEMATICS EDUCATION

INTRODUCTION

AGENCY IN MATHEMATICS EDUCATION

Paul Ernest, Brian Greer, and Bharath Sriraman

The word "critical" in the title of this collection has several meanings. One meaning, as applied to a situation or problem, is "at a point of crisis." A second meaning is "expressing adverse or disapproving comments or judgments." A third is related to the verb "to critique," meaning "to analyse the merits and faults of." All are contemporarily relevant to mathematics education.

Few would question that the world is in a critical state; the connection with mathematics and mathematics education has been most forcefully delineated by Ubiratan D'Ambrosio (2007):

> It is widely recognized that all the issues affecting society nowadays are universal, and it is common to blame, not without cause, the technological, industrial, military, economic and political complexes as responsible for the growing crises threatening humanity. Survival with dignity is the most universal problem facing mankind.

> Mathematics, mathematicians and mathematics educators are deeply involved with all the issues affecting society nowadays. But we learn, through History, that the technological, industrial, military, economic and political

complexes have developed thanks to mathematical instruments. And also that mathematics has been relying on these complexes for the material bases for its continuing progress. It is also widely recognized that mathematics is the most universal mode of thought.

Are these two universals conflicting or are they complementary? It is sure that mathematicians and math educators are concerned with the advancement of the most universal mode of thought, that is, mathematics. But it is also sure that, as human beings, they are equally concerned with the most universal problem facing mankind, that is, survival with dignity.

At the time of writing, the United States is a few weeks into a financial crisis with world-wide ramifications. It is a reasonable question to ask whether this crisis is not only due, mainly, to a failure to apply elementary mathematical principles to financial government but also, partly, to a lack of fiscal sense among many citizens that can be seen as an indictment of their education, including their mathematical education.

Critical judgments on mathematics education are nothing new, but, alongside the maturation of the field of mathematics education itself, they are now concerned with more than internal issues about the nature of teaching and what it is that students take from their mathematics education in terms of technical competence and higher-order thinking. As interest in, and attention to, the historical, cultural, social, and political contexts in which mathematics education is situated have developed, external issues about the circumstances in which children are living, and about the relationships between mathematics education and society have become prominent. The distinction between internal and external is neatly made by Skovsmose's (2006) question as to whether the leak in the ceiling of a classroom does not constitute a learning obstacle, just as much as some point about, say, the epistemology of rational numbers.

The development of mathematics education as a field has been marked by a broadening from an initial concentration in mathematics itself and psychology to a much broader interaction with human sciences, such as sociology, anthropology, ethnomathematics, gender issues, critical race theory, social history of mathematics, philosophy of mathematics, sociolinguistics, semiotics, and so on. Together with this enrichment of theoretical frameworks has come methodological diversity and liberation.

Critical mathematics educators across the world voice concern about the perceived lack of connection between school mathematics and students' lives, leading to lack of interest and alienation. They challenge the notion of mathematics and mathematics education as being morally and politically neutral. They question the role of mathematics as a gatekeeper, limiting economic and social advancement. They point to the increasingly nationalistic, and indeed militaristic, tone of much governmental rhetoric (Gutstein, 2009). All of these major concerns, and many more, are discussed by the contributors to this collection.

The emerging group of practitioners, researchers, and scholars that is evolving critical mathematics education is, above all, characterized by its recognition of the necessity of critiquing the institutions, pedagogical and research practices, and political embeddedness of mathematics education. Such a position is aligned with many other developments, including a new philosophy of mathematics (Ernest, 1991; Tymoczko, 1986) that takes as central the notion of fallibility.

In this brief introduction, we do not attempt to summarize the diversity, depth, and richness of the chapters that follow, but rather, with illustrative examples, we try to convey a sense of the common threads of criticality, in all of the above senses, that they share.

SOURCES AND ORGANIZATION OF THIS BOOK

The book is organized in four sections, namely:

1. Mathematics education: For what and why?
2. Globalization and diversity
3. Mathematics, education, and society
4. Social justice in, and through, mathematics education

The papers in the first section were prepared for Discussion Group 3 at the 11th ICME, organized by Brian Greer and Claude Gaulin. These papers were deliberately kept short, in order to maximize the chances of participants reading them in advance, and related to the specific questions for discussion, namely:

- What are the most productive ways of characterizing "mathematical literacy"?
- Should school mathematics education be dominated by the discipline of mathematics, rather than reflecting the diversity of mathematical practices?
- Can a balance be achieved between a homogeneous, monolithic, globalized curriculum and the diversity of people and forms of knowledge construction and use?
- How should mathematics education prepare people for technology?

The papers in the remaining sections are republished from the electronic *Philosophy of Mathematics Education Journal* created and maintained by Paul Ernest (http://people.exeter.ac.uk/PErnest/).

Like the papers in the Philosophy of Mathematics Education Journal, the chapters in this book are freely available on the Internet, which is the policy adopted by Bharath Sriraman for the journal The Montana Mathematics Enthusiast (www.math.umt.edu/TMME). Entirely voluntary work of this nature, making material available in this way instead of published in overly expensive books, may be considered a form of academic activism. (And see the first monograph in the series to which this book belongs (Sriraman, 2007) on social justice in mathematics education).

MATHEMATICS EDUCATION: FOR WHAT AND WHY?

The International Congress on Mathematical Education is a major quadrennial conference that brings together mathematicians and mathematics educators, between which groups exist considerable tensions. As a generalization, it is probably not unfair to say that mathematicians would tend to answer the question "What is mathematics education for?" by reference to traditional motivations such as the need to produce another generation of scholars to continue developing the discipline of mathematics; the supply of a cadre of scientists and others such as engineers who need strong mathematical competence; as a training in logical thinking and problem solving; as exposure to what is as much a part of cultural heritage as literature or music.

Critical mathematics educators, on the other hand, would go beyond these reasons and argue that mathematics education should provide people with tools to analyse, and act upon, issues important in their lives, in their communities, and to society in general. Further, it is desirable that mathematics education should give people some understanding of the complex, often hidden, roles that mathematics plays in society, "formatting" many aspects of our lives.

The discussion group was planned to confront these different perspectives and open up the issues, and the collection of papers that resulted, and is reproduced here, was assembled accordingly. In addressing the themes of the remaining three sections, they also serve as an introduction to those sections.

GLOBALIZATION AND DIVERSITY

Mathematics and mathematics education are carried on against the background of geopolitical events—globalization, growth of the knowledge economy and the attendant commodification of knowledge and corporification of schools, post-colonialism, struggles over finite resources...not to mention wars.

Academic mathematics as a discipline is an international endeavor, with well-developed institutions, publications, and networks of communication and collaboration. To a degree, the same can be said of mathematics education (Bishop, 1992), increasingly so. Atweh (Section 1, this volume) suggests that globalization may not be intrinsically good or evil, but is certainly not value-free or beyond ethical considerations, and presents both challenges and dangers. An obvious danger is that globalization, in effect, becomes Westernization (Sen, 2004, cited by Ernest, Section 2, this volume). Thus, some "Western" scholars share a concern voiced by Clements (1995, p. 3) that: "Over the past 20 years I have often had cause to reflect that it is Western educators who were responsible not only for getting their own mathematics teacher education equation wrong, but also for passing on their errors to education systems around the world." Bishop (1990) referred to "Western" mathematics as "the secret weapon of cultural imperialism."

Counter-narratives are being developed. For example, the history of the development of mathematics as written by "the winners" has been challenged by documenting, often uncovering, the contributions of non-European cultures (Joseph, 1992; Powell & Frankenstein, 1997). If mathematics education were just the transmission of the academic canon to a new generation, then world-wide homogeneity of mathematics education would be even greater than it is (Usiskin, 1999). Examples of resistance to homogeneity include post-Apartheid South Africa, where serious efforts have been made to devise mathematics education aims to contribute to improvement of the social and political conditions of the people. Brazil is another country in which mathematics educators have been heavily involved in ethnomathematics (D'Ambrosio, 2006) and in applications of mathematics as activists for social justice (Knijnik, Section 2, this volume).

MATHEMATICS, EDUCATION, AND SOCIETY

A major theme of the turn from the internal problems of learning and teaching mathematics to consideration of the external historical, cultural, social, and political contexts in which mathematics education takes place is consideration of how analysis of the roles of mathematics in society should guide the education of the children who will become citizens of that society. Thus, in Discussion Group 3 at ICME11, the final question "How should mathematics education prepare people for technology?" was not aimed at standard discussions about how the use of computers can help mathematics teaching/learning but rather to raise issues about the use of technology in contemporary society through the complementary processes of mathematization (application of mathematical models in the control of situations) and demathematization (embedding of mathematics within physical and symbolic artifacts such that it is hidden in the use of those artifacts) (Gellert and Jablonka, 2007, this volume).

SOCIAL JUSTICE IN, AND THROUGH, MATHEMATICS EDUCATION

A common theme of the quest for social justice within mathematics education has taken the form of analyzing inequities based on gender, class, and ethnic differences (though rarely their interactions). Thus mathematics often acts as a gatekeeper to educational and economic advancement, as the civil rights activist, Robert Moses, has emphasized in characterizing quality mathematics education as a civil right (Moses & Cobb, 2001) (and see Appelbaum & Davila, Section 4, this volume).

Such questions could be considered as internal. A more recent development has been concentration of external issues—in other words, social justice *through* mathematics education, providing people with tools for critiquing their own situations and acting on them. In Freire's terms (adapted for the title of his book by Gutstein (2006)) that means not just "reading" the world but also "writing" it.

Numbers can tell a story very starkly and simply. A fine book for children (and adults) called "If the world were a village" (Smith and Armstrong, 2002) provides proportional statistics that would apply if the world were a village of 100 people. Only 24 always have enough to eat, while 60 are always hungry (26 being severely undernourished), and 16 go to bed hungry at least some of the time.

FINAL COMMENT: ON BEING SELF-CRITICAL

Teachers, scholars, and researchers who consider themselves critical do not forget to turn their critical gaze on their individual and collective beliefs, attitudes, and actions and to remain open to critique by others. We are aware that as academics located in wealthy Western countries we are privileged both materially and in the freedom and autonomy that academic work still grants us (if only just!). Nevertheless, we hope that by addressing these issues we are helping to raise awareness among mathematics educators and teachers in a way that contributes a little bit to improvements in the teaching and learning of mathematics. We passionately believe in the importance of the enterprise of mathematics education worldwide, without overestimating the impact that we three editors and the score or so contributors can make.

REFERENCES

Bishop, A. J. (1990). Western mathematics: the secret weapon of cultural imperialism. *Race and Class, 32*(2), 51–65.

Bishop, A. J. (1992). International perspectives on research in mathematics education. In D. Grouws (Ed.), *Handbook of research on mathematics teaching and learning,* (pp. 710–723). New York: Macmillan.

Clements, K. (1995). Restructuring mathematics teacher education: Overcoming the barriers of elitism and separatism. In R. Hunting, G. FitzSimons, P. Clarkson, & A. Bishop (Eds.), *Regional collaboration in mathematics education,* (pp. 1–10). Melbourne: Monash University.

D'Ambrosio, U. (2003). The role of mathematics in building up a democratic society. In Madison, B. L., & Steen, L. A. (Eds.), *Quantitative literacy: Why numeracy matters for schools and colleges. Proceedings of National Forum on Quantitative Literacy, National Academy of Sciences, Washington, DC, December, 2001.* Princeton, NJ: National Council on Education and the Disciplines. (http://www.maa.org/ql/qltoc.html)

D'Ambrosio, U. (2006). *Ethnomathematics: Link between tradition and modernity.* Rotterdam, The Netherlands: Sense Publishers.

Ernest, P. (1991) *The Philosophy of Mathematics Education,* London: Falmer Press.

Gellert, U., & Jablonka. E. (Eds.) (2007). *Mathematisation and demathematisation: Social, philosophical and educational ramifications.* Rotterdam: Sense Publishers.

Gutstein, E. (2006). *Reading and writing the world with mathematics: A pedagogy for social justice.* New York: Routledge.

Gutstein, E. (2009). The politics of mathematics education in the US: Dominant and counter agendas. In B. Greer, S. Mukhopadhyay, S. Nelson-Barber, & A. B. Powell (Eds.), *Culturally responsive mathematics education* (pp. 137–164). New York: Routledge.

Joseph, G. G. (1992). *The crest of the peacock: Non-European roots of mathematics.* London: Penguin.

Moses, R. P., & Cobb, C. E. (2001). *Radical equations: Civil rights from Mississippi to the Algebra Project.* Boston: Beacon Press.

Powell, A. B., & Frankenstein, M. (1997). *Ethnomathematics: Challenging Eurocentrism in mathematics education.* Albany, NY: SUNY Press.

Sen, A. (2004) How to judge globalism, In F. J. Lechner and J. Boli, *The Globalization Reader* (Second edition), pp. 17–21. Oxford: Blackwell.

Skovsmose, O. (2006). *Travelling through education: Uncertainty, mathematics, responsibility.* Rotterdam: Sense Publishers.

Smith, D. J, & Armstrong, S. (2002). *If the world was a village*. Tonawanda, NY: Kids Can Press.

Sriraman, B. (Ed.) (2007). *International perspectives on social justice in mathematics education. The Montana Mathematics Enthusiast Monograph 1*. Missoula, MT: University of Montana Press.

Tymoczko, T. (Ed.) (1986). *New directions in the philosophy of mathematics*. Boston: Birkhauser.

Usiskin, Z. (1999). Is there a worldwide mathematics curriculum? In Z. Usiskin (Ed.), *Developments in mathematics education around the world, Volume 4* (pp. 213–227). Reston, VA: National Council of Teachers of Mathematics.

SECTION 1

MATHEMATICS EDUCATION:
FOR WHAT AND WHY?

CHAPTER 1

WHAT IS MATHEMATICS EDUCATION FOR?

Brian Greer
Portland State University

Consider the way in which the National Council of Teachers of Mathematics (NCTM) begins its draft of Standards 2000. No Socrates-like character asks "And shall we teach mathematics?" Even if the answer is a preordained "Of course, Socrates," asking the question raises a host of others: To whom shall we teach mathematics? For what ends? Mathematics of what sort?" In what relation to students's expressed needs? In what relation to our primary aims? And what are these aims? (Noddings, 2003, p. 87)

Many answers to these questions have traditionally been, and continue to be, advanced—the need to produce another generation of scholars to continue advancing the discipline of mathematics; the supply of a cadre of scientists and others such as engineers who need strong mathematical competence; as a training in logical thinking, problem solving, and creativity (Leikin, Section 1, this volume); as part of cultural heritage as much as literature or music.

In the face of change, we need to avoid what Ubi D'Ambrosio calls "the trap of the same." Computers have changed the nature of mathematics. Technological developments have radically altered the flow of information and communication in our lives, and are creating simulated hyperrealities.

Critical Issues in Mathematics Education, pages 3–6
Copyright © 2009 by Information Age Publishing

3

The amount of systematized mathematics has increased hugely, with the result that designing a curriculum is no longer easy (if it ever was) and choices have to be made. Here is a provocative question—does every student need to learn substantial amounts of mathematics (notably algebra), as is declared both possible and essential in many national documents? The mathematician Phil Davis (1999) wrote as follows:

> What is necessary is to teach enough so that the commonplace diurnal mathematical demands placed on the population are readily fulfilled. What is also necessary is to infuse sufficient mathematical and historical literacy that people will be able to understand that the mathematizations put in place in society do not come down from the heavens: that they do not operate as pieces of inexplicable ju-ju, that mathematizations are human cultural arrangements and should be subject to the same sort of critical evaluation as all human arrangements.

> At the risk of sounding like a traitor to my profession, I would say that high school algebra or beyond is not necessary to achieve this goal.

Given that a small percentage of students will continue to high-level mathematics in their careers, it is appropriate to consider what might be a suitable mathematics education to prepare the majority of students for intellectual fulfillment and as future citizens. This consideration needs to go a long way beyond compilation of a list of mathematical content. It is surely essential to make students aware of the implications of mathematization in their societies. As argued by Gellert and Jablonka, there is a further aspect, which they label "demathematization," by which they mean the invisibility of the mathematics that has been incorporated into physical and cultural artefacts. For citizenship with critical agency, an understanding of the mechanisms and effects of these related processes is needed.

In the United States in particular, but by no means uniquely, there is more and more nationalistically phrased emphasis on the importance of mathematical training of a nation's students to maintain economic competitiveness (in the case of the US, global dominance—see Gutstein, 2009). Such an attitude contrasts with the call by Ubi D'Ambrosio (2003) for mathematicians and mathematics educators to accept their ethical responsibilities for addressing the world's most universal problem, survival with dignity.

Within recent decades in mathematics education as a discipline there has been a fundamental shift to what may broadly be characterized as a humanistic view, manifested in numerous ways, including:

- rejection of a Platonist conception of mathematics, in recognition that mathematics is a human activity

- acknowledgement of cultural diversity within both academic mathematics and the diversity of other forms of mathematical practice, as emphasized in the Ethnomathematical perspective
- a broadening of the influences on the field from the dominance of psychology to include a wide range of humanistic disciplines, and a related expansion of methodological tools
- recognition that mathematics education is historically, culturally, socially, and politically situated
- more weight in curricula on making connexions between school mathematics and people's lives. Munir Fasheh (2000, p. 5) declared that: " I cannot subscribe to a system that ignores the lives and ways of living of the social majorities in the world; a system that ignores their ways of living, knowing and making sense of the world."
- teaching mathematical tools of modelling and data analysis that can be used to critique society

It was clearly fitting at an international conference to consider the globalization of mathematics education, as discussed in the paper by Atweh. To highlight that there are exceptions to the tendency towards homogenization of curriculum and pedagogy, we consider the very special case of South Africa (see the paper by Graven and Venkat), where, following the gaining of freedom, a concerted effort has been made to design a mathematics curriculum for citizenship as well as for global competitiveness. Ths debate within South Africa exemplifies a tension evident in many post-colonial societies between teaching mathematics that valorizes diverse cultures and mathematical practices, and teaching mathematics for technical advance with the associated economic benefits.

The deliberations of scholars will remain academic if there is not a collective will to provide sufficient resources for schools, and, above all, teachers. The paper by Agudelo-Valderrama describes the situation in Colombia, but a broadly similar story could be told for many countries. We need to heed Freire's declaration (1987, p. 46):

This is a great discovery, education is politics! After that, when a teacher discovers that he or she is a politician, too, the teacher has to ask, "What kind of politics am I doing in the classroom?"

However, as Apple (2000, p. 243) has pointed out:

It is unfortunate but true that there is not a long tradition within the mainstream of mathematics education of both critically and rigorously examining the connections between mathematics as an area of study and the larger relations of unequal economic, political, and cultural power.

The relationships between three aspects—mathematics as a discipline, mathematics as a school subject, and mathematics as a part of people's lives—need serious analysis. To promote their vision of what mathematics education should be for, mathematics educators need to engage politically.

REFERENCES

Apple, M. W. (2000). Mathematics reform through conservative modernization? Standards, markets, and inequality in education. In J. Boaler (Ed.), *Multiple perspectives on mathematics teaching and learning* (pp. 243–259). Westport, CT: Ablex.

D'Ambrosio, U. (2003). The role of mathematics in building up a democratic society. In Madison, B. L., & Steen, L. A. (Eds.), *Quantitative literacy: Why numeracy matters for schools and colleges. Proceedings of National Forum on Quantitative Literacy, National Academy of Sciences, Washington, DC, December, 2001*. Princeton, NJ: National Council on Education and the Disciplines. (http://www.maa.org/ql/qltoc.html)

Davis, P. (1999). Testing: One, two three; Testing: One,... can you hear me back there? *SIAM News*, March 22.

Fasheh, M. (2000, September). *The trouble with knowledge*. Paper presented at: A global dialogue on "Building learning societies—knowledge, information and human development." Hanover, Germany.

Freire, P. (with Shor, I.). (1987). *A pedagogy for liberation*. Westport, CT: Bergin & Garvey.

Gutstein, E. (2009). The politics of mathematics education in the US: Dominant and counter agendas. In B. Greer, S. Mukhopadhyay, S. Nelson-Barber, & A. B. Powell (Eds.), *Culturally responsive mathematics education* (pp. 137–164). New York: Routledge.

Noddings, N. (2003). *Happiness and education* [sic]. New York: Cambridge University Press.

CHAPTER 2

ETHICAL RESPONSIBILITY AND THE "WHAT" AND "WHY" OF MATHEMATICS EDUCATION IN A GLOBAL CONTEXT[1]

Bill Atweh
Curtin University of Technology

Discussions at international gatherings of mathematics educators such as ICME necessarily raise questions about the similarities and differences of issues as they apply to various countries and cultures. In this paper, I will discuss some implications for an increasingly globalised world of questions as to the *what* and *why* of Mathematics education raised in Discussion Group 3 during the most recent ICME gathering at Monterrey, Mexico. In particular, I will focus on the implications for curriculum development and cooperation between educators from different countries. While I acknowledge that mathematics education also occurs outside the realm of traditional formal schooling, for space considerations, I will focus on the latter. This paper commences by unpacking the concept of globalisation and illustrat-

Critical Issues in Mathematics Education, pages 7–17
Copyright © 2009 by Information Age Publishing
All rights of reproduction in any form reserved.

ing some factors in the discipline that promotes its globalisation. The complexity of issues is then illustrated by presenting different voices from industrialised and less industrialised countries about a global curriculum. Finally, I will argue for increased collaboration between educators from around the world based on ethical responsibility, one towards the other.

WHAT IS GLOBALISATION?

During the past few decades, mathematics educators have reflected an awareness of the trends of international activities in their discipline. For some, these trends are seen as great opportunities, while others regard them with great caution. Atweh and Clarkson (2001) note that the two terms *globalisation* and *internationalisation* are at times used by different authors to mean the same thing and also different authors have used the same term to mean different things. Perhaps the distinction made by Falk (1993 in Taylor, Rizvi, Lingard & Henry, 1997) which identifies two forms of globalisation processes is useful here. He calls them globalisation "from above and from below." Globalisation from above is understood as

> [t]he collaboration between leading states and the main agents of capital formation. This type of globalisation disseminates a consumerist ethos and draws into its domain transnational business and political elites. (p. 75)

On the other hand, globalisation from below

> [c]onsists of an array of transnational social forces animated by environmental concerns, human rights, hostility to patriarchy and a vision of human community based on the unity of diverse cultures seeking an end to poverty, oppression, humiliation and collective violence. (p. 75)

Atweh, Clarkson and Nebres (2003) used the term "globalisation" to refer to the shrinking world and the increasing awareness of issues and practices that affect the whole globe. Globalisation is not used here, as often it is in public media and political discourse, solely to refer to neoliberal economics of free markets, privatisation, and transnational merger trends in the late modernity. Arguably, that use of the term is what Falk might have called globalisation *from above*. Globalisation also includes the increase of collaborations between grass root movements, increased awareness of the concerns of the least disadvantaged around the globe, and the rise of multiculturalism in most countries. These might fall under the globalisation "from below" as discussed by Falk.

Understood in this way, globalisation is different from *homogenisation* (Henry & Taylor, 1997) for, as we are becoming more aware of, and having more esteem for, our similarities, we are also becoming more aware of, and having more esteem for, our differences. Nor it is *inevitable* and out of control, for the nation state remains retains considerable power to direct and control the patterns of globalisation (Henry, Lingard, Rizvi & Taylor, 1999). Similarly, globalisation is not a *utopia* (Derluguian & Greer, 2000) for it can result in imperialism and exclusion, can lead to increased gaps between the rich and the poor, ecological degradation, not to mention global terrorism. Nor is it a dreadful evil to be opposed by any means, for it can lead to collaboration at the grassroots level to oppose injustice and promote human rights. In short, I do not understand globalisation as either good or evil by itself. This is not to say, however, that it is value-free and beyond ethical considerations. On the contrary, its processes and outcomes should be carefully scrutinised as to the benefits and losses that might arise from them. This aim can only be achieved through deliberate and targeted research, reflection, and debate. Further, as I will argue below, such actions need to be done in collaboration among stakeholders from around the world.

GLOBALISATION FACTORS IN MATHEMATICS EDUCATION

Arguably, mathematics education is the most globalised discipline in education. This situation is partly due to its perceived importance for economic and technological development and to the (wrongly) perceived objectivity of the discipline that transcends national and cultural boundaries (Kuku, 1995). This globalised status is reflected by the increasing number of international publications, conferences, research and professional development activities, and most importantly, the convergence of curricula around the world (Oldham, 1989 cited in Clements & Ellerton, 1996). Moreover, these similarities have proven to be rather stable across the years; changes in curriculum in one country or certain region (mainly Anglo-European) are often reflected in other countries within a few years. Note for example, the wide acceptance of the New Mathematics movement in the 1960s, and the more recent widespread "assessment driven reforms" (Hargreaves, 1989) based on standards and profiles. Similarly, in the area of research in mathematics education, Bishop (1992) argues that similarity is a feature of many research traditions evolving in different countries around the globe. Although research in mathematics education is a relatively recent phenomenon in many countries, research questions, methods, practices and publications are becoming more standardized. Bishop concludes that these similarities have led to difficulties in identifying a national perspective of mathematics education research in any country.

In this context I will address two factors of globalisation in the discipline, the role of international organisations and regimes of international testing, and consider their implications for globalisation of the discipline.

Role of International Organisations

To illustrate the possible and diverse effects of international organisations on the global status of mathematics education, I will discuss the role of ICME and the World Bank. The role of international organisations such as the World Bank in assisting less industrialised countries in the implementation of policies of universal primary education and later for the elimination of illiteracy is undoubtedly a great achievement. According to Jones (1992), the Bank started its educational programs in the early 1960s. In the early 1990s it supported about 90 education programs in 59 countries to the total of US$8 billion with half from bank loans and half from participating local governments. However, often this assistance comes with strings attached. For example, the Bank has been the major promoter of ideas of connecting education to economic growth and the model of human development as a priority of economic development. Likewise the Bank has been the major promoter of an emphasis on primary education at the expense of secondary and higher education. Structural reforms in the recipient countries based on decentralisation, "user pays" schemes, credit system, and support to private education have often become conditions of its loans. The work of the Bank has not been without its critics both at government and public levels. In 1996, the fiftieth anniversary of the Bank, world-wide petitions and demonstrations were mixed with the celebrations of its accomplishments. International protest movements such as *50 Years Is Enough* have called for more international debate about the Bank's policies and procedures and a change in its funding into more participatory and sustainable projects. Similarly, the Bank has often been criticised for its insistence on the universal adoption of these policies irrespective of the local context (Nebres, in Atweh, Clarkson & Nebres, 2003).

While the World Bank and similar organisations may illustrate globalisation from above, professional organisations such as the International Commission of Mathematics Instruction may illustrate forces of globalisation from below. Undoubtedly the ICME conferences are the largest gatherings of mathematics educators around the world. For many educators from developing countries, they are the primary, and, in some instances, the sole contact that they have with the international scene in mathematics education. Such contact could possibly lead to further collaboration between educators outside the boundaries of the organisation itself. In this role ICME has played a crucial role in grass root globalisation. However, in spite of attempts to facilitate the participation by educators from developing countries, the congress remains dominated by educators and issues from Anglo-European countries. For many educators from developing countries, which are mainly from the Southern Hemisphere, the cost of travel, not to mention the language barriers, prohibits participation. Arguably a major limitation to communication available within the existing forums of international conferences is the limited possibility of deep dialogue due to the conferences' format. Conferences are restricted in space and/or time. Concern has been raised as to their ability to provide for deep analysis of the contexts behind the research reported in them (Silver & Kilpatrick, 1994). Some participants find international conferences too busy, large and hectic to establish meaningful contacts (Johnston, 1992) or to keep sight of the big picture on problems of mathematics education (Usiskin, 1992).

Role of International Comparisons

Perhaps there are only a few issues in mathematics education that attract more public debate from the media, politicians, and even parents, than international comparisons such as TIMMS and PISA. Several such studies have identified huge gaps in achievement between students from different countries with gaps in achievement estimated to be up to 3 years of schooling (Glewwe & Kremer, 1995). Undoubtedly, these differences reflect, among other things, the different resources available to different education systems around the world; however they do raise serious social justice issues.

This type of study has generated a considerable amount of controversy within the mathematics education literature. Kaiser, Luna and Huntley (1999) have edited a book that deals with the topic from a wide range of perspectives. The book contains discussion from both sides of the debate as well as illustration of the findings from some of these studies. The book consists of sixteen chapters with contributions from the USA, Australia, UK, Germany, Japan and the Philippines. Advocates of international comparisons have argued that such studies offer a better understanding of one's own educational system, identifying its strengths and weaknesses, in their attempts to identify approaches to reform mathematics education. For some educators these studies provide unique opportunities for a massive amount of investigation of factors that may be hard to control in a single country, such as class size, single-gender classes, and out-of-class tutoring.

On the other side of the debate, critics of international comparisons (eg. Keitel & Kilpatrick, 1999) have raised questions about the benefit of these studies to provide useful findings towards the improvement of education systems in any country. These studies have limited pedagogical benefit, since a pedagogy that might work in one country may not work in another context. Concerns were raised on whether the outcomes of these studies are perceived as biased towards the host country; that is, of those who do the data collection, the analysis and the funding. They claim that the mathematical tasks do not represent the curricula taught in many schools, teachers' questionnaires do not represent the whole range of teaching practices, and the results do not offer valid comparisons between the various countries' curricula with their divergent cultural and social contexts. Finally, Clarke (2003) identifies a major abuse of international comparative study as the imposition on participating countries of a global curriculum.

A GLOBAL CURRICULUM?

There is a great unease expressed by many English-speaking researchers about the dominance of Anglo-European thinking about mathematics education for countries around the world. Commenting on the 7th ICME conference in Canada, Usiskin (1992), perhaps summarising the feeling of many participants, notes "the extent to which countries have become close in how they think about their problems and, as a consequence, what they are doing in mathematics education" (p. 19). Yet he goes on to hope "that the new world order does not result in a common world-wide curriculum; our differences provide the best situation for curriculum development and implementation" (p. 20). This concern about uncritical globalisation of issues is shared by Rogers (1992) who, commenting on the same conference, laments that "all our theories about learning are founded in a model of the European Rational Man, and that this starting point might well be inappropriate when applied to other cultures" (p. 22). He goes further to assert that "the assumptions that mathematics is a universal language, and is therefore universally the same in all cultures cannot be justified. Likewise, the assumptions that our solutions to local problems . . . will have universal applications is even further from the truth" (p. 23).

This unease about the dominance of Western mathematics is quite strongly expressed in a keynote address to the ICME Regional Collaboration conference held in Melbourne, Australia, where Clements (1995), a leading Australian mathematics educator with extensive international experience, outlined his concerns in the following manner.

> Over the past 20 years I have often had cause to reflect that it is Western educators who were responsible not only for getting their own mathematics teacher education equation wrong, but also for passing on their errors to education systems around the world. (p. 3)

However, often these concerns do not match some voices from developing countries. At the same ICME regional conference, the president of the African Mathematical Union (Kuku, 1995) warned against the over-emphasis on culturally oriented curricula for developing countries that act against their ability to progress and compete in an increasingly globalised world. He called for "a global minimum curriculum below which no continent should be allowed to drift, however under-developed" (p. 407). Some of the reasons he presented are very relevant to the discussion here. The phenomenon of dropping out of mathematics is not restricted to developing countries. Hence, he argued, cultural relevance of the mathematics content to the culture of the student is not the only consideration in determining participation and success. Kuku expressed concern that the over-emphasis on ethnomathematics may be at the expense of "actual progress in the mathematics education of the students" (p. 406). Presumably this mathematics education is the mathematics education that is needed for economic and technological progress within their countries. Further, within each third world country there are many different cultural groups. There are no resources for implementing an appropriate ethnomathematics program for every student group. He concluded by citing examples of Asian countries that were able to achieve huge leaps in economic development through their use of "imported curricula" (p. 408).

Also at the same conference, a similar call was given by Sawiran (1995), a mathematics educator from Malaysia. Sawiran based his comments on the belief that "our experience shows that mathematics is an important ingredient of technology and therefore is a key element to 'progress'" (p. 603) (quotes in original). He concluded his address by saying that "[t]he main thrust in enhancing better quality of education is through "globalisation" of education. In this respect, it is proper to consider globalisation in mathematics education" (p. 608) (quotes in original). He added that the most important step in globalisation is through "collaborative efforts" (p. 608). In the following section I will argue that such collaboration should be based on mathematics educators taking more responsibility, one to the other. The concept of responsibility leads us into the heart of ethics.

It would be wrong to conclude that the views expressed above are a true representation of the difference of opinions and interests between all educators in the industrialised and the less industrialised countries. However, the reasons behind such calls from some educators from countries with less developed research and theories in the discipline cannot and should not be overlooked. Jacobsen (1996) discusses the increasing gap between the rich and poor countries and the curtailing of funds from these international agencies that makes it "more difficult to look for governments for improved international co-operation in mathematics education" (p. 1253). He joins Miguel de Guzman, the past President of ICMI, in calling for an increasing role of co-operation between professional mathematics educators and their associations to work to improve mathematics education worldwide.

However, as Hargreaves (1994) reminds us, the concept of collaboration should not be taken unproblematically. The remaining task in this paper is to establish a basis of such collaboration in what I will call ethical responsibility.

ETHICAL RESPONSIBILITY

In another context (Atweh, 2007) I noted that the discourse of ethics is not often discussed in mathematics education. Arguably, this absence in mathematics education is paralleled by its absence from general discourses in education and humanities in Western culture. With the rise of scientific rationality, ethics was often associated with questions of morality, dogma, codes of behaviour and legal imperatives and often seen as belonging to the domain of metaphysics rather than philosophy proper. However, this avoidance of dealing with ethical discourse is slowly dissolving. As Critchley (2002) indicates, it was only in the 1980s that the word ethics came back to intellectual discourse after the "antihumanism of the 1970s" (p. 2). Further, the post-ontological philosophical writings of Levinas (1969, 1997) have been accredited by the re-introduction of ethics within philosophy by establishing ethics as the First Philosophy.

As Levinas argues, philosophy is mainly concerned with question of being (ontology) and knowledge (epistemology). The discussion of being and knowledge are achieved by reducing the other to the same (Critchely, 1992) and by dealing with consciousness (Bergo, 1999). For Levinas, ethics is before any philosophy and is the basis of all philosophical exchanges. It precedes ontology "which is a relation to otherness that is reducible to comprehension or understanding" (Critchley, 2002, p. 11). This relation to the other that precedes understanding he calls "original relation." Critchley goes on to point out that the original contribution of Levinas is that he "does not posit, *a priori*, a conception of ethics that then instantiates itself (or does not) in certain concrete experiences. Rather, the ethical is an adjective that describes, *a posteriori*, as it were, a certain event of being in a relation to the other irreducible to comprehension. It is the relation which is ethical, not an ethics that is instantiated in relations" (p. 12, italics in original). Using a phenomenological approach, Levinas argues that to be human is to be in a relationship to the other, or more accurately, in a relation *for the other*. This relation is even prior to mutual obligation or reciprocity. Roth (2007) argues that this original ethical relationship discussed by Levinas consists of an "unlimited, measureless responsibility toward each other that is in continuous excess over any formalization of responsibility in the law and stated ethical principles."

What do we gain by adopting this ethical responsibility as a basis for international collaboration?

First, an ethical responsibility stance requires awareness that collaboration between mathematics educators from around the world is particularly problematic when it occurs between players with different needs and differing access to resources. The limited resources in some countries imply that they are more likely to copy or import ideas from the more developed regions or countries rather than to critically and empirically reflect on their appropriateness to their local context. Collaboration without such an awareness might run into the danger of becoming neo-colonialist with further draining of resources from the poor towards the rich.

Second, in late-modern and globalised times with the lack of certainty and an awareness of the complexity of the issues (Skovsmose, 2006), it may be neither desirable nor possible to establish a set of guidelines for ethical international contacts that apply to all situations. International collaborations based on ethical responsibility are necessarily transparent, reflective and accountable in examining their own rationale, aims, processes and outcomes. Questions of voice and power should always be up front. Ethically responsible collaboration should be constructed to empower individual countries to be self-reliant rather than to increase their dependency on ideas from more developed nations.

Third, collaborations that are simply based on "helping" developed countries (to become like us?) are often based on paternalistic colonial assumptions and do not contribute to ethical collaboration. Further collaborations based on ethical responsibility are based on mutual respect and trust in the ability of the different partners to contribute different types of learning to the collaborative enterprise.

FINAL COMMENT

Perhaps these disparate views are not completely incongruent in that most mathematics educators share a commitment for the development of mathematics education for empowerment of *all* students everywhere and to increasing the participation of *all* mathematics educators in international dialogue and debates. The concern is not about whether a standardised or an ethnomathematical curriculum is the solution to inequality, but about the resources and expertise that are necessary to increase the participation of students and academics both at local and global levels. This can be achieved through genuine and critical collaboration between academics internationally. Such collaboration aims to contextualise the curricula (and research findings) to local contexts without falling into a void where mutual learning is not possible. From this perspective contextualisation and decontextualisation are two interactive and complementary rather than contradictory endeavours.

REFERENCES

Atweh, B. (2007). Pedagogy for socially response-able mathematics education. Paper presented at the Annual Conference of the Australian Association for Research in Education, Fremantle, Western Australia. http://www.aare.edu.au/07pap/atw07600.pdf

Atweh, B. & Clarkson, P. (2001). Internationalisation and globalisation of mathematics education: Towards an agenda for research/action. In B. Atweh, H. Forgasz, & B. Nebres (Eds.), *Sociocultural research on mathematics education: An international perspective* (pp. 77–94). New York: Erlbaum.

Atweh, B. Clarkson, P. & Nebres, B. (2003). Mathematics education in international and global context. In A. Bishop, M. A. Clements, C. Keitel, J. Kilpartick, & F. Leung (Eds.), *The second international handbook of mathematics education* (pp. 185–229). Dordrecht: Kluwer Academic Publishers.

Bergo, B. (1999). *Levinas between ethics and politics: For the beauty that adorns the earth.* Dordrecht: Kluwer Academic Publishers.

Bishop, A. J. (1992). International perspectives on research in mathematics education. In D. Grouws (Ed.), *Handbook of research on mathematics teaching and learning,* (pp. 710–723). New York: Macmillan.

Clarke, D. (2003). International comparative research in mathematics education. In A. Bishop, M. A. Clements, C. Keitel, J. Kilpartick, & F. Leung (Eds.), *The second international handbook of mathematics education* (pp. 143–184). Dordrecht: Kluwer Academic Publishers.

Clements, K. (1995). Restructuring mathematics teacher education: Overcoming the barriers of elitism and separatism. In R. Hunting, G. FitzSimons, P. Clarkson, & A. Bishop (Eds.), *Regional collaboration in mathematics education,* (pp. 1–10). Melbourne: Monash University.

Clements, M. A. & Ellerton, N. (1996). *Mathematics education research: Past, present and future.* Bangkok: UNESCO.

Critchley, S. (1992). *The ethics of Deconstruction: Derrida and Levinas.* Oxford: Blackwell.

Critchley, S. (2002). Introduction. In S. Critchley & R. Bernasconi (Eds.), *The Cambridge companion to Levinas.* Cambridge: Cambridge University Press.

Derluguian, G. M. & Greer. S. (Eds.). (2000). *Questioning geopolitics: Political projects and organizational aspects of globalisation.* Westport, CT: Praeger.

Glewwe, P. & Kremer. M. (1995). *Schools, Teachers, and Education Outcomes in Developing Countries. Centre for International Development Working Paper No. 122.* Cambridge: Harvard University.

Hargreaves, A. (1989). *Curriculum and assessment reform.* Milton Keynes: Open University Press.

Hargreaves, A. (1994). *Changing teachers, changing times: Teachers' work and culture in the postmodern age.* London: Cassell.

Henry, M. & Taylor, S. (1997). Globalisation and national schooling policy in Australia. In B. Lingard & P. Porter (Eds.), *A national approach to schooling in Australia: Essays on the development of national policies in school education,* (pp. 46–59). Canberra: Australian College of Education.

Henry, M., Lingard, B., Rizvi, F. & Taylor, S. (1999). Working with/against globalisation in education. *Journal of Education Policy, 14,* 1, 85–97.

Jacobsen, E. (1996). International co-operation in mathematics education. In A. Bishop, et al. (Eds.), *International Handbook of Mathematics Education* (pp. 1235–1256). Dordrecht: Kluwer.

Johnston, B. (1992). Walled City. *For the Learning of Mathematics, 12(3),* 23–24.

Jones, P. (1992). *World Bank financing of education: Lending, learning and development.* London: Routledge.

Kaiser, G., Luna, E. & Huntley, I. (Eds.). (1999). *International comparisons in mathematics education.* London: Falmer Press.

Keitel, C. & Kilpatrick, J. (1999). The rationality and irrationality of international comparative studies. In G. Kaiser, E. Luna, & I. Huntley (Eds.), *International comparisons in mathematics education* (pp. 241–256), London: Falmer Press.

Kuku, A. (1995). Mathematics education in Africa in relation to other countries. In R. Hunting, G. Fitzsimons, P. Clarkson, & A. Bishop (Eds.), *Regional collaboration in mathematics education* (pp. 403–423). Melbourne: Monash University.

Levinas, E. (1969). *Totality and Infinity: An essay on exteriority* (A. Lingis, Trans.). Pittsburgh, PA: Duquesne University Press.

Levinas, E. (1997). *Otherwise than being or beyond essence* (A. Lingis, Trans.). Pittsburgh, PA: Duquesne University Press.

Rogers, L. (1992). Then and now. *For the Learning of Mathematics, 12(3),* 22–23.

Roth, W. R. (2007). Solidarity and the ethics of collaborative Research. In S. Ritchie (Ed.), *Research collaboration: Relationships and Praxis* (pp. 27–42). Rotterdam: Sense Publishers,

Sawiran, M. (1995). Collaborative efforts in enhancing globalisation in mathematics education. In R. Hunting, G. FitzSimons, P. Clarkson, & A. Bishop (Eds.), *Regional collaboration in mathematics education* (pp. 603–609). Melbourne: Monash University.

Silver, E. & Kilpatrick (1994). E pluribus unum: Challenges of diversity in the future of mathematics education research. *Journal for Research in Mathematics Education, 25,* 734–754.

Skovsmose, O. (2006). Research, practice, uncertainty and responsibility. *Journal of Mathematical Behavior 25,* 267–284.

Taylor, S., Rizvi, R., Lingard, B. & Henry, M. (1997). *Educational policy and the politics of change.* London: Routledge.

Usiskin, Z. (1992). Thoughts of an ICME regular. *For the Learning of Mathematics, 12,* 3, 19–20.

NOTE

1. Adapted from Atweh, Clarkson and Nebres (2003) and Atweh (2007).

CHAPTER 3

THE DEMATHEMATISING EFFECT OF TECHNOLOGY
Calling For Critical Competence

Uwe Gellert
Freie Universität Berlin

Eva Jablonka
Luleå Tekniska Universitet

Mathematics has penetrated many parts of our lives. It has capitalised on its abstract consideration of number, space, time, pattern, structure, and its deductive course of argument, thus gaining an enormous descriptive, predictive, and prescriptive power. A process of mathematisation has taken place in many areas of science; quantitative studies are highly valued in the humanities; it is almost impossible to understand any modelling in economics without a solid mathematical background. In all these fields mathematics can be regarded as the grammar of the particular scientific discourse. However, mathematics being the grammar implies that the characteristics of this grammar strongly influence the development of the fields in which the use of mathematics is made. It turns out to be difficult to integrate any

Critical Issues in Mathematics Education, pages 19–24
Copyright © 2009 by Information Age Publishing
All rights of reproduction in any form reserved.

idea that cannot be formulated in mathematical terms into an accepted body of mathematically formulated theories.

The impact of mathematics is by no means restricted to scientific activity. Mathematics-based decisions affect the social interactions in technological societies on many levels. On the level of the national policy, decisions about the distribution of state salaries, pensions, and social benefits rely on mathematical extrapolations of demographical and economic data provided by experts, the results of which often are communicated by means of formulae and diagrams. On the level of interpersonal relations, mathematics-based communication technologies have already changed the habits and styles of private conversations. Of course, the mathematics is mostly invisible, as in mobile phones and internet chat forums, or it is just recognised on the surface as a medium of presentation.

Why is mathematics so powerful? Mathematical thinking has the power of hypothetical reasoning: It is possible to calculate some consequences of different scenarios before the corresponding actions are carried out. Nobody needs to be afraid of any immediate consequences of mathematical thought. In the long run, however, the world of mathematical thinking transforms back to what Keitel, Kotzmann and Skovsmose (1993) describe as a *system of implicit knowledge*. In many cases, we are neither aware of the circumstances under which a particular mathematical model has been processed, nor of the purposes for its initiation. The social origins and the history of many mathematisations are immersed. Technology, including social technology, functions as a black box—and the constitutive mathematics needs not to be reflected upon anymore. The substitution of abstraction processes by black boxes produces what Keitel et al. call *implicit mathematics*.

In order to stress the point that mathematics shapes the technology with the help of which we organise much of our life, Keitel et al. introduce the notion of *realised abstraction*. Mathematical thinking becomes materialised, it becomes part of our reality, and most of the time we do not ask where it comes from or what it is—there is no necessity for doing so. Our time-space-money-system is a striking example for the implicitness of the underlying abstraction processes.

The concept of realised abstraction may make us understand that the mathematisation of our world is just one side of the coin. The existence of materialised mathematics in the form of black boxes reduces the importance of mathematical skills and knowledge for the individual's professional and social life. A demathematisation process is taking place:

This term [demathematisation] also refers to the *trivialisation* and *devaluation* which accompany the development of materialized mathematics: mathematical skills and knowledge acquired in schools and which in former time served as a prerequisite of vocation and daily life lose their importance, and become superfluous as machines better execute most of these mathematical operations. (Keitel et al. 1993, p. 251)

The process of demathematisation affects strongly the values associated with different kinds of knowledge and skills. For the user of technology it becomes more important to, first of all, simply trust the black box and, then, to know when and how to use it—for whatever purpose.

Chevallard (1989/2007) draws attention to the importance of a process he describes as follows:

> Implicit mathematics are formerly explicit mathematics that have become "embodied," "crystallized" or "frozen" in objects of all kinds—mathematical and non-mathematical, material and non-material—for the production of which they have been used and "consumed." (Chevallard 2007, p. 58)

For Chevallard (2007, p. 60), the dialectic between implicit and explicit mathematics resides in a "never-ending, two-fold process of (explicit) de-mathematising of social practices and (implicit) mathematising of socially produced objects and techniques."

It is indeed this process, which has to be the starting point of any discussion of the value of mathematical skills for an individual.

MATHEMATISATION/DEMATHEMATISATION THROUGH TECHNOLOGY

Keitel (1989) illustrates the role and possible effects of technology by the example of the mechanical clock. The construction of the clock is based on the perception of the movement of the planetary system:

> This approach is generalized and condensed to a mathematical model, transformed into a technological structure, and as such installed outside its original limited realm of significance. Earlier human perceptions of time, which had grown out of both individual and collective experiences and remained bound and restricted to these, were now rivalled and ultimately substituted for by the novel way of perceiving time. (Keitel 1989, p. 9)

The first effect of this technology is a mathematisation that makes it possible to measure time precisely and independently from the quality of the processes measured. The abstraction of comparability is presupposed. The objective character of the mechanical clock denies subjective experience of time. The specific (subjective) situation, in which time is measured, has lost its relevance. A formalisation has taken place. Time is no longer valid as a concrete sensory experience.

This objectification and formalisation still has tremendous implications: Time is regarded as the sum of arbitrarily regular units. Mathematics as the grammar of science is reinforced:

> The mechanical clock extends the domain of quantification and measurability. Applying measure and number to time means measuring and quantifying all other areas, in particular those where time and space relate to one another. The measurability of time pushes forward the development of the natural sciences as (empirical) sciences of measurement (and hence objective sciences) and mathematics as the theory of measurement. (Keitel 1989, p. 9)

Equally important, mathematics serves as the grammar of social order and social coordination. Keitel et al. (1993) refer to F.W. Taylor's introduction of "scientific management": Every complex work process can be broken down into elementary components; the time necessary to carry out these elementary components can be measured; the time in which a complex work process should be finished is the sum of the small but many "pieces of time" needed for the elementary components. Here, the measurement of time "objectively" determines work organisation. It appears as if mathematically conceptualised time were the most natural thing in the world. The mathematical abstraction, which is encapsulated in the clock, has vanished from the surface—but it nevertheless continues to be effective.

Technology can be characterised by its effect of making the underlying (mathematical) abstraction processes invisible. At the same time, technology facilitates the use of mathematics in social or technical situations precisely by liberating the user from the details of the mathematics involved. A curious correlation can be observed: Whereas the flexibility and potential of mathematical thought lies in its harmlessness—there is no immediate threat of changing the physical world by carrying out mathematical abstractions and calculations—the materialised mathematics of technologies has lost its innocence. Whereas mathematics offers hypothetical explorations into new problem solutions, the use of "frozen mathematics" in the form of technology might restrict the scope of problem solutions that could have been imagined.

MATHEMATISATION/DEMATHEMATISATION AND POWER

Skovsmose (1998) regards mathematics as an essential instrument for exercising technological power. He sees an increase in the range of applications of mathematics linked to modern information technology. Mathematics has not only become an integrated part of technological planning and decision making but also an invisible part of social structuration, encapsulated in political arguments, technologies and administrative routines. Citizenship presupposes the excavation of "frozen" mathematics.

Following this line of argument, demathematisation excludes citizenship, and development of appropriate excavation tools becomes a central issue. Skovsmose introduces positions of social groups who are involved in or affected by mathematics in action in different ways. The "constructors" are those who "develop and maintain the apparatus of reason" (Skovsmose 2006: 140). In constructing mathematics based technology, this group exercises power over "operators" and "consumers" of this technology.

Whereas the constructors are involved in developing mathematical technology, the operators are those who work in jobs, in which they have to make decisions on the input and then decisions based on the output of this technology. These job situations can be called "rich in implicit mathematics" (Skovsmose 2006, p. 142). Those who are listening to a range of offers, statements and reports containing figures and numbers, are called slightly ironically, "consumers" of mathematics. They could "vote, receive services, fulfil obligations, be citizens." Consumers are confronted with justifications of decisions based on complex models.

There is a threat to democracy because of a widening gap of mathematical knowledge between constructors and consumers. The constructors not only provide the technical knowledge for developing solutions but also have the power to define the problems and to initiate new questions. The forming of opinions and political decisions become more and more dependant on their expertise.

Skovsmose sees as one of the essential problems of democracy in a highly technological society, the development of a critical competence, which can match the actual social and technological development. If the interpretation of democracy is not restricted to formal procedures of electing a body of representatives, but also includes participation and elements of direct democracy, the status of the constructors has to be scrutinised. Decisions made on the ground of mathematical models may be inaccessible to demathematised consumers. However, citizenship includes providing a "talking back" to authority (Skovsmose 1998, p. 199). This presupposes a wider horizon of interpretations and pre-understandings of mathematical knowledge than passive consumption of offers, statements and reports. Technological competence is not sufficient for predicting and analysing results and consequences of mathematisations. Reflections building upon different competencies are needed. As Skovsmose illustrates, the competence in constructing (or using) a car is not adequate for the evaluation of the social consequences of car production.

Skovsmose (1998) identifies three groups of questions related to reflective knowledge, which focus (i) on the relationship between mathematics and an extra-mathematical reality, (ii) on mathematical concepts and algorithms, and (iii) on the social context of modelling and its implications in terms of power.

The distinction between technological and reflective knowledge, which also resembles the distinction between operators and critical consumers (in opposition to demathematised consumers), is fruitful, but still has to be further elaborated with respect to its consequences for a conceptualisation of content and forms of mathematics education. Especially with respect to those groups of people who are deprived of any kind of formal education, the tension between functional and critical education seems to be exacerbated.

ACKNOWLEDGMENT

This short paper, prepared specially for Discussion Group 3 at ICME11, is based on Jablonka & Gellert (2007).

REFERENCES

Chevallard, Y. (1989/2007). Implicit mathematics: Its impact on societal needs and demands. In U. Gellert & E. Jablonka (Eds.), *Mathematisation and demathematisation: Social, philosophical and educational ramifications* (pp. 57–66). Rotterdam: Sense Publishers.

Jablonka, E., & Gellert, U. (2007). Mathematisation–Demathematisation. In U. Gellert & E. Jablonka (Eds.), *Mathematisation and demathematisation: Social, philosophical and educational ramifications* (pp. 1–18). Rotterdam: Sense Publishers.

Keitel, C. (1989). Mathematics education and technology. *For the Learning of Mathematics*, 9(1), 103–120.

Keitel, C., Kotzmann, E., & Skovsmose, O. (1993). Beyond the tunnel vision: Analysing the relationship between mathematics, society and technology. In C. Keitel & K. Ruthven (Eds.), *Learning from computers: Mathematics education and technology* (pp. 243–279). Berlin: Springer.

Skovsmose, O. (1998). Linking mathematics education and democracy: Citizenship, mathematical archaeology, mathemacy and deliberative education. *Zentralblatt für Didaktik der Mathematik*, 98(6), 195–203.

Skovsmose, O. (2006). *Travelling through education: Uncertainty, mathematics, responsibility*. Rotterdam: Sense Publishers.

CHAPTER 4

MATHEMATICAL LITERACY

Issues for Engagement from the South African Experience of Curriculum Implementation

A core paper prepared for Discussion Group 3:
Math education: for what and why?

Mellony Graven and Hamsa Venkat
Wits University

INTRODUCTION

Due to the restricted length of this paper we will focus our attention on giving a very brief story of what our research is indicating in relation to the implementation of Mathematical Literacy as a new subject in the SA curriculum, and the issues arising within this implementation. For further reading we suggest reference to Graven & Venkat (2007) and Venkat & Graven (2007).

This is followed by introducing some key insights and issues which our research has highlighted and we believe require further exploration, discussion and investigation. Thus each insight/issue raised is accompanied by a leading question and subset of questions which require further engagement.

Critical Issues in Mathematics Education, pages 25–31
Copyright © 2009 by Information Age Publishing

MATHEMATICAL LITERACY—THE SOUTH AFRICAN CASE

What the Curriculum Says

Mathematical literacy (ML) was introduced in schools in the Further Education and Training (FET) phase (grades 10–12, learners mainly aged 15–18) in South Africa in January 2006. The subject is structured as an alternative option to mathematics, and all learners entering the FET phase since January 2006 are required to take one or other of these two options. ML is defined in the curriculum statement in the following terms:

> Mathematical Literacy provides learners with an awareness and understanding of the role that mathematics plays in the modern world. Mathematical Literacy is a subject driven by life-related applications of mathematics. It enables learners to develop the ability and confidence to think numerically and spatially in order to interpret and critically analyse everyday situations and to solve problems. (DoE, 2003, p. 9)

While the phrase 'applications of mathematics' allows for an interpretation where mathematical language, conventions, algorithms, theorems and practices can be learnt first and then applied to life related problems and everyday situations, the post-amble headed 'Context' following the Learning Outcomes, emphasizes that contexts should be engaged with in a way which enables and drives mathematical learning:

> The approach that needs to be adopted in developing Mathematical Literacy is to engage with contexts rather than applying Mathematics already learned to the context. (DoE, 2003a, p. 42)

This is re-emphasised in the Teachers' Guide although here there seems to be an attempt to highlight the importance of striking a balance between contextual and mathematical learning and an emphasis on the dialectical relationship between the two:

> the challenge for you as the teacher is to use situations or contexts to reveal the underlying mathematics while simultaneously using the mathematics to make sense of the situations or contexts... (DoE, 2006, p. 4)

Our Research

Our work within the Mathematical Literacy (ML) thrust in the Marang Centre at Wits University involves a range of strands—research, lecturing and teacher development, and raising public awareness. Our research work centrally involves a longitudinal case study, now in its third year, tracing the experiences of educators and the first cohort of learners taking ML in one inner city Johannesburg school. This work has involved weekly visits to the three ML classes in this cohort across grade 10, grade 11 and now grade 12, as well as questionnaire and interview data from learners and teachers. Additionally, we have conducted a series of classroom observations, learner and teacher interviews in three other Johannesburg schools. We also draw upon feedback from the teachers we interact with as part of our lecturing, supervision and thrust work.

Our Findings

We have identified a spectrum of agendas (Table 4.1) that teachers work with as they navigate their teaching of Mathematical Literacy across the FET band (see Graven & Venkat, 2007). While the spectrum is likely to be refined and reviewed over time it has proven useful as a tool for analysing mathematical literacy practices.

Our documentary analysis of the curriculum suggests that in intention at least agenda 2 is foregrounded.

Our classroom observations and interviews to date seem to suggest that learners' have tended to be largely positive about ML. We are well aware of the fact that this may well not be the case more broadly and that in the classrooms that we have observed teachers have tended to work with agendas towards the left hand side. In these classrooms both learners and teachers have noted substantive differences (compared to mathematics experiences)

TABLE 4.1   A Spectrum of Agendas

Context driven (by learner needs)	Content and context driven	Mainly content driven	Content driven
Content in service of context	Content and context in dialectical relationship	Context in service of content	No clear need for context
Driving agenda: To explore contexts that learners *need* in their lives (current everyday, future work & everyday, and for critical citizenship) and to use maths to achieve this.	Driving agenda: To explore a context so as to deepen math understanding and to learn maths (new/GET) *and* to deepen understanding of that context.	Driving agenda: To learn maths and then to apply it to various contexts.	Driving agenda: To give learners a 2nd chance to learn the basics of maths in GET band (grades 0–9)

in: the nature of tasks in ML (engagement with a scenario rather than application of maths in 'word problems') and the nature of interaction in ML (much slower pace, more discussion and group work). See Venkat & Graven (2007) for elaboration and evidence of such changes.

Some teachers on our ACE and postgraduate courses have reported low levels of motivation and lack of interest amongst learners in their ML classes. Anecdotal evidence from educators suggests that this kind of ongoing negativity is associated with a lack of substantive change in pedagogic practice. In some cases this is due to an interpretation by the teacher of ML as involving 'basic maths' (towards the right of the spectrum), and consequentially, teaching that incorporates the kinds of tasks and pedagogic practice that have predominated within learners' earlier experiences with mathematics.

Our table highlights various issues that are experienced by teachers when working with a particular agenda. In particular we point to the issues of authenticity of context, development of mathematical progression and discrepancies of continuous and summative assessments. These issues are experienced by teachers in different ways depending on their primary driving agenda. Below we summarise these issues and raise some questions that merit further discussion.

ISSUES FOR DISCUSSION

Authenticity and Mathematical Progression

Agendas on the left hand side demand a certain degree of authenticity in the tasks and scenarios that learners engage with. This can sometimes raise tensions in relation to mathematical progression. Thus many teachers might wish to trim off certain aspects of a context as they deal with mathematics which they feel learners are not yet ready to deal with. Trimming such contexts affects the authenticity of the task. Other teachers feel that mathematical progression is less of an issue in mathematical literacy and that one draws on the maths you need when and where you need it (and then only as far as you need it and not further).

Curriculum documents struggle to build mathematical progression into assessment standards from one grade to another—so for example assessment standards from one grade to the next simply say "in more complex contexts" or provide what they believe are more complex contexts from one grade to the next.[1]

CRITICAL QUESTIONS

- In what ways might teachers and materials developers work with such tensions between authenticity and mathematical progression?
- How should mathematical literacy year to year planning deal with progression (contextual and/or mathematical)?

Discrepancies in Performance between Continuous and Summative Assessments

In ML there is an increased amount of project work or extended activity work which is often done over an extended period of time, is mediated by the teacher and draws on other learners ideas (often in group work). Success in such work sometimes contrasts with weaker performance in time constrained, unmediated and individualized assessments.

CRITICAL QUESTIONS

- What do such discrepancies tell us about learner ML competences? Are such discrepancies a problem?
- Is there a place for individualized, time constrained, summative assessments in ML? If so, what is their place and how should performance on these be weighted against performance on other types of activities?
- What about issues of assessment validity? [e.g., in SA the externally set and marked Grade 12 examination carries the most weight—seen to have greater validity and reliability—but such assessments sometimes reduce the scope of ML]

ML in Its Own Right versus ML in Contrast Mathematics

Much of the data that we have received to date on ML is in contrast to learner and teacher experiences of Mathematics. Thus many of the successes for learners are often in contrast to their largely negative experiences of learning mathematics. While this provides interesting comparative data it skews the data towards aspects which contrast to mathematics. ML is now in its third year of implementation. We are beginning to get comments from learners which critique the issues arising in ML classrooms in their own right without comparison to Maths—comments about ML involving the need to make sense of situations for example. As researchers too, we are aware of the mathematical lenses that we bring to our observations. Depending on how one views ML such lenses can be obstacles in the way of being able to see the range of non mathematical outcomes achieved in classes.

CRITICAL QUESTIONS

- In what ways do our mathematical experiences/lenses through which we view ML an influence what we see and how we interpret what we see? How do we reflect on the influence of these experiences?
- How do varying definitions of ML affect the appropriateness of the influence of mathematical lenses on its research? [e.g., If ML was instead incorporated into or renamed "Life Orientation" would we research it through a different lens?]
- How (if we see this as desirable) do we find ways to research ML in its own right without ongoing comparison to mathematics?

Language Issues

In our research to date there is a conspicuous absence of comments by learners and teachers noting any language difficulties (note in SA the majority of learners are taught in English which is not their first language). Much has been written on the difficulties of learning mathematics in this context (see Setati, 2005), and many SA educators have noted that integration with contexts could be problematic due to the increased English language demands.

In contrast learner interviews have highlighted accessibility in relation to language demands in ML with comments such as 'It's in English' unlike Mathematics which was compared to a language they did not speak or understand (e.g., Latin). Our hypothesis is that changes in both the nature of activities and mediation of those activities (increased discussion in class with learners around scenarios and contexts used) are supporting learners in the language demands of ML and the use of a more everyday, rather than more technical, register alleviates some of the language demands that are noted in mathematical classes.

CRITICAL QUESTIONS

- What is the nature of the language issues that emerge in the teaching and learning of mathematical literacy? How are these similar or different from those that emerge in the teaching of mathematics?
- How should language issues be researched in ML? How (if at all) should such research differ from research on language issues in mathematics classes?

REFERENCES

DoE. (2003). National Curriculum Statement Grades 10–12 (General): Mathematical Literacy: Department of Education.

DoE. (2006). National Curriculum Statement Grades 10–12, Teacher Guide, Mathematical Literacy: Department of Education.

Graven, M. & Venkat, H. (2007) Emerging pedagogic agendas in the teaching of Mathematical Literacy. AJRMSTE, Vol 11 (2), pp. 67–86

Venkat, H. & Graven, M. (2007) Insights into the implementation of Mathematical Literacy. Proceedings of the Thirteenth Annual National Congress of the Association for Mathematics Education of South Africa (AMESA), Uplands College, Mpumalanga, Vol 1, pp 72–83.

Setati, M. (2005) Teaching Mathematics in a Primary Multilingual Classroom. JRME. NCTM, USA. 36(5), 447–466.

NOTE

1. Many assessment standards (ASs) are the same for all three grades with different contextual examples given in each grade. No detail is given on what makes one context more complex than another, or what progression within a context might entail.

CHAPTER 5

THE PURPOSE OF SCHOOL MATHEMATICS

Perspectives of Colombian Mathematics Teachers

Cecilia Agudelo-Valderrama
Universidad Distrital F. J. C., Bogotá, Colombia

For decades, the goal of educating *critical citizens* has been the flag of the Colombian educational system. In the specific area of school mathematics this goal has been underlined as fundamental in curriculum guidelines, emphasising the use of *real-life* situations as the centre of classroom activity. Teaching school mathematics for "real life," however—according to the findings of this study carried out with the participation of thirteen mathematics teachers with a range of teaching experiences—was making sure that the list of content items included in schools' syllabuses was covered, for a prime concern for the schools was to secure a good place in the league tables produced by the National External Examination body. The teachers' conceptions (encompassing knowledge, beliefs, and attitudes) of the requirements of the external examination had a strong influence in their

Critical Issues in Mathematics Education, pages 33–37
Copyright © 2009 by Information Age Publishing

thinking about *what mathematics education is for, and why*. These findings highlight the difficulty of moving away from a school mathematics dominated by the *academic* curriculum designed with the needs—as established by the education system—of the few moving on to higher education. It is argued that the mathematics education for "critical citizenship" put forward by Colombian policy needs to emphasise a diversity of situated, social, cultural, and political mathematical practices, giving important attention to the mathematical theories that help to describe and explain those practices—with mathematics as a critical tool to analyse and develop awareness of reality. Clarification of the tasks of those responsible for the implementation of such curriculum (which includes external assessment) is needed.

Here are two statements from teachers:

> ... the teacher has to aim at having his/her pupils prepared in the best possible way for the ICFES [Colombian Institute for the Promotion of Higher Education] examination. So one knows ... one cannot stay too long in one topic because, even if these pupils are not moving on to higher education, they themselves, their parents, the Ministry, everybody! measures the school and the teachers by the ICFES results. (Juan, Int. 1)

> ... the classroom activities discussed in the focus group session are very good to motivate the pupils ... they are examples of the type of work which can help pupils to build the path to algebra ... but I have to be careful not to expend too much time in that sort of activity ... pupils should work on the formalisations, the universal language of mathematics so that they are prepared for life, for the ICFES. (Nora, Focus group)

With a focus on the perspectives of teachers, the study on which this paper is based investigated the relationship between Colombian mathematics teachers' conceptions of beginning algebra and their conceptions of their *own* teaching practices. The teachers' understandings of their teaching practices (i.e., their conceptions of their roles as teachers, their explanations of why they teach in the way they teach) were explored with a view to unravelling their conceptions of change in their teaching. The vignettes above, provided by two of the participating teachers, are representative of the perspectives, of the majority of participants, on *the purpose* of the teaching of beginning algebra. Further, they are representative of the teachers' justifications for classroom incidents, and of their answers regarding the questions that guided the analysis of the teachers' conceptions of their own teaching practices—"why do they teach in the way they do" and "why it is difficult to incorporate alternative teaching approaches in their classrooms."

The study was conducted with the participation of thirteen qualified secondary school mathematics teachers who taught at six different (state and private) schools in Bogotá, during the academic year of 2002. After an exploration of the conceptions of the initial group of thirteen, nine teachers with a variety of conceptions were selected in order to conduct case studies. The participating teachers varied greatly in age and teaching experience; some of them were young beginner teachers, some were experienced teachers who had started their teaching at the primary level and some others, beside their secondary school job, were teaching at universities. Some of the teachers had done postgraduate courses in Education, and had participated in classroom-based professional development programs. Information about the methodology and significant findings can be found in Agudelo-Valderrama (2006, in press), Agudelo-Valderrama, Clarke and Bishop (2007).

Eight out of the nine case study teachers justified their teaching decisions, and explained their teaching acts, by the need to prepare the pupils for higher mathematics levels and, ultimately, for the external examination. The teachers' knowledge and beliefs about the social and educational systems had a strong structuring power in their thinking. In general, while the teachers' conceptions of beginning algebra played a strong role, their conceptions of the social/institutional factors played a stronger role in their conceptions of "what school mathematics is for."

SOME BACKGROUND INFORMATION ON MATHEMATICS EDUCATION IN THE COLOMBIAN CONTEXT

One of the fundamental aims of education—which has been highlighted as crucial in the General Law of Education, issued in 1994 and currently in force—is "the development of the individual's *critical, reflective and analytical* mind... needed by individuals to participate *actively* in a *democratic* society... the search for solutions to problems... and to contribute to the social and economic progress of the country" (Articles 5 and 21, italics added). In the specific case of mathematics education, "*problem-solving and learning with meaning*" have been recommended by the National Curriculum Guidelines as "the centre of classroom activity" (Ministerio de Educación Nacional, 1998).

The main objective of "any mathematical work is to help the learners find meaning of the world that surround them.... Through the learning of mathematics students should acquire a set of powerful tools to explore, represent, explain and predict reality... to act in and for a real world" (p. 35).

Before the issuing of the General Law of Education, the Mathematics National Curriculum—a prescriptive and *academic* curriculum, designed for those moving on to higher education—was based on the formalistic aspect of a hierarchically organised list of topics where algebra was a packaged course to be taught in Grades 8 and 9. The standard National External Examination at Grade 11 has always been used to control admission to higher education and to judge the academic quality of schools.

The Challenge of Educating Critical Citizens through School Mathematics in Colombia

The mathematics curriculum guidelines encourage the provision of learning environments that promote the exploration of "real-life" situations in order to educate students "to act in and *for* a real world." Within the broader frame of educating *critical citizens*, this evokes a curriculum that includes a diversity of mathematical practices, giving importance to specific aspects highlighted by Jablonka (2003) in her different conceptualisations of *mathematical literacy* (e.g., mathematical literacy for cultural identity, for social change, for environmental awareness, for developing human capital).

(Note that there are no literal translations of the terms "numeracy" and "mathematical literacy" into Spanish, so with the term "mathematical literacy," in Colombia, I refer to "mathematics education for citizenship.") Indeed, all these conceptualisations are of necessary inclusion in a "mathematical literacy" curriculum like the one put forward by the Colombian curriculum guidelines. However, the majority of teachers in this study saw little opportunity to incorporate teaching approaches which created space for the consideration and analysis of real-life situations and the promotion of meaningful learning. Paraphrasing the teachers: "to engage the students in such learning activities, takes too much time which is not available in the school setting, and that type of work or learning is not recognised in the external examination."

The influence of external assessment on classroom practices is not a new finding of research, but these findings call attention to the great power that the teachers' conceptions of social/institutional factors of teaching play on their conceptions of *the purpose of school mathematics*, and highlight the difficulty, in Colombia, of moving away from a school mathematics dominated by the *academic* curriculum designed with the needs in mind of the small elite moving on to higher education. These findings have important implications for the education programmes of those responsible of "mathematical literacy" education in Colombia, and call for a *systemic* change strategy. How can we, as educators, work for the empowerment of teachers so that they start to move towards the *type of teacher* (i.e., "internal attribution teacher," see Agudelo- Valderrama, in press; Agudelo-Valderrama, et al., 2007) and *teaching* that is exemplified in the rhetoric of Colombian policy?

Schools are responsible for the mathematics education of all students. So the curriculum needs to be developed to satisfy the needs of a democratic, and "today's global and *knowledge-based* society" (World Science Forum, Budapest, 2003 at www.sciforum.hu), for the education of critical citizens as well as the mathematical needs of those moving on to higher (mathematical) education. This means that the curriculum needs to reflect a diversity of mathematical practices, carrying clear purposes of promoting critical discussion and development of awareness of real-life situations and, at the same time, giving important attention to the mathematical theories (including those related to an identified and agreed core mathematical content) that help describe and explain those practices. As Bishop (2006) argues:

> The ethnomathematics literature indicates that numeracies are all about particular practices, and they also include meanings, beliefs and conceptualisations. Learners bring many numeracy, and ethomathematical, practices to their education…but the naive learner lacks any understanding of the mathematical theories that help to explain those practices. Without these tools they become trapped by those practices, just as most adults currently are, without the understanding to question them and, perhaps, even develop alternative practices....

We need research and conceptualisations of "mathematical literacy" in Colombia, "which clarify the educational task facing those responsible for [mathematical literacy] education" (Bishop, 2006), *their educators,* and those interested in mathematical literacy external assessment.

REFERENCES

Agudelo-Valderrama, C. (2006) The growing gap between policy, official claims and classroom realities: Insights from Colombian mathematics teachers´ conceptions of beginning algebra and its teaching purpose. *International Journal of Science and Mathematics Education, 4*, 513–544.

Agudelo-Valderrama, C., Clarke, B., & Bishop, A. (2007). Explanations of attitudes to change: Colombian mathematics teachers´ conceptions of the crucial determinants of their teaching practices of beginning algebra. *Journal of Mathematics Teacher Education, 10*(2), 69–93.

Agudelo-Valderrama, C. (2008) The power of Colombian mathematics teachers' conceptions of social/institutional factors of teaching. *Educational Studies in Mathematics, 68,* 37–54.

Bishop, A. J. (2006). Culture, numeracy and values in relation to mathematics education. XXII Coloquio Distrital de Matemáticas y Estadística, Universidad Distrital FJC, Universidad Nacional y Universidad Pedagógica Nacional, Bogotá. (www.coloquiodistritaldematematicasyestadistica.com)

Jablonka, E. (2003). Mathematical literacy. In A. J. Bishop (Ed.), *Second International Handbook of Mathematics Education* (pp. 75–102). Dordrecht: Kluwer.

Ministerio de Educación Nacional (1998). *Matemáticas—Lineamientos Curriculares.* Bogotá.

CHAPTER 6

TEACHING MATHEMATICS WITH AND FOR CREATIVITY

An Intercultural Perspective

Roza Leikin
University of Haifa

CREATIVITY AND GIFTEDNESS

Creativity is typically used to refer to the act of producing new ideas, approaches or actions, while innovation is the process of both generating *and applying* such creative ideas in some specific context (Horowitz & O'Brien, 1985; Piirto, 1999; Davis & Rimm, 2004; Sternberg, 1999). Creativity is manifested in the production of a creative work (for example, a new work of art or a scientific hypothesis) that is both *novel* and *useful.*

Unlike many phenomena in science, there is no single, authoritative perspective or definition of creativity. The association between high IQ and giftedness and creativity is not simple. O'Hara and Sternberg (1999) provided some support to Torrance (1974) suggestion which is termed "the threshold hypothesis" which holds that a high degree of intelligence appears to be a necessary but not sufficient condition for high creativity. An alternative perspective, Renzulli's three-ring hypothesis, sees giftedness as

Critical Issues in Mathematics Education, pages 39–43
Copyright © 2009 by Information Age Publishing
All rights of reproduction in any form reserved.

based on both intelligence and creativity (e.g., Davis & Rimm, 2004; Renzulli, 2002; Sternberg, 1999).

Though most researchers agree that giftedness and creativity are associated, clearly not all gifted children are creative. Clearly, cultures vary considerably in how they view these issues, the importance they ascribe to creativity, the measures they use to identify it, the domains which seem to them important, and the ways they employ to foster creativity in the different domains. The following question remains quite obscure to this date:

- How to identify and assess creativity, and especially how to foster creativity?
- How creativity giftedness and relationships among them are perceived in different cultural contexts?

CREATIVITY AS RELATED TO MATHEMATICS EDUCATION

Research literature distinguishes between general and specific giftedness, and general and specific creativity (e.g., Piirto, 1999). *Specific giftedness* refers to clear and distinct intellectual ability in a given area, for example, mathematics. It is usually reflected in socially recognized performance and accomplishment. *Specific creativity* is expressed in clear and distinct ability to create in one area, for example, mathematics.

Torrance (1974) defined *fluency, flexibility* and *novelty* as the main components of creativity. Krutetskii (1976), Ervynck (1991), and Silver (1997) explored the concept of creativity in mathematics in the context of multiple-solution tasks. In this context (Silver, 1997, Ervynck, 1991, Leikin & Lev, 2007), *flexibility* refers to the number of different solutions generated by a solver, *novelty* refers to the conventionality (relatively to a specific curriculum) of suggested solutions, and *fluency* refers to the pace of solving procedure and switches between different solutions.

There are many open questions about the development of creativity which arise in at least two contexts in mathematics education.

1. Fostering Student Creativity

It has become commonplace for educators to accept that the goal of mathematics education is more than the mastery of a body of algorithms and methods, and that mathematics is an ideal training ground for the development of logical reasoning in students. It is less widely accepted in the field that mathematics is also a training ground for creativity. Some questions associated with the notion of creativity in mathematics might be:

- Can creativity be actively fostered in the classroom?
- What classroom techniques lead students to exercise their minds creatively? What areas of mathematics lend themselves most to students' independent exploration?
- What must a teacher know, or know how to do, in order to support students in developing their creativity in mathematics?

These are questions that can be asked with regards to students and classrooms on any ability level and any age level

2. Fostering Creativity in Teaching

A second context in which we may discuss the phenomenon of creativity within mathematics education is in the context of teaching. The act of teaching may itself be ordinary or creative, standard or novel. A teacher may simply follow the textbook. A more creative teacher might give her own examples that illustrate the points in a textbook. A still more creative teacher might invent his or her own explanations or activities to convey a concept or method. And there are certainly levels of creativity above this one. Questions about creativity in teaching might be the same as those about creativity among students. But additional issues come up in this wider context:

- How does creative teaching interact with the need to standardize, to measure, to be held accountable?
- How do we train teachers to be creative?
- How do we manage teachers or schools, to allow for creativity to emerge?
- What is the role of creative teacher in the classroom? In the school? In the profession?

CREATIVITY AND CULTURE

The phenomenon of culture, of ethnicity, of belonging to a group, runs very deep. All humans have culture, just as all humans have language or music or dance. However, the specific nature of culture varies widely. The multicultural approach to creativity is especially relevant in multicultural countries which are characterized by great diversities of languages, cultures, traditions, mentalities, and educational achievements. The cultures in different countries have characteristics which can be used to help develop the creative abilities.

To increase students' creative potential, a variety of methodologies may be employed, based on the ideas of Vigotsky (1984), Freudental (1977), Davydov (1996) and Polya (1945/1973). An example of the important source for creative mathematics education is the experience of schools established approximately 30 years ago in Russia at the initiative of the famous Soviet mathematician Kolmorogov (1965). In this context we may ask the following questions:

- How artifacts of local culture may be used in stimulating creativity?
- How is the creative student (or teacher) regarded in different cultural contexts by peers? By authorities? What role does such a person play in the society?
- How is cultural borrowing viewed by the host culture? Is there resistance? Attraction to the exotic? When cultural practices change, what are the forces at work to change it?

ACKNOWLEDGEMENTS

The paper addresses some issues discussed at the international workshop "Intercultural aspects of creativity " held in Haifa, March, 2008 with support of Templeton Foundation. The author thanks Dr. Mark Soul for his contribution in preparation of the earlier version of the article.

REFERENCES

Davis, G. A. & Rimm, S. B. (2004). *Education of the gifted and talented* (5th ed.). Boston, MA: Pearson Education Press.

Davydov V. V. (1996). *Theory of Developing Education,* Moscow: Intor (in Russian).

Ervynck, G. (1991). Mathematical creativity. In Tall, D. (Ed), *Advanced Mathematical Thinking* (pp. 42–53). Netherlands: Kluwer

Freudental H. (1977). *Mathematik als padagogische aufgabe,* Stuttgart: Ernst Klett verlay (in German).

Horowitz, F. D. & M. O'Brien (Eds.), (1985). *The gifted and talented: Developmental perspectives.* Washington, DC: American Psychological Association.

Kolmorogov A. N. (1965). On the content of mathematics curricula in the school: *Mathematics in the School,* 4 (in Russian).

Krutetskii, V.A. (1976). *The Psychology of Mathematical Abilities in Schoolchildren.* (Translated by Teller, J.; edited by J. Kilpatrick and I. Wirszup). Chicago, IL: The University of Chicago Press.

Leikin, R. & Lev, M. (2007). Multiple solution tasks as a magnifying glass for observation of mathematical creativity. In the *Proceedings of the 31st International Conference for the Psychology of Mathematics Education.* Pólya, G. (1945/1973). *How to solve it.* Princeton, NJ: Princeton University.

O'Hara, L. A., & Sternberg, R. J. (1999). Learning styles. In M. Runco & S. R. Pritzker (Eds.), *Encyclopedia of creativity* (Vol. II) (pp. 147–153). San Diego: Academic Press.

Piirto, J. (1999) *Talented children and adults: Their development and education.* Upper Saddle River, N.J.: Prentice Hall.

Renzulli, J. (2002). Expanding the Conception of Giftedness to Include Co-Cognitive Traits and to Promote Social Capital. *Phi Delta Kappan,* 84(1), 33–58.

Silver, E. A. (1997). Fostering creativity through instruction rich in mathematical problem solving and problem posing. *ZDM,* 3, 75–80.

Sternberg, R. J. (Ed.). (1999) Handbook of creativity. New York: Cambridge University Press

Torrance, E. P. (1974). *Torrance tests of creative thinking.* Bensenville, IL: Scholastic Testing Service.

Vigotsky L. (1984). *Psychology of Children. Collected works,* v.4, Moscow: Pedagogy (in Russian).

CHAPTER 7

WHOSE MATHEMATICS EDUCATION?

Mathematical Discourses as Cultural Matricide?

Fiona Walls
James Cook University

"Mathematics education, for what and why," begs an additional question: "Whose mathematics education?" Among the ideas for discussion in this group is the concern that "*school mathematics has scant relevance to the personal and collective lives of the students or the adults they will become.*" This paper takes a step beyond "*reactions against Eurocentric narratives of the history of mathematics*" to the question of the sexualization of mathematical discourses (Irigaray, 2002), including notions of mathematical literacy. Within such discourses, women appear only by virtue of their invisibility.

A globally recognised phenomenon in mathematics education is the differing ways in which boys and girls participate and achieve in their learning of mathematics. According to the recently publicised research of Guiso, Monte, Sapienza and Zingales (2008) based on the PISA analysis, the "gen-

Critical Issues in Mathematics Education, pages 45–52
Copyright © 2009 by Information Age Publishing
All rights of reproduction in any form reserved.

der gap," long perceived to exist between girls and boys in mathematics, disappears in societies that treat the sexes equally and in which men and women have access to similar resources and opportunities, suggesting that boys are not innately better at mathematics than girls, and any difference in test scores is due to nurture rather than nature. When girls have equal access to education and other opportunities they do just as well as boys in mathematics tests.

Despite such findings, persistent beliefs in the existence of biological cognitive differences between boys and girls, that have traditionally advantaged boys' performance in mathematics, continue to be expressed. The following analysis of the Guiso et al. research findings is a prime example:

> New studies may be shedding light on the issue. In a nutshell, some of the latest research points to three conclusions that offer something to satisfy both sides—but overall paint a bright picture for those eager to see more women enter mathematics and sciences. The key findings:
>
> - Girls are as good at math as boys given the proper environment.
> - Males may have an edge in spatial thinking abilities, which are useful in math—and this advantage may be very ancient, evolutionarily speaking.
> - Deep-rooted though this difference may be, females can surmount it with just a little work.[1]

What is overlooked in both the Guiso et al. research, and its widely publicised reactions such as the one above, is the question of whose mathematics education we are talking about. It is accepted without question that girls must study and succeed at a subject that is masculine, that is, a discursive formation that has been, and continues to be, socially constructed through the life experiences, world views, practices and priorities of men. Women must "enter" mathematics and science as invited guests, as privileged participants, not as architects or founding mothers. With "just a little work" they can "surmount" their "deep-rooted" difference/disadvantage that goes with being female.

Luce Irigaray describes the widespread exclusion of women's discourses in Western society as *monosexuality* or *sexual indifference*. Even within mathematical discourses of diversity and inclusion this indifference can be discerned as a pervasive *negligence*, a systematic *oversight*, an exclusory *deafness* that relentlessly fails to substantiate female/feminine/woman/girl mathematical narratives and contributions to the discipline of mathematics as distinct (not to be confused with singular and essential as Diana Fuss (1990) describes), as sexually nuanced, and as situated in women's and girls' contextualised, gendered, lived experiences in society. The female register is scripted as echo, mimicry, and ventriloquism within malecentric mathematical discourses.

The (mono)sexuation of mathematics education discourses is particularly apparent when we examine the choice of plenary conference speakers, the choice of "pressing" (gender blind) issues on which to focus our collective attention, the choice of theories by which we *examine, interrogate,* and *inform* our practice, or the method and form of what counts as acceptable/believable (male sanctioned) research, consistently revealing a marginalisation, minimisation, and dematerialisation of the *female* in mathematics. The mathematics that has been adopted worldwide as *the* mathematics children should learn, has a patriarchal genealogy. Margaret Wertheim's penetrating examination of the cultural/historical construction of Western mathematics (Wertheim, 1997) shows how the tessellating/intersecting discourses of *mathematics, physics,* and *God* can be viewed as a *gender war* that consciously and explicitly constructs the feminine out of mathematics through a binary oppositional alignment of female with evil, ungodly, impure, irrational, and illogical. This is reinforced by the research of Valerie Walkerdine (1998) which demonstrates how the *feminine* is *subtracted* from mathematics, as science, reason, and the male mind are coproduced through the psychological discourses of education.

The pervasive, global, exclusion of girls and women from schools, from universities, from the field of mathematics, and from equal *and* different standing in mathematics education policy formation, implementation, research, and debate, particularly in developing countries, reveals connections between masculinity, mathematics, and social power which cannot be ignored. Currently, for women to engage in mathematics education they must wear a cloak of invisibility that affords them temporary status as honorary males in a male domain, speaking a male mathematical language.

If we *acknowledge mathematics as a human activity*, then mathematics that can be seen as distinctly *created and used by women and girls*, should appear as often, as prominently and as equally valued, as that of men and boys in our mathematical discourses. What might a *feminine mathematics* include? Ethnomathematical studies have gathered data about the ways in which the construction of tangible women's artefacts such as woven baskets, mats and quilts, engage methods of thinking relationally, algebraically, and spatially. A feminine mathematics must also take account of how women's mathematical thinking is shaped by their social positions and everyday existences. Women within most cultural groups are cast as nurturers, caterers, carers, and managers of people within family, community, workplace and institutional structures. In these roles, women develop sophisticated cognitive matrices by which they map, order, and seek optimal fit among such competing variables as time, location, space, movement, resources, com/demands and emotions.

Women's mathematical catering skills and wisdoms have traditionally been passed from woman to woman across generations, communicated as an oral body of knowledge generated and transmitted through women's working and talking together. For example, my mother taught me the science/art of sewing clothes by engaging with me in an active process. This involved:

(a) selecting the pattern and fabric, as constrained by budgets, personal preferences, and fashion
(b) placing and cutting pattern pieces in the most efficient way. This required highly developed spatial sense, since the fabric might be doubled, pattern pieces folded along lines of symmetry, aligned with the grain of the fabric, following the nap, or cut on the bias for binding, without waste. Calculation of expanding seams to accommodate children's growth, the creation of darts, rouching and clipped curves to fit flat surfaces to the human form, and matching checks or stripes were also taken into account, and
(c) sewing, fitting, and finishing the garment. Seams were sewn a consistent distance from the raw edges in parallel lines, darts as precise, gently tapering triangles, gathers as the drawing of a larger length of fabric into even folds (scaling) to match a smaller piece, buttons and button holes aligned and spaced precisely and evenly.

It took me many years to become an accomplished seamstress. Through my engagement in this socio-symbolic work, I was constituted as *female* rather than as *mathematical.*

Another example of women's mathematics embodied in cultural technologies can be found in the basket weaving of the women of the Pacific islands. I bought the basket shown in Figure 7.1 from a market in Port Vila, Vanuatu. I was told that this design is known by the women who weave such baskets as "Stars." Its Escher-like tessellation is not only visually striking, but mathematically sophisticated. The pattern is produced by weaving coloured strands of dried pandanus leaf in a particular repeating order, with, as far as I can ascertain, a base unit of six strands. The basket is a 3-dimensional hollow shape with a square base. Not only is the pattern of stars repeated flawlessly around the continuous surface of the basket's sides, but also across its base. The design and production of such an intricately patterned object requires deep spatial and algebraic "sense," yet the women of Vanuatu have at best received only the most basic formal schooling. Taught by their mothers and mothers-in-law, the women carry their weaving patterns in their minds. They are continually producing new designs, one woman describing how her design came to her in a dream. When she awoke she rushed off to convert it into a woven reality. The women often weave in groups with young children by their sides, and I have spoken with Pacific adults who describe how, through the distinctive swish of the

Figure 7.1 'Stars' pattern on Vanuatu women's hand-woven pandanus basket.

pandanus strands as particular designs are created, they can recall becoming rhythmically engaged with pattern from a young age. To discount such mathematics as somehow inferior to the "Western" mathematical systems of symbols, abstraction, logic, and proof is to ignore the social and cultural origins of mathematics in general, and the origins and "rootedness" in everyday life, of women's mathematics in particular. Beliefs about women's inabilities to reasoning spatially fail to take account of the complex geometric dimensions of women's technologies.

Garment and basket construction, as with many other women's cultural practices such as catering for family and community events, maintaining agricultural plots, and trading and exchanging goods and services to sustain family and community economies, engages thinking, reasoning, and working mathematically in complex spatial, numerical, and relational networks, simultaneously engaging the socially situated mathematics of production and use. By contrast, mathematics of the Western male lineage is purged of human context, distanced, disembodied, and detached from everyday purpose, formulaic, and presented as an unchallengeable immutability. Thus purified, (masculine) mathematics has been socially exalted in its exclusion of *embeddedness* in life in general, and women's lives in particular. Because all children, male and female, are expected to learn only male versions of mathematics as a "basic" of their schooling, mathematics education can be viewed as a form of cultural matricide, that is, a muzzling, negating, extinguishing practice.

To explain why women's mathematics have been counted out of the mathematical content and processes that now constitute an almost universal school mathematics curriculum, we need to look to women's social positioning. Bourdieu (2001) takes a socio-cultural approach to understanding the ways in which the foundational and defining "symbolic violence" of male domination has become so deeply embedded in the collective human unconscious that we are barely aware of its omnipresence, let alone able to perceive or question its profound impacts and implications, particularly for women. He notes how social institutions such as family, school, church and state reify, sanctify, and eternalise the power of men. He concludes that male domination can be reduced only through political intervention that takes account of the effects of domination exerted through the objective complicity between the structures embodied in women and men as social actors, and the structures of the major institutions through which not only the masculine order but the entire social order is enacted and reproduced (Bourdieu, 2001, p. 117).

Judith Butler (1990, 2006) warns that traditional feminist approaches that have attempted to draw together and strengthen women under a unifying sense of "we," have perpetuated the underlying "truth" that produces and sustains male domination—the idea of *natural* or *essential* characteristics or qualities of "masculine" and "feminine" in an epistemologically oppositional frame. She looks instead to social and socialised structuring or "practices of signification" (p. 197) of these concepts, to describe gender as performative, as things we do, rather than what we "are." She offers the view that what is necessary to move from essentialist positions that maintain male dominance is a deconstruction of masculine/feminine identification and its accompanying gender performances. If this were enacted within mathematics education, we might expect to see girls and boys behaving in gender-unexpected ways, for example, boys and girls struggling or succeeding at spatial problems in equal proportions. But unless performative change extended to the ways that teachers behave, the ways that curricula are developed, the ways that politicians profile and prioritise mathematics education, and even the ways that mathematics education research is valued, legitimised, or rejected, little change would be possible.

So what of approaches that attempt to represent the female in mathematics? Through her examination of the "masculine" in mathematics education, Mendick (2006) expresses unease about the notion of a female or feminised mathematics, suggesting instead an *opening up* of mathematics to unfix gender from its binary frame. Such re-framing might acknowledge and embrace mathematics as lived, as creatively imagined, as socially and culturally derived and constructed, as connected and connecting, and as complex and challenging in its demand for social aptness, timeliness, practicability, and sustainability, thus embracing (women's) multiple ways of working mathematically. But there is a continuing danger that in uncoupling the binary, women and girls remain invisible.

Using Irigaray's (2002) argument that sexual difference is inscribed at all levels of the socio-symbolic exchange, and her more recent work that offers a vision of *sexual* "difference" as something to be productively embraced as *human* difference, a reinscribed sexual intersubjectivity in mathematics education would require firstly an acknowledgement that women's mathematical world-views, wisdoms, voices and practices exist, as diverse and situated as they may be. Once recognised, women's mathematics might be made visible as valuable and necessary in our societies. Thirdly, multiple feminine (*and* masculine) reconstituted and reconstitutive mathematical discourses and discursive practices might claim and be given equal standing in mathematics education everywhere, from classroom cultures to academic production and exchange, embraced as strengthening and enriching, rather than weakening the discipline of mathematics, and in turn the field of mathematics education. To continue to ignore or denigrate the mathematics of women is to perpetuate a form of educational discrimination at best, or educational misogyny at worst.

This paper provides a starting point for reopening and redeveloping discussion around an issue that has failed to disappear, despite the ongoing efforts of women in mathematics education and despite the so-called "success" of girls and women in the few societies where women are encouraged to participate and achieve in areas that have traditionally been considered masculine domains.

REFERENCES

Bourdieu, P. (2001). *Male domination*. Palo Alto, CA: Stanford University Press.

Butler, J. (1990) (2006). *Gender trouble*. New York. Routledge Classics.

Guiso, L., Monte, F., Sapienza, P., & Zingales, L. (2008). Diversity: Culture, gender and math. *Science*. 320 (5880), 1164—1165.

Irigaray, Luce (2002). *To speak is never neutral*. New York: Routledge.

Irigaray, Luce (2006). *The way of love*. London: Continuum.

Fuss, Diana (1990). *Essentially speaking*. New York: Routledge.

Mendick, Heather (2006). *Masculinities in mathematics*. Milton Keynes, UK: Open University Press.

Walkerdine, Valerie (1998). *Counting girls out: Girls and mathematics*. London: Falmer.

Wertheim, Margaret (1997). *Pythagoras' trousers: God, physics and the gender wars.* London: Fourth Estate.

NOTE

1. World Science, May 2008 http://www.world-science.net/othernews/080531_math.htm

CHAPTER 8

THE TENSION BETWEEN WHAT MATHEMATICS EDUCATION SHOULD BE FOR AND WHAT IT IS ACTUALLY FOR[1]

Alexandre Pais
University of Lisbon

This article is fuelled by two aspects of my life: my work as a mathematics teacher, and my research in the academy. By describing the sociological problems I feel as a teacher and confronting them with the research made in mathematics education, I will try to emphasise the tension that exist between what we, as teachers, researches, parents, think math education should be for and the real role performed by math education in our society. In order to exemplify that tension I will present some considerations about my participation in Discussion Group 3 in ICME11.

1. SOCIOLOGICAL PROBLEMS OF A MATH TEACHER

I have been a mathematics teacher for the last eight years, in several Portuguese public schools. During my first years my ideas about education were

Critical Issues in Mathematics Education, pages 53–60

oscillating between didactic knowledge and pedagogical knowledge, which provided me a very professional but narrow view of my role as a teacher. I felt comfortable. I was enjoying my first years of teaching, with lots of new things to do, and plenty of didactic ideas to implement in the classroom, in order to allow success to my students.

But problems and contradictions started to arise. For example, the fact that some of the nice, carefully worked out, plans that I prepared for my classes failed because of strange things that aren't supposed to happen in a classroom. Things that had to do with the presence of thirty children with wills, fears, desires, problems, families, that definitely weren't the ones I imagined when preparing the class in the comfort of my home.

I started to realize that school was a place of conflict, where knowledge was just an alibi to the practice of *educare*.[2] There are many examples of this disciplinary strategy. Like, for instance, all the bunch of disciplinary mechanisms deployed by teachers and others school functionaries to control children. Like the advice to only smile at students after the first trimester, to site the students in a way that makes conversations between them impossible, to have them doing routine exercises just to avoid the noise and the confusion that occur when trying to develop more problematic situations, the use of evaluation as a device for controlling behaviour, the use of blackmail by saying things like "if you don't be quiet and study you will become a vagabond," or "if you don't study you will be the last in the class," or even using the parents: "if you don't behave I will talk to your father," knowing that his or her father will beat him or her. The examples could continue, and show how teachers incorporate, with an incredible sense of mission, the role of transforming a child into a student, and, in the process, transforming themselves into a teacher. Although I felt uncomfortable with that type of behaviour, in school, in my classroom, in my role as teacher, I was not an exception and a feeling of disillusion was progressively taking the place of the optimism of the first years.

Naturally, contradictory feelings started to appear about my role as a teacher, and awareness of school's role in society. By that time I had started a master's degree course in university, with allowed me the contact with ideas from critical theory, in particular critical mathematics education. Although the Portuguese curriculum explicitly mentions the importance of working with student on topics of mathematics and society, it is content-oriented and the high-stakes tests are always present, putting pressure on teachers and students to be glued to specifically mathematical content. That corrupts any possible change. As a teacher I feel that the only thing I can do is to confront students with the reality, with the contradictions I feel, and, in doing so, contribute to politicizing mathematics education in particular and school in general. But I am conscious that little transformation has been made.

I was not making any change as a teacher, although I had the will to do so and was developing some activities in my school with that purpose. I simply felt that all those enterprises were suppressed by a school structure that has other purposes for education visible in mechanisms like the high-stakes tests and selection role taken care of by school; the culture of individualism that students felt, and the appeal to be competitive; the central mathematical curriculum. The contradictions arose when I started to criticize, in an academic field, the same practices I developed as a teacher: I continue to propose lots of routine exercises, preparing kids for exam, reproving children, conveying the idea that mathematics is just for some, taking an authoritarian role as a teacher, controlling students, etc.

So, although some classes went really well, with the application of all my didactical knowledge, I started to felt more and more as someone being used to perform someone else's job. This feeling and the opportunity to get back to university to study (taking the master's degree) allowed me the opportunity to start thinking more clearly about the entire educational dimension that surpasses didactic knowledge and school. I started to situate my work as a teacher in the world, in a society, which I progressively started to look at with a critical stance. I started to be politicized about education.

2. UNDERSTANDING SCHOOL

Most of the difficulties I felt as a teacher were related not with didactical issues (how to motivate students, how to teach algebra in a significant way, how to evaluate oral communication of mathematics contents, etc.), but with non-didactical aspects, like the contradictions I mentioned before. Contradictions about the way I understand mathematics, school and society, and the practices, discourses, and regimes of truth that circulate in school, and for which I was just another link in the chain. I realized that in order to understand those contradictions I needed to go beyond the classroom, beyond school, to have a bigger picture of the role of school in society and my role as a teacher. So I started to research about education, using Foucault as a guide.

As is mentioned by Stoer, Cortesão, and Correia (2001), "schoolarization" has become the only legitimate modality of thinking about education. My main goal with that research was to illuminate justifications and explanations that to us, in our "schoolarized" society, appear as natural. These hidden assumptions include the idea that school is a necessary institution, that people should spend all their childhood and adolescence in such institutions, and the idea that mathematics is one of the most important achievements of mankind, and one of the most important school subjects.

By reading Foucault, and the archaeological and genealogical analysis of formations like the human sciences, and institutions like the prison, the asylum, the hospital, I started to look at school not as a necessary, or natural, institution of humankind, but a rather specific and contingent institution, with specific purposes that have less to do with the transmission of knowledge, than with the governance of people.

Foucault goes against the humanistic position that separates knowledge from power. According to Foucault, they are intrinsically linked, and could be seen as productive forces. Taking school as an example, power is exercised through mechanisms like the curriculum, the placement of students in a classroom, the techniques deployed by the teacher to teach, the examination, the registration techniques, that simultaneously are effects of power and create knowledge about the students. This knowledge is then use to justify the exercise of power.

In school, education is a disciplinary device (Foucault, 2003), which fabricates the individual. Following Foucault's thought, school has become one of the major modern disciplinary centres of the body. Obligatory in modern societies, in school we are introduced to the disciplinary society, via academically recognized knowledge, by the way we submit our bodies and minds to the training devices. There exists in school a huge amount of corporal discipline, whether it is in the space organization, or in rules as norms about what is considered to be good and bad behaviours. It is in school that the human being, no longer a person but a student, starts to understand the hierarchy of behaviours and knowledge, by means of the creation of classificatory systems that limit, integrate, and exclude them. School plays the role of an apparatus to govern a population by fabricating the kinds of subjects that hegemonic society stipulates as normal, through the dissemination of norms that function as calibration devices.

Education is a state question, under the influence of national political projects and international rules that are being disseminated with more or less resistance. The goal of education is not just to educate students in the sense of intellectuality, behaviour, and citizenship, but also to implant the desire for education, because the job market demands motivated and intelligent people, people with attitude and lots of self-esteem. For instance, in the Portuguese mathematical curriculum we can read that one of the main goals of mathematics education should be the development in students of a *taste* for mathematics. That is, students must not just *know* mathematics, they must also *like* mathematics. According to Fendler (1998), the purpose is to govern the soul: teachers have not only the responsibility to govern the moral aspect, but also the feelings, the desires, anxieties, in order to produce the desired citizen: "becoming educated, in the current sense, consists of teaching the soul—including fears, attitudes, will, and desire" (p. 28).

This fabrication of subjectivity, by means of mechanisms, techniques, norms, rules, discourses, positions us in order to become knowledgeable and capable of being administered. Universal and compulsory schooling catches up the lives of all young children into a pedagogic machine that operates not only to impact knowledge but also to instruct in conduct and to supervise, evaluate, and rectify pathologies.

So, school is far from being the place of education. On the contrary, it has developed the function of reducing, dominating, and suffocating education by the ways of re-inscribing it within the structure of the state. Knowledge in school can be understood as an alibi, an alibi for the formation of the subject needed by the modern society.

3. FROM SCHOOL TO ACADEMY: STUDIES ON CRITICAL MATHEMATICS EDUCATION AND THE ICME DISCUSSION GROUP

After eight years of teaching I decided to stop, and went back to university, to do a PhD course on mathematics education, with the question in mind: how to develop a research that explicitly addresses all the sociological problems I felt as a teacher?

At that point, I had already realized that the field of mathematics education research is a field stemming from psychology, where the main focus of research has been the classroom, with all the knowledge of how to better teach, learn, and evaluate teachers and students in there school tasks. I realized that most of the mathematics education research takes school and mathematics for granted. The focus is not in the "why" but in the "how." And huge amounts of research have been produced based in that unquestionable certainty.

So I found refuge in a marginal part of mathematics education research that progressively was taking steps in order to implement a sociological approach to the problems of the field. We could call those studies critical mathematics education, although they are very different from each other, and some of them are far from being critical.

When I was choosing the discussion group to go to in ICME11, I didn't find any specific mention of critical mathematics education. But that doesn't mean that concerns related with critical mathematics education didn't appear in the discussion groups of ICME. For instance, DG13 addressed the challenges posed by different perspectives, positions, and approaches in mathematics education research, and DG18 was about the role of ethnomathematics[3] in mathematics education. But since I first took a look into the DG proposals I immediately got seduced by DG3. The title of the group was very explicit about the radicalism of the discussion: *Mathematics Educa-*

tion: for what and why? So there I was, trying to understand my sociological problems in the field of academic research, by participating in a radical discussion group.

However, during the first readings of the discussion paper and the others papers that fuelled the discussion, I become suspicious about the radicalism of the discussion. For instance, in the introductory text of the group (see Greer, this volume) the aspects listed to justify the teaching of mathematics, this is, to answer to the question *What is mathematics education for?*, are all of them a decoy. In a rough way we can group them in four categories: because it is useful to the student in her/his daily life; because is an important cultural heritage that must be preserved; the argument that the study of mathematics develops psychological skills; and final, the most propagated in the recent years, because it contributes to the development of participative citizens in a socio-political perspective. All those arguments are a decoy. It is easy to deconstruct them and show how fragile and incoherent they are, specially the socio-political argument (Pais, 2005).

But that doesn't mean, as is mentioned in the same introductory text that "school mathematics has scant relevance to the personal and collective lives of the students or the adults they will become." It's exactly the opposite! It's because mathematics in school fails to promote all the aims that are proudly announced by educators, that mathematics is the most relevant school subject in governing people's life. Mathematics appears in school with other surreptitious functions.

If we take from the work of Foucault the notion of normalization device, then we can say that school mathematics acts like a device that makes behaviour elements to be reinforced or punished, as they adhere to the rule or not. One of the effects of teaching mathematics is to approximate the student to the norm, by putting in motion mechanisms that penetrate their bodies, their gestures, and their behaviour. Normalizing is associated with the governance of people, by "controlling their multiplicities, using them at maximum, maximizing the utility of their work and activity, thanks to a system of power susceptible of controlling them" (Foucault, 2004, p. 105). Basically, "making grow at the same time the docility and the utility of all elements of the system" (p. 180). As stated by Popkewitz (2002), who takes into account the work of Foucault on education:

> [the mathematical curriculum] is an inscription devise that makes the child legible and administrable. The mathematics curriculum embodies rules and standards of reason that order how judgments are made, conclusions drawn, rectification proposed, and the fields of existence made manageable and predictable. (p. 36)

Looking with this perspective, the questions "What should mathematics education be for?," and "Why mathematics education?" gain a completely different array of answers. The turning point here is the fact that we aren't trying to answer those questions based on what we wish, as mathematics educators, mathematics education to be, but, instead, what mathematics education is: a normalization device. So, ignoring the real role performed by mathematics in school, and continuing to stipulate desirable goals to mathematics education, could work as a decoy, as something that by obscuring the role of governance taking place in school serves to support an ideal (sometimes identified as being critical) that makes invisible others ways of seeing reality.

4. THE TENSION BETWEEN WHAT WE WANT AND WHAT IS

So even in the field of so called critical mathematics education, and in a radical discussion group, traps in understanding the aims of mathematics in school are present.

As researchers, we want mathematics education to be a school subject important to student's lives. We convey, to teachers and public opinion, the idea of mathematics education as a central discipline in a more and more mathematical knowledge-based society. In some sense, we reinforce the importance of mathematics education, and by doing that, we contribute to the role of governance taking place at school.

Should our focus of research be not only what we imagine that mathematics education should be for, but also what role mathematics is asked to carry out in schools? How is the teaching of mathematics contributing, in a disguised way, to the fabrication of the kind of people that our society needs? What does it means to educate people to be participative in a more and more market-orientated society? How are we researchers concerned with the development of an idea of citizenship, and encouraging a mathematics education that makes accessible powerful tools to analyze, critique, and act upon, social and political issues, problematizing the role of school in our society?

Those are some of the questions that pump into me during the exercise of confronting my experience as a teacher and the experience of doing research. There is obviously a fissure between teacher practice and researcher practice. In some way, that could be explained by the fact that a teacher is, in fact, in school, every day, dealing with contradictions, mechanisms, and discourses than compel him/her to exercise power, to change people's lives (mostly the student's life). She/he will, sooner or later, in a superficial or profound way, have to deal with those contradictions. From my experience, most teachers get refuge in educational dogmas to avoid dealing with those contradictions, and to be able to continue their work. But when you start to erase all the dogmas and certainties about education, you get lost. And that is an important step to becoming critical. You start looking for alternatives.

REFERENCES

D'Ambrosio, U. (2006). Etnomatemática e Educação. In G. Knijnik, F. Wanderer, & C. Oliveira [Eds.], *Etnomatemática: currículo e formação de professores*. Santa Cruz do Sul: EDUNISC.

Fendler, L. (1998). What is it impossible to think? A genealogy of the educated subject. In T. Popkewitz, & M. Brennan, [Eds.], *Foucault's Challenge: Discourse, Knowledge, and Power in Education*. New York: Teachers College Press.

Foucault, M. (2003). *Vigiar e punir*. (27ª ed.). Petrópolis: Editora Vozes.

Foucault, M. (2004). *Microfísica do Poder* (20ª ed.). Rio de Janeiro: Edições Graal.

Pais, A. (2005). *Uma abordagem crítica das relações entre sociedade, ciência e matemática num contexto educacional (tese de mestrado)*. Lisboa: DEFCUL.

Popkewitz, T. (2004). The alchemy of the mathematics curriculum: Inscriptions and the fabrication of the child. *American Educational Research Journal, 41*(1), 3–34.

Stoer, S., Cortesão, L., & Correia, J. (2001). *Transnacionalização da Educação: Da crise da Educação à «Educação» da Crise*. Porto: Afrontamento.

NOTES

1. This paper was prepared within the activities of Project LEARN: Technology, Mathematics and Society (funded by Fundação Ciência e Tecnologia (FCT), contract no. PTDC/CED/65800/2006.
2. Latin word that means to put right what is twisted or not conformed.
3. An important part of the research on ethnomathematics has developed a strong critique of school. Ubiratan D´Ambrosio (2006), for instance, clearly pointed out the role of governance performed by school in our society, which he calls pastoral education. The program of ethnomathematics as he proposes implicates, at least, a radical change in school.

CHAPTER 9

MATHEMATICS EDUCATION: FOR WHOM?

Mônica Mesquita
University of Lisbon

To rethink our role as researchers of the mathematics education process could be a way to think about the relation between for what and why mathematics education exists. Some thoughts, that grew from my inner dialogues as a researcher, teacher, student, and mother that I am, were developed within practices inside multiple systems in which I was engaged, bringing some questions that became a paper from the necessity for sharing them in the Discussion Group 3 of the ICME environment.

RESEARCHERS AND EDUCATION: OUR SITUCIONALIDADE

The relation between researchers and education is much more than a simple professional connection. The praxis of researcher in the educational process goes beyond research practices—we have other roles in the world as educators: the roles of parents, of teachers, and as students in our own researcher's practices. I take as my task to focus on the role of the mathematics education process in relation to our multiple identities, questioning ourselves, mainly searching to realize the strong relations that exist between mathematical education and school:

Critical Issues in Mathematics Education, pages 61–64
Copyright © 2009 by Information Age Publishing
All rights of reproduction in any form reserved.

In my case, the appearance of contradictory feelings about my role as a teacher, and the awareness of school's role in a hegemonic society. The fact that I was not making any change as a teacher, although I had the will to do so and was developing some activities in my school, with that purpose. I simply felt that all those enterprises were suppressed by a school structure that has other purposes for education visible in mechanisms like the high-stakes tests and selection role taking care by school; the culture of individualism that students felt, and the appeal to be competitive; the central mathematical curriculum. The contradictions arose when I started to criticize, in an academic field, the same practices I developed as a teacher: I continue to propose lots of routine exercises, preparing kids for exams, reproving children, conveying the idea that mathematics is just for some, taking an authoritarian role as a teacher, controlling students, etc. I felt the need of understanding the role of school in our society. (Pais, 2008)

In my case, I am reacting to the appearance of feelings about my roles as teacher but still as researcher and mother, and the awareness of the role of school in society at large, planting and cultivating the hegemony in the actual societies. I could feel some changes as teacher, researcher, and mother in the relation between school and me but these changes are at a level to contribute different strategies to maintain the order that exists in our hegemonic society. The lack of connection between the school and the process of knowledge, as a collective work, that I could feel as teacher, researcher, and mother gave me support to ask myself about my active role as a critic of the system that I am constructing and that constructs me. Do I want it? What does education mean? Do I agree with this order of the power of corporification[1] and of the ethic of identification maintained by the school institution?

RESEARCHERS AND SCHOOL: A PATHWAY TO CORPORIFICATION AND IDENTIFICATION

School, as a modern institution, has the purpose of governing. Governance refers to the principles of classifications that differentiate and order who we are, should become, and those who are not "capable." In school we learn to be governed. Which behaviours are right, which things can and can't be said? This is how school appears in modernity, as an apparatus to govern the population by fabricating the kinds of subjects that hegemonic society stipulate as normal, by the dissemination of norms that work as calibration/standardization devices.

Corporification acts by means of power, mechanisms, techniques, norms, rules, discourses that position us in order to become knowledgeable and capable of being administered. Universal and compulsory schooling catches up the lives of all young children into a pedagogic machine that operates not only to impact knowledge but also to instruct in conduct and to supervise, evaluate, and rectify pathologies.

Intrinsic in this process of corporification is the simultaneous process of identification with the other, in the multiple systems in which human beings are engaged, reinforced by the schooling process, which promotes better things in their lives. These identifications are sustained by the ethic of schooling. One strong focus of the school is to maintain the elitist ethic, although some activities have been developed to bring the voice of the other trying to resist the hegemony. The rules exist in the schooling process to support the "normal" . . . normalizing societies.

RESEARCHERS AND MATHEMATICAL KNOWLEDGE: OUR FOCUS

Except for the name of the discipline and the words of disciplinary knowledge, school subjects have little relation to the intellectual fields that bears their names in schooling (Popkewitz, 2002). In the corporification and identification sense, mathematical knowledge is only an alibi to the learning of rules and norms, through mechanisms and techniques like the type of activities proposed in a mathematics class by the discourse that embodies mathematical knowledge, the exams, etc. The education of mathematics is only recognized in a schooling sense.

The power and the ethic of mathematical knowledge in the schooling process bring not only the necessity of students, teachers, researchers, and parents to have success but also the necessity to be happy as a part of it. The mechanisms and the rules construct and are constructed in the scholar's body by means of obligation and desire. Intrinsic in this body, the relations of power and ethic appear linked with the process of pleasure and submission, and with the political correctness where, respectively, the duty becomes pleasure and the pleasure becomes duty.

To view mathematical knowledge as a closed system, as a divine thing that we need to obey, breaks with its possibility of exchanging energy in interaction with others, with the non-schooling mathematical knowledge; breaks with the way of maintaining it as an opening tool of systems through voices that come from all the identities that we have inside us as researchers.

MATHEMATICS EDUCATION: FOR WHOM?

Totally inserted inside this obligation and desire "game" some thoughts came to me as, for example, *for whom are we playing this game?* My feeling is that I am being cynical, because I know the ideology that is inside my roles as mother, teacher, student and researcher... I criticise it, but I continue living it, planting and cultivating the hegemonic scholar system.

ACKNOWLEDGMENT

This paper was prepared within the activities of Project LEARN: Technology, Mathematics and Society (funded by Fundação Ciência e Tecnologia (FCT), contract no. PTDC/CED/65800/2006. A more extended version of the ideas discussed here is presented at http://www.mes5.learning.aau.dk/Papers/Pais_Mesquita.pdf.

REFERENCES[2]

Baldino, R. (1998). Assimilação Solidária: escola, mais-valia e consciência cínica. *Educação em Foco*, *3*(1), 39–65. Minas Gerais: Universidade Federal de Juiz de Fora Editor. http://www.gritee.com/participantes/cabraldinos/Escola%20 e%20mais%20valia.pdf

Foucault, M. (2003). *Vigiar e punir.* (27ª ed.). Petrópolis: Editora Vozes.

Foucault, M. (2004). *Microfísica do Poder* (20ª ed.). Rio de Janeiro: Edições Graal.

Pais, A. (2008). Paper given at Fifth Mathematics Education and Society Conference, Albufeira, Portugal.

Popkewitz, T. S. (2002). Whose heaven and whose redemption? The alchemy of the mathematics curriculum to save. *Proceedings of the Third International Mathematics Education and Society Conference.* (pp. 34–57). Aalborg: Centre for Research in Learning Mathematics.

Zizek, S. (2003). *Bem-vindo ao deserto do real:cinco ensaios sobre onze de Setembro e datas relacionadas.* São Paulo: Boitempo.

Zizek, S. (1998). *UMBR(a): From "Passionate Attachments" to Dis-Identification.* http://www.lacan.com/zizekpassionate.htm

NOTES

1. I developed the concept of "corporification" to categorise the action that determines the place of the physical (collective or individual), mind (knowledge), or institutional (juridical or political) bodies
2. Key references are included here that are not specifically cited in the text, but represent major influences on my thinking.

SECTION 2

GLOBALIZATION AND CULTURAL DIVERSITY

CHAPTER 10

MATHEMATICS EDUCATION IDEOLOGIES AND GLOBALIZATION

Paul Ernest
University of Exeter

In this paper I tell two tales. One is a tale of the role of ideology in the globalization of mathematics, science and technology education research, and its social and political implications. The other thinner tale is the story of my personal situatedness within the intellectual and material worlds I inhabit. To critique the globalization of educational research in these domains without acknowledging my situatedness and the boundaries of my complicity would be only half of the story, and would lack reflexivity and the necessary acknowledgement of the complexity of the issues involved.

Postmodernity has adopted Bacon's (1597/1997) insight that *knowledge is power* and exploited it in the knowledge economy. My aim is to partly subvert this order and through providing a lens that reveals some of the ideologies at work in educational research to offer knowledge workers a tool for carrying their endeavours forward and resisting or harnessing some of the forces at play in globalization.

Critical Issues in Mathematics Education, pages 67–110
Copyright © 2009 by Information Age Publishing
All rights of reproduction in any form reserved.

GLOBALIZATION AND THE KNOWLEDGE ECONOMY

One of the defining characteristics of postmodernity is the dramatic emergence of globalization across the domains of industry, commerce, technology (including information and communication technologies), culture and education. Globalization is the social change brought about by increased connectivity among societies and their elements, including the merging and convergence of cultures. The principal means is the dramatic enhancement of transport and communication technologies used to facilitate international cultural and economic exchange. This has led to the formation of a 'global village' (McLuhan 1964), with closer contact between different parts of the world, providing increasing possibilities of personal exchange, mutual understanding and friendship between 'world citizens', which is especially notable and important in the worlds of education and research.[1]

However, the principal driver is economic globalization. This consists of the opening up of international markets and the freedom to trade in them, and the multinational location of corporations controlling workers in several countries and marketing products and services in many, possibly other, countries (Hobsbawm 1994). The outcome is the corresponding erosion of national sovereignty in the economic sphere as profit-making multinational or transnational corporations circumvent the bounds of local laws and standards through moving their operations from country to country. Reinforcing this principal driver is the promotion of the culture-ideology of consumerism. In this the world media act as purveyors of the "relatively undifferentiated mass of news, information, ideas, entertainment and popular culture to a rapidly expanding public, ultimately the whole world" (Sklair 2004: 74). This is at one and the same time one of the major products for sale on international markets, and also the principal component of the culture-ideology of consumerism that creates and expands the markets. Thus globalization is often seen as global Westernization (Sen 2004)

A central dimension of globalization is the new role of knowledge, and in particular its commodification and exploitation in the global knowledge economy (Peters 2002). The knowledge economy differs from the traditional economy in that it is knowledge, rather than products or services, that is treated as the primary saleable and exploitable commodity. In addition to knowledge-products, human capital in the form of human knowledge and competencies are the key component of value in a knowledge economy and knowledge-based organizations. Through the focus on knowledge, the use of appropriate technologies is able to diminish the effect of geographical location (Skyrme 2004).

Both governments and corporations have become aware of the tremendous power and profitability of knowledge and information. Knowledge is not only *power*, as Francis Bacon (1597/1997) said, but is also *money*. Hence the capitalist principles of ownership, investment, production, marketing and profit maximization have been utilized in the domain of knowledge. The consequent application of business models and policies in this area results in the increasingly tight control and hierarchical management of knowledge. As capitalism colonizes the knowledge domain, and applies the 'science of production' to it, there has been a policy transfer into education. "At international level there has emerged a coherent set of policy themes and processes through which policy makers (at national, international and trans-national) levels are reshaping education systems" (Ozga 2005: 118). The traditional humanistic values of education policy have been "replaced by a totalizing and unreflective business-orientated ideology expressed through a discourse based on markets, targets, audits, 'quality performance' and human resource management" (Avis *et al.* 1996: 20).

The commodification of knowledge has led to performativity and managerialism in schools, universities and throughout education and its policy drivers and management. From this new perspective the success of education is measured in terms of the achievement of numerical targets. The value of teachers and academics is defined in terms of performance measures. Educational institutions and structures are managed as systems with resource inputs and performance outputs. According to this system the underlying values are those of efficiency, 'value-for money' and productivity, underpinned by the profit motive and its analogue, the maximization of outputs. Teaching and research are viewed as mechanical processes, a means to the end of producing knowledge and human capital.

From an ethical standpoint the transfer of the values of the corporate world into education is deeply problematic. Bakan (2004) has described the behavior of the corporation as psychopathic. "As a psychopathic creature, the corporation can neither recognise nor act upon moral reasons to refrain from harming others. Nothing in its legal makeup limits what it can do to others in pursuit of its selfish ends, and it is compelled to cause harm when the benefits of doing so outweigh the costs." (Newton 2004: 52). Thus to apply the values and ethics of the corporation to social and human undertakings, such as education and research, not to mention governance, medicine and other professions whose ultimate focus is human well being, is a travesty of their underlying purposes. While economics (money and its equivalents) plays a part as a *means* of improving human well being, happiness, social justice, etc., it should never become an *end* in itself as it is in the corporate world. By applying the values of corporations and institutions to human beings the incongruity and inappropriate nature of these values for the world of human-centred concerns is highlighted.

At first glance the attribution of human psychological characteristics like psychopathy to social and political entities seems far-fetched, even though in law the corporation (named after its analogue, the body) is treated in many ways like a person, with rights and responsibilities. However this analogy goes back a long way. "By art is created that great Leviathan called a commonwealth, or state, in Latin Civitas, which is but an artificial man" (Hobbes 1651/1962: 59). Hobbes goes on to develop in some detail the analogy between the powers and functions of the state and the human body. More recently critical theorists of the Frankfurt school applied Freudian and other concepts from depth psychology to modern society. Adorno *et al.* (1950) studied the origins of fascism in the psychology of the authoritarian personality. Miller (1983) locates it in the psychological damage resulting from abusive child-rearing practices. Others, such as Marcuse (1964) continued to explore the psychological roots of social organizations and ideologies, including some of the pathologies of modern society, similar to the characterization of the corporation as psychopathic. Anthropologists have also utilized the correspondence between the psychological and the social. Douglas (1966) offers a bold theoretical analogy between purity, ritual cleansing and concern with individual's body boundaries and orifices, on the one hand, and social group membership, structures and group actions, on the other. In each case these and other theorists (e.g., Vygotsky 1978, Lasch 1984) argue for the existence of a strong relationship between characteristics on the psychological plane and the social or group plane. There is controversy over in which direction, if either, the causal links flow, although my own view is that social and psychological characteristics are linked in an endless mutually constitutive cycle.

However, in the present context, the key point is the dramatic juxtaposition of, and contrast between, the values of corporations and social groups on the macro-level and the widely accepted personal values of caring and connection on the micro-level, as they apply to personal behaviour and inter-personal relationships (Noddings 1984, Gilligan 1982). Foregrounding this contrast serves to emphasize the inappropriateness of applying a business-orientated ideology with its knowledge-commodification, performativity and managerialist values to education. The potential damage is greater than simply the degradation of the employment conditions and lives of educators, researchers and knowledge workers, as bad as this might be. By degrading the contexts and conditions of learning there is also the risk of damage to the nurture and growth of children. Throughout their years of education learners are vulnerable, but especially so in the early years of schooling. At this stage much of the child's basic social and moral growth, is taking place, as well as the foundations for their overall intellectual and identity development. Threats to the health of these areas could have serious unanticpated negative consequences for society as well as for the individuals involved. But to the corporate and materialist mentality this is an irrelevant 'externality', Milton Friedman's term for the external effects of a transaction or activity on a third party or any other who is not involved in the transaction (Newton 2004). Unlike the military term 'collateral damage', horrific as it might be, the term 'externality' does not even acknowledge the negativity of such incidental impacts or by-products.

THE KNOWLEDGE ECONOMY
AND MATHEMATICS EDUCATION

In this paper my aim is to explore the relevance and impact of globalization and the knowledge economy for research in mathematics education, taking particular cognizance of the 'developing' country perspective.[2] Mathematics education is both a set of practices, encompassing mathematics teaching, teacher education, curriculum development, researching and research training, as well as a field of knowledge with its own terms, concepts, problems, theories, subspecialisms, papers, journals and books. Likewise, educational research is both a process and a product. Looking at these fields as processes and practices, since they are geographically located and embodied in organised social activities, is more immediately revealing of the role they play in the knowledge economy than starting with the objectivized products of research, although the latter will rapidly figure in my account too. My primary expertise lies in the area of mathematics education, but there are several aspects of the following account that extend more widely to incorporate science and technology education as well, and sometimes further afield, so I shall broaden my claims where this seems appropriate.

There are a number of ways in which the effects of the global knowledge economy impacts on mathematics, science and technology education. I have identified four main dimensions or areas of impact as follows, although there may well be more. First of all, there is the export of university education from Western countries to Eastern and 'developing' countries. Students are recruited internationally to engage in educational study and research both through international franchising and distance learning courses in their home countries, and through study at universities in the exporting countries. This is supported by scholarships financed by the students' countries of origin, by self-funding for those from wealthy backgrounds, or less often, by grants from 'developed' countries such as Commonwealth Scholarships. The result is a net inflow of funds to 'developed' countries (the *asymmetric economic effect*) and an inflow of knowledge and expertise to 'developing' countries. However, there are two subsidiary effects, which I shall term the ideological and recruitment effects. The *ideological effect* is the ideological orientation or saturation that accompanies the flow of knowledge and expertise to 'developing' countries. For the intentional importation of knowledge, skills, expertise and research methodologies is always accompanied by a set of implicit values, together with epistemological and ideological orientations. These may replace or co-exist with the recipient student's own orientation, but cannot be wholly rejected if the recipient is to be successful in acquiring and applying the knowledge and expertise. The *recruitment effect* concerns the recruitment of the most able personnel of 'developing' countries to work for knowledge organizations (academic or commercial) based in the 'developed' countries, including what is traditionally known as the 'brain drain'.

Second, there is the recruitment and mobility of educational researchers and academics for employment, consultancy and research projects internationally. This includes the importation of expertise from the West or 'developed' countries in the form of permanent or fixed term contracted staff, including researchers and project staff, as well as bought-in consultants for the faculty of higher education institutions, acting as external examiners, staff trainers, and so on. One outcome of this inflow of expertise and knowledge to 'developing' countries is the ideological effect noted above, because overseas trained experts must, inescapably, bring their ideological orientations with them. The movement of personnel also includes visits by academic research staff funded from 'developed' countries to conduct research in 'developing' countries. Such projects may be focused on specific features of the local culture such as the gathering of ethnomathematical field data, typical of research in the interpretative paradigm (e.g., Saxe 1991, Lave and Wenger 1991). They can also be focused on applying some predetermined framework as in international assessment and comparative studies (e.g., SIMS, TIMSS, PISA) typical of scientific paradigm research. In both of these types of research there is what I shall term the *appropriation effect*. In this, locally gathered knowledge from 'developing' countries is appropriated for academic and other uses in 'developed' countries.[3]

Third, there is the international regulation (and promotion and marketing) of the products of educational research via international bodies, conferences and associated publications, discussed further below. In mathematics education there are international coordinating bodies such as the International Commission on Mathematical Instruction (2001) organising conferences and study projects, including the ICME, PME, HPM, IOWME organizations and conferences, and independent series of conferences (e.g., MES, ALM, CERME) bringing together researchers from many countries.[4] These often make concerted attempts to include representation of researchers from 'developing' countries and hold conferences in such locations as well. However, due to the underlying economic inequalities and the high costs of international travel (as well as the institutional biases) the representation from 'developing' countries is limited. The number of international conferences held in 'developing' countries is also very limited (ICME 11 in Mexico in 2008 will be the first in this series, and only 10% of the approximately thirty PME conferences to date have been held in 'developing' countries (Mexico, Brazil and South Africa). The locations are primarily chosen to suit the convenience of mathematics education researchers in 'developed' countries, who make up the largest attendance group. Because of this location bias an outcome is another instance of the asymmetric economic effect, the net inflow of funds to 'developed' countries.

There also is an increasing number of regional international conferences such as in the Caribbean and Latin America, South East Asia (EARCOME), and Southern Africa (SAARMSTE), organised and run primarily for regional participants and benefits. However, these are internationally perceived to be of second rank in importance, value and prestige.[5] This is due to what might be termed the *dominance effect*. Research and researchers from the Northern and 'developed' countries who communicate and publish in English dominate the international research community in mathematics, science and technology education, in terms of both power and prestige. Second in order of dominance come Northern and 'developed' country researchers who communicate and publish in other European languages, e.g., French, German and Spanish. Last come the researchers in 'developing' countries communicating and publishing in local, i.e., non-European, languages.

Fourth, there is the primary source of the dominance effect, the international journals and other research publications in mathematics, science and technology education. This research literature, which incorporates the full range of academic publications including journals, texts, handbooks, monographs, and web sources is largely based in Northern and 'developed' countries, and is largely Anglophone at the high prestige end. Although journals, publishers and conference committees reach out to many countries for their editorial panels and members the locus of control remains firmly Eurocentric. This leads to the intensification of the ideological effect, as does the Eurocentricity of international research organizations and conferences mentioned above. In addition, the research literature is marketed internationally adding to the asymmetric economic effect, an inflow of funds to 'developed' countries. The high prestige end of the research literature market is in many cases matched by high prices (e.g., the expensive Kluwer/Springer books), although prices are even higher for some other scientific and medical publications. Some of the literature originating in not-for-profit organizations, such as the Journal for Research in Mathematics Education (USA) and For the Learning of Mathematics (Canada), is sold at moderate prices by Western standards. Nevertheless, it remains expensive for 'developing' countries when the costs of foreign currency conversion and international postage are included.

These four dimensions of the knowledge economy in mathematics, science and technology education combine to give an asymmetric set of flows. The primary activity is that of the sale of knowledge and expertise by 'developed' to 'developing' countries. This leads to the asymmetric economic effect, the inflow of money from 'developing' countries to 'developed' countries. This is no surprise as the knowledge economy is all about the commodification and sale of knowledge. The role of 'developed' countries as the source of knowledge and expertise also leads to the dominance effect, in which Western and 'developed' countries dominate the production and warranting of high value knowledge through control of the high prestige publications and conferences and impose Eurocentric epistemologies, methodologies and standards on it in their gatekeeper roles. It also leads to the ideological effect, whereby researchers in 'developing' countries are subject to and internalize the ideological and epistemological presuppositions and values of this dominant research culture. For to fail to do so is to be excluded from the high prestige channels for knowledge publication and dissemination. A further outcome of the ideological effect whereby researchers subscribe to Eurocentric research standards and values, is the recruitment effect. This is the 'brain drain', the migration of some of the most skilled researchers and knowledge workers from 'developing' to 'developed' countries. The acceptance of and admiration for Western academic standards makes the temptation of improved personal economic standing, as well as improved conditions of work and career opportunities, almost irresistible.

One final asymmetry arises from the appropriation effect. This is where 'developed' country researchers capture local knowledge and make representations of local practices and take this knowledge home with them. There it is reconfigured in a high prestige way such that it has significant value in the 'academic market'. The career value of such data in fields such as anthropology has long been a feature of the academic scene, and this effect has also emerged in mathematics and science education in the last two decades or so, with growing attention to ethnomathematics and ethnoscience. International assessment and comparative data in education is similarly imported, possibly after collection with local help, and is a marketable and prestigious academic product in 'developed' countries, where it features in many leading research publications. It also has extra value because it feeds into the managerialist mechanisms of education policy and control. Governments and other agencies take international comparison data as a key measure of performance and use it to evaluate and judge educational systems (analogous to the role of economic indicators). As an indicator of success in education it is a source of national pride (or shame) and it is also selectively used in the media to influence political and popular opinion.

Making Space for the Personal

In the above account I have used the voice of an impersonal structural analyst to describe the role of the knowledge economy in the managerialization of education, and its role in the globalization and internationalization of mathematics, science and technology education. Conducting an analysis from this perspective reveals important imbalances and dimensions of exploitation, but it obscures two features. The first is the agency, and at times the resistance and countervailing intentions of the actors in play. Second it is an outward gaze that hides the identity of the commenting observer.

My structural analysis of the knowledge economy in education and educational research is painted in broad brush strokes that fail to show in fine detail the varied roles, motivations and expressions of agency among the individuals directly involved, as well as those on the sidelines. Ideological orientations are by no mean monolithic in any research culture, whether in 'developing' or 'developed' countries. Educational practice and research are primarily vocational undertakings which are usually motivated as ends in themselves, dedicated to the improvement of education and the enhancement of understanding, rather than driven by ulterior motives. While most individuals combine the goals of contributing to education and research with expectations of their own personal material and esteem enhancement, it does not seem inappropriate that individuals should be rewarded and recognised for their contributions, if they are not profiting unduly or at others' expense. Likewise, individuals involved in the frontline of the knowledge economy, marketing higher education and research to 'developing' countries, or cooperating at the receiving end, are entitled to feel they are performing a useful and valuable function for all of the parties concerned, provided, in my view, that this service is non-exploitative. This raises the issue of how to characterize exploitation, and I suggest that some of the criteria that might be used to distinguish useful from exploitative aspects of the knowledge economy are as follows:

1. The service provided should be beneficial for the individuals involved, such as learners, students, and local researchers.
2. The education and research activities should be undertaken in a way that is cognizant of the local context of the recipient country and should be suited or tailored to local needs and local perceptions of needs.
3. The provision should enhance self-reliance and economic and educational development in the recipient country.
4. The underlying economic arrangements should not be an excessive drain on the resources of the recipient country.[6]

Of course the decision as to whether any particular knowledge provision activity meets these conditions depends on the perceptions and values of whoever is making the judgement. So no such justifications can be absolute or even persuasive to everyone involved.

Likewise, educational researchers visiting 'developed' countries and collecting and taking local knowledge home with them need not be seen as primarily exploitative. What I have described as the appropriation effect can be motivated by the wish to seek recognition and prestige for culturally embedded knowledge and practices for the benefit of both 'developing' and 'developed' countries. The conceptualization and recognition of ethnomathematics, for example, contributes both to a positive revaluing of culturally embedded modes of thinking and the reconceptualization of mathematics as a less Eurocentric field of knowledge. It also provides a set of multicultural resources for the teaching of mathematics to diverse groups as a global resource.

The inclusion of a more representative set of researchers in the organisation and management of international research associations, conferences and editorial boards, as well as among the contributors to conferences and publications, need not be viewed merely as tokenism. It is beneficial because it enables the voices and concerns of researchers from 'developing' countries to be heard and for them to contribute to academic decision-making within the international community of researchers. This can broaden the research agenda and also make research more accountable to the social concerns of a broader range of constituents and stakeholders. In addition, it can have an impact, albeit small in the first instance, in reducing the dominance effect described above via the reduction of the knowledge and expertise imbalance and the shifting or broadening of the underpinning ideological orientations.

Countries whose participation in educational research is recent or on a relatively small scale can gain and probably have already benefited from the traditions and activities in 'developed' countries, especially in mathematics, science and technology. For many would regard educational research as currently being in a healthy state worldwide. There are strong links between theory and practice, and space is made for constructive dialogue and critique. A wide variety of research methods and methodologies are disseminated and legitimated worldwide. A broad range of issues including social issues have been problematized as suitable topics and questions for research. Furthermore, 'developed' countries no longer control and impose their research agendas and practices on 'developing' countries as they once did in the immediate post-colonial eras. Thus researchers in mathematics, science and technology education in 'developing' countries are more epistemologically empowered, and increasingly set their own research agendas, as this special issue illustrates in a modest way. They have more space for resistance and critique, for intellectual self-determination and autonomy, and increasingly have access to the necessary concepts and tools for conducting high quality research. Some of this is home-grown, but much of the knowledge, expertise and epistemological self-confidence originates through knowledge links with 'developed' countries on a personal, institutional or national scale.[7]

The second aspect of the personal I want to treat is my own identity as a mathematics education researcher. This text does not issue from some objective logical space or an idealized rational being. It is voiced by an embodied and culturally situated human being. I, the author, am an academic who has been permanently employed in a university in Great Britain, a 'developed' country and an ex-colonial power.[8] My direct experience of 'developing' countries is limited, comprising short visits to most continents for academic work or leisure purposes, as well as a two year stint as a locally employed academic in the Caribbean region (Jamaica) in the early 1980s. Thus my intellectual outlook on the issues of globalization and education, my 'gaze', is that of a 'developed' country intellectual. This position gives me the autonomy and the intellectual and material freedom to consider, critique and theorize. But it also means that I do not have the direct experience of oppression and the anger it kindles as resources to sublimate into my intellectual work.[9] My academic identity is a result of enculturation into the practices, modes of thinking and presuppositions of my Anglocentric cultural milieu. Presumably my gaze, voice, and academic output of talks and writings are emanations and productions of my academic identity and underlying presuppositions. But, as we have learned from post-structuralism, individuals have multiple identities which are produced and elicited in different contexts (Henriques *et al.* 1984). So the question of authenticity arises: to what extent am I able to comment authentically on issues concerning 'developing' countries? There is also an ethical question: to what extent is it morally legitimate for me, given my positioning, to comment on issues concerning oppression, social justice, the problems of 'developing' countries, and those of minorities or oppressed groups defined by sex, race, class, or disability?

On the one hand, I can feel, in a miniature and fleeting way, through the pan-human faculty of empathy, the pain of excluded and oppressed peoples. I believe that all human morality is founded on the principle that you and I are the same, and that but for luck your pain and suffering could just as easily be mine. So I should help you overcome what I would not like to experience myself. But on the other hand, is my academic work on social issues part of the appropriation effect described above? Am I intellectually colonizing the problems and issues that belong to others for my own academic benefit and enhanced cultural capital?

I reject these (self-directed) charges put in such stark terms. For my view is that the problems of any oppressed group do not belong solely to that group but are problems of humankind. Therefore the ethical issue is not one of the ownership of social problems but that of their exploitation. If academics like myself use social problems and issues in academic writing for personal benefit without contributing in some direct or indirect way to the clarification or solution of the problems, or to consciousness raising about their existence, then this risks being exploitative.

This discussion raises again the interesting philosophical question of the extent to which any individual's epistemological and ideological outlook is a function of their personal psychology, their upbringing, and their experience of the surrounding socio-cultural milieu. From the perspective of social constructivism and other perspectives including constructivism, depth psychology, and anthropology, it is clear that an individual's worldview is in some sense a function or product of their experiences, both private (psychological) and public (social/cultural). Thus, one can sometimes see links between an individual's experiences and their worldviews, and sometimes also see the link between specific events and changes in their outlooks. Actually the word 'see' is a misleading metaphor here, because often it will be biographical or autobiographical narratives that *tell* us about these changes and their imputed causes. Valerie Walkerdine (Walkerdine and Lucey 1989) has written about how being a woman of working class origins in the context of her family background expectations meant that her academic advancement and success was perceived as a greedy, hubristic overreaching of herself, epitomized in the critical dictum "much wants more" frequently thrown at her. She makes the point that her persistent internal anger at this negation of her aspirations, and her related identity conflicts, help to fire her critical academic work.

However, it is also clear that the 'functionality' described is not mechanically causal. Individual humans construct their own identities in complex, unpredictable and ultimately not fully knowable ways, and their course of (self)development may include qualitative changes in both consciousness and outlook triggered internally by relatively small experiences. Womack (1983) illustrates this when he describes how an interest in pursing a single mathematics problem while a student in school led to success, teacher encouragement and the growth of a fascination with mathematics which resulted ultimately in his choice of a career in mathematics education. But others might well have responded differently, for there is no simple causal link between experiences and personal outlook and identity. Furthermore, as noted above, individuals do not end up, at any given moment, with a single essential identity but construct multiple identities which are produced and elicited in different contexts (Henriques *et al.* 1984). The central issue is that the key intervening processes between the social context, background, and experiences and the personal identity and worldview of an individual are the idiosyncratic meaning and interpretation functions of that individual (Schütz 1972). Identity formation is a recursive process, whereby personal development at any given time depends on preceding personal development. This provides an interpretative 'lens' through which all experiences are refracted and made sense of, which holistically leads to further personal development and growth, including growth and development of the interpretive lens itself.

With all these caveats, it remains possible to see (read) individuals whose outlooks, worldviews and intellectual productions reflect their origins and personal experiences, as well as others who transcend their origins. These two descriptors can even apply to the same individual. In the present era of globalized educational outcomes comprising marks and grades it is easy to forget that one of the traditional aims of education is to facilitate and enable people to transcend and overcome their origins. The German concept of Bildung refers to the process of education as spiritual, intellectual and character formation, that is the growth, development, enhancement and fulfillment of an individual's human potential through the experience of education. From this perspective the main goal of education is to enable the student to transcend the limited outlooks of their childhood and social origins.

Dewey argues that a key purpose of education is to take "the child out of his familiar physical environment, hardly more than a square mile or so in area, into the wide world–yes, and even to the bounds of the solar system. His little span of personal memory and tradition is overlaid with the long centuries of the history of all peoples." (Golby *et al.* 1975: 151). Dewey uses geography and history as metaphors for the boundedness of the child's experience and understanding, limits imposed by their origins, something that education should enable the child to transcend. Another metaphor from my own childhood and adolescence spent immersed in reading is that it allows one to see the world through a thousand pairs of eyes. Books open up the life narratives of many others to the reader, and good fictional narratives contain the lived truth of human experience, as the interpretative research paradigm acknowledges (Ernest 1994). In this respect I feel that literacy can potentially play a larger part in Bildung than numeracy, although this latter undoubtedly has an important part to play in the development of critical citizenship (Ernest 1991, 2000, Frankenstein 1983).[10]

As a knowledge worker, I am not simply a researcher and writer. A key dimension of my academic role, one that brings funds in more directly to my employer, is that of teacher. I no longer teach children or train/educate beginning teachers, which I did for more than 20 years. My primary role now is to teach and supervise mid-career teachers and other education professionals on masters and doctoral programmes, both Doctor of Education (EdD) and Doctor of Philosophy (PhD). Since 1994 most of this teaching and supervision has been distance learning based, both for British and overseas students. A large part of this involves participation in the global export of education, with my masters and doctoral students spread across the world, including North and South America, the Caribbean, the

Bahamas and Bermuda, mainland Europe, Africa, and the Middle and Far East. Costs rarely permit the students travel to me, or me to travel to them, so communication is by post, telephone and currently mostly by electronic means such as e-mail. I sit like a spider at the centre of a web disseminating knowledge of mathematics education and research methodologies and methods. Initially, this takes the form of handbooks on selected central themes in mathematics education.[11] However, the transactional style I employ is to elicit from the students a choice of research topics related to their own professional context and then help them to refine, shape and make feasible their own chosen areas of investigation. Then pairwise we engage in an extended 'conversation' as they conduct their research and write up their reports. Both masters and taught doctorate (EdD) students have to work through smaller assignments, each self-selected and conducted in this way, before extending their reach to a final larger study for the dissertation or thesis. In addition, a part of the students' developing research skills is to identify an appropriate literature base themselves as is needed for their specialist, self-chosen investigations.

In terms of content, students are exposed to an overview of themes in mathematics education selected by me (with peer approval in the original accreditation processes), as indicated above. However they explore self-selected topics relevant to their own professional contexts and interests in depth for their assignments, both empirically and in terms of the literature. Thus they apply the theory they meet on the courses and in the literature to their own professional practice. In terms of pedagogy, there is a combination of exposition via the handbooks, and a negotiated self-directed investigational style, as they pursue their own assignment topics. In terms of assessment, students work in progress is formatively assessed and then their completed research reports summatively assessed. The whole assessment load is on the project reports which combine a learning and research exercise with an assessment exercise.

In terms of the global knowledge economy, my teaching activities in mathematics education contribute to a number of the dimensions identified above. First of all, through recruiting international students I am exporting university education to both 'developed' and 'developing' countries from the UK, contributing to the asymmetric economic effect. Secondly, there is an accompanying ideological effect. For as I have argued, the communication of knowledge, skills, expertise and research methodologies is accompanied by epistemological and ideological presuppositions and values. This also reinforces the dominance effect, because my language of instruction is English, and the vast preponderance of papers, books and researchers I cite are from the Europe and North America. This communicates a Eurocentric view of mathematics education research, which is further reinforced by the assessment standards and procedures I apply. Here, of course, there are institutional checks and balances at work, so that even if I tried to apply 'rogue' anti-Eurocentric assessment standards, whatever these might be, they would be challenged and reformatted in the accepted mode.

It is possible to partially justify these practices, even if they sustain the ideological and dominance effects, by an argument that can be applied analogously to the goals of schooling. In school teaching it is necessary to address society's assessment targets to provide the basis for learner progress, both in terms of certification and applicable knowledge. Likewise, education courses (in mathematics, science and technology education) need to provide both the accepted forms of certification (higher degrees) and the high prestige knowledge and skills for expert professional functioning and career progression.

This argument provides necessary but not sufficient conditions for the selection of content for the programmes I teach. In addition to the brief justification provided above with regard to the content, I would strongly defend the importance of the skills targeted for development, including the central one of criticality. Briefly, this is the ability to engage in the careful formulation, analysis and evaluation of claims so that they are linked to an evidential basis and a carefully reasoned argument or narrative. Wielded knowledgeably, this skill enables students to make social and philosophical critiques and fosters their autonomy and independence of thought. Indeed, it is a central element of critical citizenship (Ernest 1991, 2000, Frankenstein 1983). This, together with other skills involved in research, is central to those elements of Western academic values that I would wish to defend. However, there are further covert values and epistemological assumptions making up the ideological effect, which I analyze more closely and critique in the next section.

However, before moving on to this I feel I ought to account for my own critical 'gaze' and research methodology. I have argued that as a situated 'developed' world academic I have been encultured into one of the dominant research ideologies, which in my case is critical academic liberalism. This is reflected in my choice of teaching topics and emphasis on guided autonomy and criticality as aims and values in my teaching. But this does not reveal the sources of my research methodology and methods. For this I need to give a brief account of my intellectual development.

I was trained in mathematics, logic and philosophy, especially the philosophy of mathematics. However, to make a living I became an untrained secondary school teacher and then teacher trainer in the area of mathematics. Through these practices I became encultured into the worlds of mathematics teaching and later mathematics education. Initially my research as a mathematics educator was on pedagogical issues, the primary problematic issues for teaching and teacher education. However, it was not very long before I realized that my intellectual resources as a philosopher of mathematics were applicable to mathematics education. At first I strove to find direct links between the philosophy of mathematics and mathematics teaching (Ernest 1985). But in pursuing these ideas more deeply I was led into considering the implicit philosophies of teachers (their beliefs and belief systems) and those underpinning mathematics curricula (their covert ideologies and epistemologies) (Ernest 1991). The research methods I used were to look beneath the surface of teachers' actions and utterances for deeper meanings, and likewise to analyze curriculum texts for indicators of their covert values, ideologies and epistemologies. This has led to my characteristic research methods and methodology, namely to analyze texts of different kinds for covert meanings, somewhat in the style of literary criticism. This is the method that underpins my claims to discern covert ideologies and values in varying domains of mathematics education, such as in this paper.

Of course this research methodology is risky. It depends in part on my personal intuition and discernment, and is hard to validate empirically. But my research does not rely solely on my powers of discernment. Through acquaintance with a broad range of literatures across the sciences, social sciences, humanities and arts I am able to apply a variety of concepts and methods of analysis that are novel or less frequently used in mathematics education research. To strengthen my claims I utilize this literature to cite supporting insights, theories and analyses. Beyond this I rely on the critical judgement of colleagues and referees to help me distinguish between analyses worth pursuing and those that are commonplace or unconvincing.

IDEOLOGY AND EDUCATIONAL RESEARCH

In this section I want to explore some of the main ideological components of educational research, primarily in mathematics education (but also in science and technology education), that are at play in the ideological and dominance effects discussed above. There are four main components I shall discuss. These are the commodification of knowledge and managerialism, the idea of progress and progressivism, individualism, and the myth of universal academic standards in educational research.

The Commodification and Fetishization of Knowledge

In order to analyze its role in the knowledge economy it is necessary to look more closely at the concept of knowledge. To philosophers, knowledge is justified true belief, although further subtleties and caveats are required for a definition that would satisfy them. Beliefs are held by people but philosophy mostly concerns itself with propositional knowledge, that is knowledge that can be represented sententially as a sequence of propositions. From this disembodied perspective, knowledge consists of justified claims. To social scientists and educational researchers, beliefs are more than claims. It is not just that beliefs are located in individuals or groups, as opposed to being impersonalised expressions of claims. Rather it is that to hold beliefs is to be committed, to a greater or lesser extent, to their contents. Beliefs are representations of information in the form of assertions or claims to which persons have some commitment or ownership. This brings in the basic 'stuff' of knowledge, namely information. Information consists of signs or semiotic expressions, some of which are in the form of representations or models of elements of external or experienced reality, thus having a modeling function. However, some information, although still representational, consists of signs at play in a semiotic field.[12] This includes all of the various media used for communication and entertainment.

Thus there is a hierarchy of forms of 'knowledge'. At the bottom level is plain information, which is the raw product of the knowledge economy. At the next level are beliefs which are more than information because there is commitment or ownership. At the top level is knowledge in the strict sense, comprising justified beliefs or claims. The ascent of these levels typically adds value or represents added value in the knowledge economy, for information is used to build beliefs (e.g., through advertising), and complex and expensive research procedures are used to validate claims (e.g., in medical research).

There is one further form of knowledge, embodied or tacit knowledge that is held by persons individually or situated in social practices. This includes all the various forms of professional expertise and skill that persons develop through their practices. Evidently, skilled personnel with this type of knowledge are an important part of the resources involved in the knowledge economy. Such humanly embodied knowledge, or to put it better— knowledge production or knowing capacities—are very important in the social use and functions of knowledge, even if backgrounded and neglected in traditional epistemology.

In any discussion of knowledge it must be acknowledged that there are opposed and competing epistemologies at play, which I shall characterize as modernist and postmodernist. Modernist philosophical perspectives see knowledge as objective, abstract, depersonalized, value-neutral and unproblematically transferable between persons and groups. Such perspectives are central to much of modern epistemology and since the time of Descartes, and can be traced back to Plato. This epistemology serves the knowledge economy well in one central respect, for it supports the view that knowledge can be disembedded from its contexts of production and is readily transferable and marketable as a commodity. It also supports the notion that the research methodologies and standards employed in mathematics, science and technology education are universal and applicable across the world, irrespective of their origins. This helps to support the domination effect discussed above.

The objectification and commodification of knowledge that follows from the modernist epistemology and is necessitated by the knowledge economy that exploits it, reflects a set of values. These valorize the measurable outputs of knowing over those that are less easily measured. They prioritize knowledge products over knowing processes, the cognitive domain over the affective domain, and value knowledge and the intellect over feeling and being. From the point of view of the knowledge economy, this is perceived to be necessary. But from the perspective of education, this is much more problematic. Both perspectives see knowledge as a means to an end, but these ends are very different. The knowledge economy aims at profit, whereas education aims at social and personal growth and development, and objectivized indicators of this should not overpower, dominate or replace what they are meant to stand for.[13]

Where do these values come from? During modernity, the scientific worldview has come to dominate the shared conceptions widely held in society and by the individual. This worldview prioritizes what are perceived as objective, tangible, real and factual over the subjective, imaginary or experienced reality, and over values, beliefs and feelings. This perspective rests on a Newtonian-realist worldview, etched deep into the public consciousness as an underpinning 'root metaphor' (Pepper 1948), even though the modern science of relativity and quantum theory shows it to be scientifically untenable. In postmodernity this viewpoint has developed further, and a new 'root metaphor' has come to dominate, namely that of the accountant's balance-sheet. From this perspective the ultimate reality is the world of finance, i.e., money, and other related measurable quantities. In particular, knowledge is commodified and fetishised as a quantifiable and marketable entity. It is no longer seen as indissolubly tied in with human knowing. Thus in education and research it is the hard measurable outputs that are valued, not the softer processes and human dimensions of knowing.

Elements of this critique of the values of modernity are well anticipated in the work of Marcuse (1964), Young (1979), Skovsmose (1994) and Restivo et al. (1993). The way the balance-sheet model and its associated mechanisms work is as follows. Much of the working of modern society is regulated by deeply embedded complex mathematized systems, including taxation, welfare benefits, industrial, agricultural and educational financing, the stock markets, banking, etc. Such systems are automated and carry out complex tasks of information capture, policy implementation and resource allocation. Niss (1983) named this the 'formatting power' of mathematics and Skovsmose (1994) terms such socially embedded systems 'realised abstractions'. The point is that complex mathematics is used to regulate many aspects of our lives, with very little human scrutiny and intervention, once the systems are in place.

Although mathematics provides the language for these systems, the overt role of academic mathematics in this state of affairs, i.e., that which we recognise as mathematics *per se*, is minimal. It is management science, information technology applications, accountancy, actuarial studies, economics, and so forth, which are the sources for and inform this massive mathematization on the social scale. Underpinning this, at both the societal and individual levels, is the balance-sheet metaphor, for economic or market value is the common unit in which virtually all of the activities and products of contemporary life are measured and regulated.

There are two overall effects. First, most of contemporary industrialized society is regulated and subject to surveillance by embedded and part-hidden complex mathematical-based systems ('black boxes'). These are automated through the penetration of computers and information technology into all levels of industry, commerce, bureaucracy, institutional regulation, and more generally, society. The computer penetration of society is only possible because the politicians', bureaucrats' and business managers' systems of exchange, government, control and surveillance were already quantified and in place before the information technology revolution. Knowledge already had a central use in controlling society and the population (Foucault 1976).

Second, individuals' conceptualisations of their lives and the world about them employs a highly quantified framework. The requirement for efficient workers and employees to regulate material production profitably has necessitated the structuring and control of space and time, and for workers' self-identities to be constructed and constituted through this structured space-time-economics frame (Foucault 1970, 1976). We understand our lives through the conceptual meshes of the clock, calendar, work timetables, travel plans, finance and currencies, insurance, pensions, tax, measurements, graphical and geometric representations, etc. This conceptual framework positions individuals as regulated subjects and workers in an information controlling society/state, as consumers in post-modern consumerist society, and as beings in a quantified universe. Knowledge frameworks are now inscribed into our very beings, rewriting our subjectivities and our existential selves. What we have seen in the past 25 years is the growth and tightening of these nets to encompass professionals, first teachers and civil servants, then doctors, lawyers and academics.

The commodification and fetishization of knowledge, and performativity and managerialism throughout education and employment in general, are just some of the products of these ideological shifts. They also have profound implications for the nature and processes of education. For the commodified view of knowledge entails that it can be disembedded from its cultural context and conveyed without loss of meaning. But to assume that signs and knowledge tokens carry their own full meanings with them has damaging consequences for education, where transmission models of teaching presume that knowledge can be 'handed over' or 'delivered' to learners (Seeger and Steinbring 1992). The power of constructivist learning theories in education is that they acknowledge that individuals must recreate meanings afresh, based on their idiosyncratic pre-existing meaning structures and experiences (Steffe and Gale 1995).

Postmodernist epistemologies extend these insights and embrace the multiplicity of coexisting meanings, perspectives and systems. Knowledge is viewed as socially and culturally embedded, value-laden, and not transferable across contexts without significant transformations and shifts in meaning. Explicitly represented knowledge is understood semiotically in terms of signs and texts that draw upon the meanings historically formed and owned by groups and individuals. As signs travel there will often be shared or preserved elements of meaning, depending on the proximity and interactions of the donor and recipient cultures, but there will also always be differences and nuances of interpretation. To create readings or interpretations of signs requires persons to invoke and marshal their own meanings, understandings and interpreted experiences.

Just as postmodernist epistemologies see the meaning of information as located in cultures, so too is the commitment to and ownership of beliefs; and the validation of knowledge claims depending on communities of knowers and researchers. Likewise, the value of experts and knowledge workers will depend on their role and positions within knowledge-orientated institutions and projects. Gibbons *et al.* (1994) describe the production of knowledge in postmodernity as falling into two modes. Mode 1 knowledge production comes from a disciplinary community and its outcomes are those intellectual products produced and consumed inside traditional research-oriented universities. The legitimacy of such knowledge is determined by the university, the academics working within the knowledge area, and the academic journals that disseminate the knowledge. Typically academic research in mathematics, science and technology education would fall in this category.

In contrast, Mode 2 knowledge is the identification and solution of practical problems in the day-to-day life of its practitioners and organizations, rather than centering on the academic interests of a discipline or community. Mode 2 knowledge is characterized by a set of attributes concerned with problem-solving around a particular application and context.

1. The different knowledge and skills of the practitioners are drawn together solely for the purpose of solving a socially (including indus-

trial, commercial and technological) motivated problem, and hence are integrated and transdisciplinary rather than restricted to single academic discipline or area of study.

2. The trajectory follows the problem-solving activity, and the context, conditions and even the research team may change over time according to the course of the project.

3. Knowledge production is carried out in an extensive range of formal and informal organizations including but extending well beyond universities.

4. The focus on socially motivated problems means there is social accountability and reflexivity built in from the outset of the project.

The key point made by Gibbons *et al.* (1994) is that the 'know how' generated by Mode 2 practices is neither superior to nor inferior to Mode 1 university-based knowledge, it is simply different. As well as different projects, there are different sets of intellectual and social practices required by Mode 2 participation compared with those likely to emerge in Mode 1 knowledge production.

Mode 2 knowledge production may be newly recognised in postmodernity, although it has antecedents stretching as least as far back as Aristotle's (1953) recognition of Techne (applied or technological knowledge). Its recognition is a central part of postmodern epistemology (Lyotard 1984, Rorty 1979). However, it is also a central part of the new knowledge economy where knowledge and experts are classified in terms of their functionality and marketability, rather than their foundational basis and disciplinary categorization. In this sense the global knowledge economy is postmodern. For in the knowledge economy one feature stands out: the marketability of knowledge is the prime factor, and this marketability does not discriminate between different forms of knowledge except in terms of value, price and ease of sale. From this perspective there is no discrimination between the qualitative results of interpretative paradigm research and the hard quantitative results of scientific paradigm research except insofar as it affects price and marketability. For example, since the 1990s politicians have utilized the qualitative results of focus group research to evaluate the impact of policies, just as they have used the quantitative results of survey research for a much longer period. The market is neutral to qualities such as these except according to how it impacts on economics. But having said this, as is indicated above, marketability has a profound impact on what types of knowledge are valued.

Thus ironically, postmodern epistemology does not provide a conceptualization of knowledge that leads it way from the knowledge economy. Instead it delivers it directly to the marketplace. However, there are ways of using the reconceptualization of knowledge to offer a more democratic and socially responsive approach to research in education. For example, the mode 2 knowledge production category provides a useful way of describing 'bottom up' projects initiated by teachers, activists and or concerned citizens. Examples can include environmental or anti-globalization activities (e.g., Klein 2000), peoples' education movements (e.g., School of Barbiana 1970, Freire 1972) and the development of ethnomathematical and multicultural materials for teachers (e.g., Mathematical Association 1988, Wiltshire Education Authority 1987). Such projects can follow from what Habermas (1971) characterizes as the critical interest, namely, the desire to change society for the better.[14] This provides the basis for the critical-theoretic paradigm in educational research, the central feature of which is the desire not just to understand or to find out, but to engage in social critique and to improve or reform aspects of social life. Thus it involves critical action research on social institutions, and interventions aimed at social reform and increasing social justice. In particular it aims at the emancipation, empowerment and the development of critical consciousness among the participants and the others targeted. What distinguishes this type of research is that social and institutional change is primary and knowledge is secondary. It is thus mode 2 type knowledge production because it does not seek primarily to satisfy academic standards, but to solve real problems.

The critical-theoretic paradigm is often associated with action research among the 'teacher-as-researcher' movement, with teachers working to change their teaching or school situations to improve classroom learning (Carr and Kemmis 1986). However, in my view such action research balks too often at addressing social justice and oppression in society to fit comfortably under the critical-theoretic paradigm. In contrast, activism by environmentalist, anti-globalization, human rights, and gay rights organizations may be a better fit. One of the most successful strategies of such groups is to construct mode 2 knowledge and deliver it to the media with sufficient impact to sway public and political opinion, and thus to help initiate social change.[15] In education, mode 2 knowledge production is typically involved in development projects that prioritize social change and empowerment over publishable research. What characterizes such projects is the deep commitment of the researchers and activists involved, who measure their successes in terms of social changes and not in terms of knowledge capture or production. Such approaches either reject the commodification and fetishization of knowledge by the knowledge economy or subvert it in order to pursue their own ends.

Progressivism and the Idea of Progress

Modernism originated with the philosophers of the 17th and 18th century Enlightenment who believed in the ultimate power of reason to reveal truth, and through rationality to advance society towards the 'good life'. This movement subscribed to the values that prioritize the cognitive and intellectual over feeling and being discussed above. But it also led to a further ideology, namely progressivism with its fetishization of the idea of progress. Rational thought, it was believed, would solve all the problems of life and lead to enlightened living and happiness for all. Reason and knowledge especially in the areas of science and technology would provide the means for this continued and continuous improvement.[16] In the 19th and 20th centuries many of the key theories that underpinned modern thought were based on the idea and assumption of continual progress. However another notion became entangled with this, namely that progress was automatic, and stemmed from the natural order of things as well as from the application of conscious thought. Hegel and Marx's theories of history both assumed the inevitability of social progress, without the application of reason. Darwin's theory of evolution is underpinned by the idea of biological progress, i.e., the survival of the more 'fit', through the elimination of the least 'fit'. The Eugenics movement of the late the 19th and early 20th century believed in the perfectibility of the human race through selective breeding.[17] Educational and psychological theorists including Herbart, Spencer and Piaget assumed that ontogenesis and phylogenesis unfold in a sequence of stages, and that later phases are in some measure more complete than or superior to earlier phases of this development.

The problem with the idea of progress is that what is perceived in some rational, scientific or technological way to be more developed is also seen as superior to, or more complete than, the less developed. Thus rural or tribal societies are deemed to be inferior to industrialized, technological or urban societies. Such perceptions provided justifications for the conquest and colonialisation of many countries, for they were being enlightened; being given knowledge to deliver them from their ignorance, their 'undeveloped state'. These perceptions continue to provide a justification for the exploitation of 'developing' countries. These ideas also underpin racism, which saw and continues to see the yellow or brown skinned peoples of countries in the East or South as less 'developed' and hence as inferior.[18]

Such conceptions persist in many overt and covert ways. The very terms I have used, 'developing' countries versus 'developed' countries have the unintended connotation that 'developing' countries are inferior and 'developed' countries superior.[19] After all, 'developed' includes in its meanings: to be further down the road of progress, and nearer to fulfillment and perfection. Several of the effects I have identified above, such as the domination and ideological effects serve to reinforce this fetishization of progress and the implicit devaluation of 'developing' countries in the East and south. Some philosophers and social theorists have gone so far as to announce 'the end of history' and 'the end of ideology', a teleological notion that human and social development were somehow destined to reach the 'final' state that we are reaching after the fall of communism, dominated by global capitalism and Western ideology (Fukuyama 1992).

But this 'final state' with its fetishization of progress is far from perfect (nor is it inevitable). The progress of rationality applied in society, technology and industry has led to a disregard for and despoliation of the environment. The underlying Western model of the good life is that of consumerism, where progress is measured in terms of the consumption and acquisition of more material goods. Furthermore, consumerism depends on "the myth of consumer inadequacy" according to which relief from this state of inadequacy can only be obtained by purchases and further consumption, and even this relief is only temporary (Collis 1999).

Even in education we see constant innovation, with bigger and better electronic resources (calculators, computers, data projectors, electronic whiteboards, etc.) and the associated software, viewed as necessary but year-on-year absorbing disproportionate financial resources.[20] These tools have undoubted benefits, but how often do we see a reasoned analysis of their costs and benefits set against the possibility of more and better paid teachers, books, broader forms of inservice training, and other education resources?

Overall, one of the by products of progressivism and the fetishization of the idea of progress is to over-valorize the products, position and power of the industrialized Western 'developed' countries. This ideology views the 'developing' countries of the South and East as primitive and inferior. This leads to a continuation and justification of the historically unequal relations between the countries on the economic, cultural and intellectual domains. It leads to hubristic over-valuation of the educational and research traditions of the anglophone West and supports the ideological and dominance effects identified above.

Individualism

One of developments of postmodernity has been a 'social turn' in the way many leading edge researchers conceptualize learning in mathematics, science and technology education (Lerman 2000, Atweh *et al.* 2001). However for all of this utilization of social concepts in educational theories, the dominant ideology of research in the 'developed' countries of the North remains firmly individualistic. Educational researchers typically choose their own personal research specialism and style, and work individually, on projects of personal preference. They may form temporary alliances or research groupings, but these usually have their own stratified hierarchy from principal researcher to research assistant, with their own incipient individualism.

In addition, educational researchers' careers are generally concerned with competing and being rewarded individually on high status written output, not on any social impacts or other measures that might be valued more highly outside of the academic world.

Most educational researchers are positioned somewhere along a scale that encompasses teachers researching for a higher degree, teacher educators conducting research, and university researchers teaching and supervising others' research. In each case the person's main function is as a teacher or supervisor, so that pursuing their own individual research interests is a subsidiary rather than their main breadwinning activity, even if they are evaluated on their research performance. Alongside this majority activity there also exists a parallel career course for contract researchers who work on funded projects full time. However, this second string of research workers is a much smaller group, and most of the researchers involved work on projects initiated by government or other agencies, or in a minority of cases, projects initiated by the principal researchers themselves, but constrained by the requirements of the funding agencies.

Thus in education research agendas are primarily set by researchers themselves, albeit with some accountability to their employers. But this is usually for the quantity and quality of their research output, and not for its content. Thus the dominant aim of researchers in mathematics, science and technology education in countries of the North is to satisfy themselves and their peers. This is a manifestation of individualism, in which individuals pursue their own agendas semi-autonomously, competing for rewards and esteem.

I should make it clear that I am both a willing participant in, and a beneficiary of, this system. I am able to pursue whatever writing and research projects take my fancy, and provided that my output is held in esteem by local and international peers, I am rewarded by my employing university for my efforts. Furthermore, I have and would strongly resist any attempts to control or direct my research efforts, because of my strong desire to pursue my own interests, preferences and research agendas. Thus in my own professional life I am a product and supporter of an individualistic work pattern. I like to think that the research directions I pursue, mostly philosophical and theoretical, are the best deployment of my skills and talents. Nevertheless, I feel no compulsion to address what some might see to be the most compelling problems or useful issues that concern the teaching and learning of mathematics in my region or country. I make no claim to be exempt from the ideology of individualism, indeed it is deeply inscribed in my personal and professional being.

Despite the so-called 'social turn' in some researchers' conceptualisation of learning, constructivism remains the leading theory of learning among education professionals. This sees students as knowing subjects who are individual sense-makers; who understand the classroom and contexts independently; who have isolated and independent inner lives of thought, which emerge into the outer public arena as speech or actions. This view of the knowing subject is an expression of the both of the ideologies of individualism and educational progressivism (Ernest 1991). But there are major weaknesses in this position.

First of all, there is the sentimentalisation of the knower. Knowing is not always the sweet sense making of untrammeled reason and intelligence that modernism and progressivism would like to portray. Knowing is also about fighting for psychic survival in the face of emotional and other forms of threat, it is about interpreting hidden coercive meanings underlying the literal spoken word of a teacher or another. Knowing is about deviant scheming to compensate for personally sensed deprivations or inadequacies. Knowing is about the desire for forbidden objects of gratification approached covertly under a patina of acceptable behaviour and discourse. Knowing is in the mind of the eagle as it sights and takes its prey. Knowing and the knowing subject should not be sentimentalized.

A second area of weaknesses concerns the politics of education. A focus on the activities and immediate surrounding context of the knowing subject ignores structural inequalities in terms of class, gender, race, etc. It ignores the dominant ideologies of schooling. Ultimately it also ignores the fact that the main obstacle to individual sense-making and self-realization is the active opposition to the political empowerment of the learner as a democratic citizen by most forms of institutionalized education.

Thirdly, there are theoretical weaknesses with an individualistic form of constructivism, as well as with other philosophies that prioritize the individual over the social. The Western ideology of individualism that characterizes modernity backgrounds the essentially social and communitarian nature of humanity. But knowledge and indeed all forms of everyday life and functioning could not exist without that quintessentially social construction of humanity, language. While it may be possible to give a plausible if superficial account of the knowing subject in individualistic terms, this is not possible to do so to any depth for the feeling subject. For our feelings are inescapably tied in with our inter-personal relations starting with our parental/caregiver experiences. Our biological beginnings are within the body of another, and our development into persons can only be fulfilled through mutual loving relations with others. We are not isolated and self-contained individuals but social beings that are part of communities and indeed are formed through our very sociality.

Both the emphasis on individual choice for researchers and the individual as the essential unit for the conceptualization of education, are expressions of the ideology of individualism. This has long dominated Western thought, and is emerging most sharply in the 'free market' or 'marketplace' metaphors of recent social policy. The dominant ideology of individualism is based on a model of the individual as an isolated rational being with an independent perception of reality, and which acts on the basis of its own rational analysis and thought. This model emerged in the 17th and 18th century enlightenment and is epitomized most starkly in Leibniz's Monadology, a philosophy of independent and unrelated coexisting entities. In addition, the values of liberty and freedom have been much vaunted from the days preceding the French Revolution to the current temporary world domination by one superpower, the United States of America. These forground and valorize not only the meaning of freedom as liberation from oppression, but also the meaning of freedom as being free from social constraints and responsibilities. Admittedly this latter freedom of individual action has limits in the relevant legal framework. However, freedom in this sense, expressed in the actions of corporations on the global scale, is able to circumvent and manipulate many of the laws and limits to actions experienced by individuals nationally.

Individualism combined with the workings of the 'free market' provides the basis for the ideology of consumerism. This sees human beings as agents in the material world working, accumulating and expending wealth, and owning and consuming material products and 'experiences'. Although a small minority of concerned and principled citizens in western industrialized countries try not to participate in this globalizing universe of consumption, most, like myself, have consumerism inscribed deeply within our subjectivity.[21] We enjoy the benefits of modern industrialized society including almost unlimited access to power, heat, water, housing, transportation, material possessions, electronic appliances, media products, shopping, food, and other luxury goods and experiences. In the postmodern era, Descartes' *cogito*, 'I think therefore I am', the dictum which helped initiate the modernist era of individualism, has been replaced by 'I shop therefore I am', as the artist Barbara Kruger vividly expresses it (Baudrillard 1988). "One's body, clothes, speech, leisure pastimes, eating and drinking preferences, home, car, choice of holidays, etc. are to be regarded as indicators of individuality of taste and sense of style of the owner/consumer." (Featherstone 1991: 83).

However, the ethos of consumerism and many of the products involved embody values that conflict with those of a range of cultures and creeds located in countries around the globe. For example, the increasing sexualization of popular music, fashions in clothing, visual media and indeed childhood, is unacceptable to many cultures valuing modesty, including some sub-cultures in the Western world. The social construction of persons as consumers through such influences is also problematized by non-traditional critical perspectives as well, which see their impact on subjectivity and personal identity as negative. Furthermore, consumerism is not sustainable, for not all world citizens can consume resources as wastefully as Western consumerism requires, which is already a major cause of the current environmental and ecological problems.

McBride (1994) has offered a powerful critique of how individualism is the hidden ideology of much of modern school mathematics. She argues how the emphasis on individual choice, and on mathematics as underpinning rational choice, permeates school mathematics texts, and represents a perspective whose strategy is to deny or conceal the historical, social, cultural, political nexus in which all knowledge making and practices take place. Thus even school mathematics is complicit in promoting and sustaining the ideologies of individualism and consumerism.

From the perspective of research, there are alternate ways of conceptualizing the issues. In a country with severe social problems, or a serious lack of resources, or both, in short, a 'developing' country, it can be argued that solving social problems is far more urgent than satisfying individualistic researchers. One answer to the problem of how to evolve a new style of researching in mathematics, science and technology education in a 'developing' country, is to work in coordinated research teams on shared problems of social significance. Rather than let an individualistic ideology drive research and education, resources can be directed at teams working on socially relevant research projects. This should take research projects closer to the critical research paradigm. For the critical research has the virtue of being concerned to improve some aspects of the social context, situation or institutions. Although most educational research is concerned to improve schooling indirectly, critical paradigm research has the advantage of specifying this goal up-front, and not being concerned to try to leave the situation being investigated undisturbed.

The disadvantage is that there are often hidden institutional sources of resistance to change, such as teacher and pupil ideologies, institutional structures, and so on, which may prevent the desired progress. If there is no progress, and there is little of the validated knowledge that research in the other two dominant educational paradigms seek to find, then there may be no worthwhile outcome for the energy and time invested. However, if social change and improvement is the overwhelming goal, then this is a gamble worth taking.

Despite the widespread ideology of individualism, the 'social turn' in research in mathematics, science and technology education (Lerman 2000, Atweh *et al.* 2001) has done more than focus on social theories of the teaching and learning. It has also foregrounded ethical, cultural and political dimensions of educational research in a way unheard of two decades ago. In mathematics education, conference series on the Political Dimensions of Mathematics Education and Mathematics, Education and Society, and on Ethnomathematics are regular features of the international research scene. These have legitimated the prioritization of social justice issues and research and education for social change, and provided vehicles for the politicization of a generation of younger researchers. Thus revealing ideological dimensions of research in mathematics, science and technology education does not serve to validate individualism but to problematize it and other hidden assumptions and values.

The Myth of Universal Academic Standards in Educational Research

In postmodernity, one of the main currents of thought is "incredulity towards metanarratives," a critique of universalist explanatory frameworks or epistemologies (Lyotard 1984). A suitable case for treatment in this respect is the myth of universal academic standards in educational research, especially in mathematics education. In this area the received view is that there are universal, reliable and consistent academic research standards applied at the highest levels, and that all of the leading refereed international journals and international conferences in education subscribe to and apply the same standards of rigour in the evaluation of submitted papers. I shall challenge this and label it a myth. It is a myth which helps foster unity and integration in the international research community, but it also sustains the domination and ideological effects discussed above.

There are, of course, dissenting voices. Most notable among these are researchers who label themselves as postmodernist, such as Patti Lather: Postmodernity "is a time of the confrontation of the lust for absolutes, [to] produce an awareness of the complexity, historical contingency, and fragility of the practices we invent to discover the truth about ourselves." (Lather 1992: 88). "Positivism is not dead. What is dead, however, is its theoretic dominance and its 'one best way' claims over empirical work in the human sciences." (Lather 1992: 90). Gergen (1999) and Denzin (1997) have referred to the 'legitimation crisis' following the rejection the absolute authority of science whose standards and epistemological modus operandi other disciplines have to aspire to and emulate. Donmoyer (1996: 19) writes about the problems of taking over the editorship of an international educational research journal and observes: "There is little consensus in the field about what research is and what scholarly discourse should look like."

It is worth noting that a similar received universalist view also prevails among researchers in mathematics, although a growing minority of mathematicians, philosophers and social researchers reject this as a myth. Those supporting the view that there are universal academic standards in mathematics often argue that mathematical papers present new mathematical truths and their warrants, usually in the form of proofs. Since mathematical truths are absolute and universal, they claim, so too are the standards for truth. Those arguing against this universalist view point to the historically shifting and incomplete standards of proof and truth in mathematics. Some also argue that mathematical truths, like the concepts of mathematics, are social constructions (Ernest 1998, Hersh 1997, Tymoczko 1986). Without reiterating the complex arguments for and against these positions, the point I wish to make is that even in mathematics, where claims to truth and objective standards for the validity of knowledge are perhaps the strongest in all fields of enquiry, there is fierce controversy over their universality. Hence the claims for the universality of standards in mathematics education must be that much weaker.

In order to question their claimed universality it is helpful to ask what academic standards in educational research might be, and which aspects of research reports they foreground and evaluate. To this end I shall distinguish three features of academic papers relevant to acceptance: their *content, knowledge production strategies*, and *textual features*. In addition, the *social and organizational features* of the refereeing process are also likely to impact on the judgements made. Further features could be distinguished but these enable some of the most salient points to be discussed.

A wide range of criteria are used for judging research papers. For example, a typical detailed scheme includes the following ten evaluation criteria: 1. Significance of themes, 2. Relevance of themes, 3. Clarity of thematic focus, 4. Relationship to literature, 5. Research design and data, 6. Data analysis and use of data, 7. Use of theory, 8. Critical qualities, 9. Clarity of conclusions, 10. Quality of communication (Learning Conference 2005). Using specific criteria shapes the evaluation of papers but it cannot eliminate the subjective, situated element in the judgements. Such a practice merely provides a structure, distributing the application of subjectivity over a range of pre-specified categories. All evaluative judgements applied to creative knowledge productions depend on the experience and expertise of the referee, acquired in the social practices of evaluation, but exercised through the application of individual agency. Even then "your colleagues are there too, looking over your shoulder, as it were, representing for you your sense of accountability to the professional standards of your community" (Wenger 1998: 57). Final judgements as to the acceptability of a paper are made by editors or committees, who coordinate several such reports and apply further procedures in cases of disagreement. Thus ultimately the rulings are intersubjective and socially constructed, building on the subjective judgements of experts but also transcending them through social agreement.

1. The *content* of a paper includes the particular theme or topic area treated and the subdomains of enquiry explored or interrogated by the research questions and objectives. It also includes the theories used, the link

to base disciplines and the interdisciplinarity of the inquiry. All academic disciplines and fields of study, including research in mathematics education, can be expected to shift their boundaries and contents to some extent over time. This will naturally impact on judgements of appropriateness and relevance of submissions, and thus will affect the standards applied as well as which research community is called on to make such judgements.

For example, the International PME Group organizes annually what is widely recognised as the leading international research conference in mathematics education.[22] Each year an international committee including local organisers is constituted to select papers for presentation at the international venue chosen for that year. In 1989 the Paris conference committee rejected all papers on teacher beliefs as not being relevant to the field of psychology of mathematics education. This led to disquiet among the delegate membership because this topic area was an emerging subfield which now is a major strand of research in the domain. From 1990 (and before 1989) such papers have been accepted, provided they passed the same 'quality threshold' procedures as other papers. This demonstrates graphically that judgements of relevance to the subfield are variable.

A rejoinder might be that quality judgements need not be affected by the relevance issue. However, in my view such judgements cannot be separated completely from issues of quality, since acceptability can only be based on the knowledge and experience of the gatekeepers. As Kuhn (1970) has argued for the physical sciences, knowledge of a field depends on familiarity with a range of paradigmatic examples, such as papers, arguments, applications of methods and problem solutions. So referees making judgements about new examples beyond their range of familiarity might well be influenced by their willingness to accept novelty. They are required to make a creative act of evaluation which may vary from person to person. Referees deeply entrenched in different external base disciplines (e.g., psychology, mathematics, philosophy, sociology) might make different judgement calls from each other, based on their experiences of these fields, as well as from referees whose discipline is mathematics education. As Wenger (1998: 254) points out: "At boundaries things can fall through the cracks—overlooked or devalued because they are not part of any established regime of accountability."

Lerman (2003) analysed changes within the field of mathematics education by looking at the papers accepted for PME conferences, and papers published in two of the leading research journals in mathematics education JRME (*Journal for Research in Mathematics Education*) and ESM (*Educational Studies in Mathematics*) over the past 15 years. Lerman *et al.* (2002) found a shift in underlying theories and methodological bases over the period in ESM research papers, with many more drawing on social theories (social constructivism, Vygotsky, social interactionism, etc). The percentage of papers employing social theories rose from 9% in the early years of the period to 34% in the later years. Lerman (2003) also notes that since 1990 there has been a marked drop from 24% to 6% in the proportion of purely 'empirical' papers for PME that do not explicitly use theoretical frameworks. Those that draw on the traditional areas of mathematics and psychology have also diminished (by about 12% for papers in both PME and ESM). Thus there are shifts in the subject area, as well as in the knowledge demands on referees in making judgements.

2. The *knowledge production strategies* in a paper include the research methodology and methods employed. In broad terms, three overall 'families' of knowledge production strategies might be distinguished, although finer distinctions would also be revealing. Papers from within the interpretative research tradition are typically exploratory, constructing an interpretation of the inferred meanings of the research subjects, as represented in qualitative data. Papers from a scientific research perspective typically are hypothetico-deductive, testing generalizations against data that is quantitative or classified according to a preconceived framework, although they may do this more or less formally. There are also theoretical, conceptual, philosophical or critical papers which clarify concepts or reflect on theories, policies and practices without presenting new empirical data. These represent three subtraditions in social science research (Habermas 1971) and there are different criteria for judging the quality of papers in each of them. Furthermore, different referees will have different levels of experience and expertise in these different families and are likely to differ in their judgement calls too. There have also been shifts in journal recognition and acceptance of these three types too, and it is noted that over the past 15 years JRME "moves from an initial emphasis on quantitative to qualitative . . . achieving a more balanced use of methods" (Lerman 2003: 6).

It might be claimed that the same criteria are employed in making judgements about research papers following all three paradigms. Gage (1989) claims that the struggle for supremacy between their proponents, the 'Paradigm Wars', are over. However, many authors argue that what is presented as consensus is in fact the domination of the field by proponents of the modernist myth of the universality of research criteria (Denzin 1997, Gergen 1999, Lather 1991, 1992). Donmoyer (1996) questions whether a consensus over research evaluation criteria exists. Anderson and Herr (1999: 15) assert that the 'New Paradigm Wars' are still being waged and complain that "we can't use current validity criteria to evaluate practitioner research." So differences in knowledge production methodologies and research paradigms seem to provide a strong argument for rejecting the myth of the universality of research evaluation criteria.

3. *Textual features* of the papers encompass the rhetorical dimension including forms of argument and persuasiveness of the discourse and the

rigour of the reasoning, as well as the organisation and presentation of the text. The structure and organisation of texts is the area that is most susceptible to explicit prescription, and there are very detailed standard guides available (American Psychological Association 2001, University of Chicago Press Staff 2003). However, judgements concerning the rhetorical dimension depends on how persuasive the reasoning is perceived to be by referees, and the same arguments about the background and expertise of referees rehearsed above apply here.

4. The *social and organizational features* of the refereeing process play a constitutive role in the social construction of acceptance decisions, as is also discussed above. For example, from the early 1990s PME conferences required that full papers be submitted in January for the following Summer's conference. This was a shift from previous practice in which only a one page abstract was required. Clearly different standards are at work in scrutinizing a finished paper from accepting a one page summary. The rigour of the process is increased, since it is not enough for authors to promise some contents in an abstract. Instead the final paper must be submitted and be judged academically worthy in terms of all of the criteria discussed here.

Overall, I am arguing for the relative and changing nature of research quality criteria in the field of mathematics education. I believe the same holds true within all disciplines and fields of study, as I suggested in the case of mathematics, but I will not pursue this generalization here. My claim is that every learned journal and international conference has different (i.e., non-identical) enacted standards for the acceptance of submissions, which share only a 'family resemblance' (Wittgenstein 1953). Although there are shared features, to a greater or lesser extent, standards differ and are a function of the different communities of scholars serving as gatekeepers with their own situated practices and expertise. In the preceding discussion I have illustrated how the experiences, expertise and disciplinary background of individual referees impacts upon and helps to shape their judgements, creating divergences and differences as well as shared features and resemblances. In particular, their membership of a range of different communities of practice, including experiencing the roles of student, teacher, researcher, author, referee, editorial board member, etc., will shape their evaluation practices. In these social situations working consensuses over acceptability are achieved, although sometimes conflicts over standards occur, and these will normally lead to new resolutions and occasionally even to shifts in standards.

The myth of the universality of research standards in mathematics education may serve a useful function, in helping to sustain the notion that there is a unique field of study called mathematics education. As such it helps to create the illusion of the existence of a unique international mathematics education research community, when what actually exists is a complex set of interlocking and interacting but different and distinct practices and communities located and dispersed globally. The myth also helps to encourage the mathematics education research community to strive towards greater consistency and reliability in its research standards and practices. In other words it expresses an aspiration which may be unachievable, rather than an actual attainment.

However, if research quality standards are function of the differing practices and needs of different research communities, contexts and countries, there is the issue of whose interests they serve. Since the standards and values are a function of their geographical and historical location, namely, the early 21st century Anglophone West, they serve the interests of this dominant research culture. Thus they are a part of the ideological effect and support the dominance effect discussed earlier. This raises a further question. Should not different criteria for judging research quality be applied as is appropriate for different contexts? Just as in education, where assessment standards vary as a function of the aims and objectives of the curriculum they assess, so too standards for research papers should reflect the aims and social purposes underlying the research and its context? While this does not mean that research standards for conferences or publications in, say, countries of the South should be set at a lower level (whatever that might mean), it might be the case that more emphasis is placed on community orientated research. Research is required to be ethical in all research traditions, and it might be that a broadened concept of ethics including community responsibility should be applied.

Such a dialogue is already taking place outside of research in mathematics education. In discussing the evaluation of educational research, Kincheloe and McLaren (1994) reject the criteria of validity, reliability and objectivity as they are understood and applied in scientific paradigm research. Instead they propose the notion of 'catalytic validity' introduced by Lather (1986) as more appropriate term for describing the criteria for establishing rigour in the study of the social world which is characterized by extreme complexity and unpredictability. This description of the contextual features resonates with the call to reconceptualize research in "mathematics, science and technology education in contexts of rapid change, conflict, poverty and violence" (Vithal *et al.* 2004: 3, Vithal *et al.* 2007). Lather defines catalytic validity as "the degree to which the research process re-orients, focuses and energizes participants toward knowing reality in order to transform it" (Lather 1991: 68). This, it is claimed, is a more rigorous test of validity, because it not only seeks to understand the world, but it also seeks to move those it studies to understand the world and the way it is shaped in order for them to transform it. Thus there more at stake in the challenge to the "'one best way' approach to the generation and legitimation of knowledge about the world" (Lather 1992: 87) than merely the issue of research criteria. It also concerns the problem of how to mobilise knowledge production, educational research in particular, so as to have a major impact on society and social structures of power and domination.

CONCLUSION

In this chapter I have explored some of the ways in which the globalisation and the global knowledge economy impacts on mathematics, science and technology education research. In particular, I have explored some of the imbalances between 'developing' and 'developed' countries created and sustained by the global knowledge economy. The major effect is economic, the inflow of funds to 'developed' countries in return for the export of knowledge and expertise. There is also a recruitment effect, i.e., the 'brain drain' of many of the most skilled personnel from 'developing' countries. I also described the appropriation effect in which knowledge gathered locally in 'developing' countries is appropriated for gain. I have also tried to show that none of these effects are simple and purely exploitative; that there are complex eddies and countercurrents as well as the main flows of globalization. By considering the subjectivities and agencies at play I have tried to reveal some of the complexities involved. I have illustrated this with reflections on my own role as an agent of the knowledge economy located in the dominant research culture.

However, my main focus has been on critiquing the dominance and ideological effects. The dominance effect is enacted through the way that research institutions, organizations and publications from Northern and 'developed' countries, typically anglophone, dominate the international research community in mathematics, science and technology education, both in terms of power and prestige. This helps to sustain the ideological effect, whereby all researchers including those in 'developing' countries are subject to and internalize the ideological and epistemological presuppositions and values of this dominant research culture. I have identified four components of the ideological effect. First, there is the reconceptualization of knowledge and the impact of the ethos of managerialism in the commodification and fetishization of knowledge. This is probably the most important ideological dimension that characterizes the postmodern knowledge economy, with clear and direct impacts in education and research. Second, there is the ideology of progressivism with its fetishization of the idea of progress. This helps sustain imbalanced and racist views of the value of different nations and the worth of different peoples and races. Third, there is the further component of individualism which in addition to promoting the cult of the individual at the expense of the community, also helps to sustain the ideology of consumerism. Fourth is the myth of the universal standards in mathematics education research, which can delegitimate research strategies that forground ethics or community action more than is considered 'seemly' in traditional research terms. This is the final capstone of the system that keeps the dominance effect in place.

In the words of Bacon (1597/1997) 'knowledge is power', and my aim in providing this critique of ideology in mathematics, science and technology education research and its globalization is to provide researchers with a means of gaining power and control over some of the hidden dimensions of social and educational research and its methodology. An awareness of the roles of the commodification of knowledge, managerialism, progressivism and individualism in the culture of educational research enables them to be revealed, scrutinized from an ethical perspective, and countered. Awareness of the lack of universal standards is also empowering in enabling researchers to develop and rely more on their own critical judgements, and to select their avenues of publication pragmatically. Thus far from showing that educational research for social change and betterment is hopeless, I believe that revealing some of the countervailing ideological currents offers a basis for wiser and more informed action.

REFERENCES

Adorno, R., Frenkel-Brunswick, E. Levinsion, D. and Sanford, R. (1950) *The Authoritarian Personality*, New York: Harper.

American Psychological Association (2001) *APA Publication Manual, Fifth Edition*, Washington, DC: APA.

Anderson, G. L. and Herr, K. (1999) The New Paradigm Wars: Is There Room for Rigorous Practitioner Knowledge in Schools and Universities? *Educational Researcher*, 28 (5): 12–17.

Aristotle (1953) *The Ethics of Aristotle* (The Nichomachean Ethics, translated by J. A. K. Thomson), London: Penguin Classics.

Atweh, B., Forgasz, H. and Nebres, B., Eds., (2001) *Sociocultural Research in Mathematics Education: An International Perspective*, Mahwah, New Jersey: Lawrence Erlbaum Associates.

Avis, J., Bloomer, M., Esland, G., Gleesons, D. and Hodkinson, P. (1996) *Knowledge and Nationhood*, London: Cassell.

Bacon, F. (1597/1997) *Meditations Sacrae and Human Philosophy*, LaVergne, Tennessee: Lightning Source Inc.

Bakan, J. (2004) *The Corporation*, London: Constable.

Bassey, M. (1990–91) On the Nature of Research in Education (Parts 1–3), *Research Intelligence* Nos. 36 (Summer 1990), 35–38; 37 (Autumn 1990), 39–44; 38 (Winter 1991), 16–18.

Baudrillard, J. (1988) 'How the West was lost', *Guardian Newspaper*, Review Section, 21 October 1988, page 1, London: Guardian Newspaper Group.

British Broadcasting Corporation (2005) *The Power of Nightmares*, 3 part TV series screened beginning Tuesday, 18 January, 2005 at 2320 GMT on BBC Two, details accessed at 1211 on 21 July 2005 at <http://news.bbc.co.uk/1/hi/programmes/3755686.stm>.

Carr, W. and Kemmis, S. (1986) *Becoming Critical*, London: Falmer, (reprinted 1988.).

Collis, D. (1999) The Abuse of Consumerism, Zadok Paper S101 Winter 1999. Consulted via <http://www.zadok.org.au/papers/collis/colliss10101.shtml> at 1219 on 6 July 2005

Denzin, N. K. (1997) *Interpretive Ethnography*, London: Sage.

Donmoyer, R. (1996) Educational Research in an Era of Paradigm Proliferation: What's a Journal Editor to do? *Educational Researcher*, 25(2): 19–25.

Douglas, M. (1966) *Purity and Danger*, London: Routledge and Kegan Paul.

Ernest, P. (1985). The philosophy of mathematics and mathematics education. *International Journal for Mathematical Education in Science and Technology, 16*, 603–612.

Ernest, P. (1991) *The Philosophy of Mathematics Education*, London: Falmer Press.

Ernest, P. (1994) *An Introduction to Educational Research Methodology and Paradigms*, Exeter: University of Exeter School of Education.

Ernest, P. (1998). *Social constructivism as a philosophy of mathematics*. Albany, NY: SUNY Press.

Ernest, P. (2000) Why teach mathematics?, in Bramall, S. and White, J. Eds. *Why Learn Maths?* London: Bedford Way Papers, 2000: pp. 1–14.

Featherstone, M. (1991) *Consumer Culture and Postmodernism*, Sage, London, p. 83.

Foucault, M. (1970) *The Order of Things: An Archaeology of the Human Sciences*, London:

Foucault, M. (1976) *Discipline and Punish*, Harmondsworth: Penguin.

Frankenstein, M. (1983) Critical Mathematics Education: An Application of Paulo Freire's Epistemology, *Journal of Education*, Vol. 165, No. 4, 315–339.

Freire P (1972) *Pedagogy of The Oppressed*, London: Penguin Books

Fukuyama, F. (1992) *The End of History and the Last Man*, Boston, Massachusetts: The Free Press.

Gage, N. L. (1989) The Paradigm Wars and Their Aftermath: A 'Historical' Sketch of Research on Teaching Since 1989, *Teachers College Record*, 91(2) 135–150.

Gergen, K. J. (1999) *An Invitation to Social Construction*, London: Sage.

Gibbons, M., Limoges, C., Nowotny, H., Schwartzman, S., Scott, P. and Trow, M. (1994) *The New Production of Knowledge*, London: Sage.

Gilligan, C. (1982), *In a Different Voice*, Cambridge, Massachusetts: Harvard University Press.

Golby, M., Greenwald, J. and West, R. Eds (1975) *Curriculum Design*, London: Croom Helm and Open University Press.

Gordon, P. (1989) The New Educational Right, *Multicultural Teaching*, Vol. 8, No. 1, 13–15.

Habermas, J. (1971) *Knowledge and Human Interests*. London: Heinemann.

Henriques, J. Holloway, W. Urwin, C. Venn, C. and Walkerdine, V. (1984). *Changing the Subject: Psychology, Social Regulation and Subjectivity*. London: Methuen.

Hersh, R. (1997). *What is mathematics, really?* Oxford: Oxford University Press.

Himmelfarb, G. (1987) *Victorian Values*, London: Centre for Policy Studies.

Hobbes, T. (1651) *Leviathan*. (Reprinted in Fontana Library, Glasgow: William Collins, 1962).

Hobsbawm, E. J. (1994) *Age of Extremes*, London: Michael Joseph.

International Commission on Mathematical Instruction (2001) About ICMI, Bulletin No. 50, Webpage Accessed via <http://www.mathunion.org/ICMI/bulletin/50/about.html>, accessed at 0736 on 20 July 2005.

Kincheloe, J. L., & McLaren, P. L. (1994). Rethinking critical theory and qualitative research. In N. K. Denzin & Y. S. Lincoln (Eds.), *Handbook of qualitative research* (pp. 138–157). Thousand Oaks, CA: Sage.

Klein, N. (2000) *No Logo*, London: Flamingo, HarperCollins publishers.

Kuhn, T. S. (1970) *The Structure of Scientific Revolutions*, (2nd edition), Chicago: Chicago University Press.

Lasch, C. (1984) *The Minimal Self*, London: Picador (Pan) Books.

Lather, P. (1986) Issues of validity in openly ideological research: between a rock and a soft place, *Interchange*, *17*(4) 1986: 63–84.

Lather, P. (1991) *Getting smart: Feminist research and pedagogy with / in the postmodern.* New York: Routledge.

Lather, P. (1992) Critical frames in educational research: Feminist and post-structural perspectives, *Theory into Practice*, 31(2): 87–99.

Lave, J. and Wenger, E. (1991) *Situated Learning: Legitimate Peripheral Participation* Cambridge, MA: Cambridge University Press

Lawlor, S. (1988) *Correct Core : Simple Curricula For English, Maths And Science*, London: Centre For Policy Studies.

Learning Conference (2005) Sample Referee Report Form for 12th International Conference on Learning, University of Granada, July 2005, and International Journal of Learning. Accessed via <http://data.commonground.com.au/forms/SAMPLE-RefereeReportForm.pdf> at 0714 on 15 July 2005.

Lerman, S. (2000). The Social Turn in Mathematics Education Research, in J. Boaler, Ed., *Multiple perspectives on mathematics teaching and learning*, Westport, Connecticut: Ablex Publishing, 2000: 19–44.

Lerman, S. (2003) *The Production and Use of Theories of Teaching and Learning Mathematics* (ESRC End of Award Report), Swindon: Economic and Social Research Council.

Lerman, S. Xu, G. and Tsatsaroni, A. (2002) Developing Theories of Mathematics Education Research: The ESM Story, *Educational Studies in Mathematics* 51(1), 2002: 23–40.

Lyotard, J. F. (1984). *The Postmodern Condition: A Report on Knowledge*. Manchester: Manchester University Press.

Marcuse, H. (1964) *One Dimensional Man*, London: Routledge and Kegan Paul.

Maslow A. H. (1954) *Motivation and Personality*, New York: Harper

Mathematical Association (1988) *Mathematics in a Multicultural Society*, Leicester: Mathematical Association.

McBride, M. (1994) The Theme of Individualism in Mathematics Education, *For the Learning of Mathematics*, 14(3) 36–42.

McLuhan, M. (1964) *Understanding Media*, New York: Mcgraw Hill.

Miller, A. (1983) *For Your Own Good: Hidden cruelty in Child-Rearing and the Roots of Violence*, New York: Farrar Straus Giroux.

Newton, D. (2004) Evil Inc, *The Guardian Weekend*, 2 October 2004: 52–53.

Niss, M. (1983) Mathematics Education for the 'Automatical Society', R. Schaper, Ed., *Hochschuldidaktik der Mathematik* (Proceedings of a conference held at Kassel 4–6 October 1983), Alsbach-Bergstrasse, Germany: Leuchtturm-Verlag, 1983: 43–61.

Noddings, N. (1984) *Caring: A Feminine Approach to Ethics and Moral Education*, Berkeley: University of California Press.

Ozga, J. (2005) 'Travelling and Embedded Policy: the case of post-devolution Scotland within the UK' in Coulby, D. and Zambeta, E., Eds., Globalisation and Nationalism in Education (The World Yearbook of Education 2005), London: Routledge Falmer, 2005.

Pepper, S. C. (1948) *World Hypotheses: A Study in Evidence*, Berkeley, California: University of California Press.

Peters, M. (2002) Education Policy Research and the Global Knowledge Economy, *Educational Philosophy and Theory*, Vol. 34, No. 1, 2002: 91–102.

Rawls, J. (1972) *A Theory of Justice*, Oxford: Oxford University Press.

Restivo, S., Van Bendegem, J. P. and Fischer, R. Eds (1993) *Math Worlds: Philosophical and Social Studies of Mathematics and Mathematics Education*, Albany, New York: SUNY Press.

Rorty, R. (1979) *Philosophy and the Mirror of Nature*, Princeton, New Jersey: Princeton University Press.

Saxe, G. B. (1991) *Culture and Cognitive Development: Studies in Mathematical Understanding*. Hillsdale New Jersey: Lawrence Erlbaum Associates.

School of Barbiana (1970) *Letter to a Teacher*, Harmondsworth: Penguin Books.

Schubert, W. H. (1986) *Curriculum: Perspective, Paradigm, and Possibility*, New York: Macmillan.

Schütz, A. (1972) *The Phenomenology of the Social World*, London: Heinemann.

Seeger, F. and Steinbring, H., Eds. (1992) *The Dialogue between Theory and Practice in Mathematics Education: Overcoming the Broadcast Metaphor*, Bielefeld, Germany: I. D. M., University of Bielefeld.

Sen, A. (2004) How to judge globalism, in F. J. Lechner and J. Boli, *The Globalization Reader* (Second edition), Oxford: Blackwell, 2004: 17–21.

Sklair, L. (2004) Sociology of the Global System, in F. J. Lechner and J. Boli, *The Globalization Reader* (Second edition), Oxford: Blackwell, 2004: 70–76.

Skovsmose, O. (1994) *Towards a Philosophy of Critical Mathematics Education*, Dordrecht: Kluwer.

Skyrme, D. (2004) The Global Knowledge Economy, consulted via URL <http://www.skyrme.com/insights/21gke.htm>, accessed at 1437 on 26 April 2004.

Sriraman, B. and Steinthorsdottir, O. (2007) Social Justice and Mathematics Education: Issues, Dilemmas, Excellence and Equity, *The Philosophy of Mathematics Education Journal* No. 21 at <http://www.ex.ac.uk/~PErnest/>.

Steffe, L. P. and Gale, J., Eds., (1995) *Constructivism in Education*, Hillsdale, New Jersey: Lawrence Erlbaum Associates.

Sundaram, K. (1999) *Herder, Gadamer, and 21st Century Humanities*, Paper from Twentieth World Congress of Philosophy, Boston 1998. Consulted via URL

<http://www.bu.edu/wcp/Papers/Educ/EducSund.htm>, accessed at 1006 on 16 July 2005.

Tymoczko, T. (Ed.). (1986). *New directions in the philosophy of mathematics*. Boston: Birkhaüser.

UNESCO (2002) Teaching and Learning for a Sustainable Future, (multimedia teacher education programme), Paris: UNESCO. Consulted via URL <http://www.unesco.org/education/tlsf/index.htm>, accessed at 1150 on 6 July 2005.

University of Chicago Press Staff, Eds., (2003) *The Chicago Manual of Style, 15th Edition*, Chicago: University of Chicago Press.

Vico, G. (1961). *The new science*. Ithaca/London: Cornell University Press. (Original work published 1744).

Vithal, R., Setati, M and Malcolm, C. (2004) *Call for papers for a new volume on: Methodologies for researching mathematics, science and technological education in societies in transition, A UNESCO-SAARMSTE Book Project*, South Africa: University of KwaZulu-Natal.

Vithal, R., Setati, M and Malcolm, C., Eds., (2007) *Methodologies for Researching Mathematics, Science and Technological Education in Societies in Transition*, Cape Town, South Africa: Heinemann.

Vygotsky, L. S. (1978). *Mind in Society, Cambridge*. Massachusetts: Harvard University Press.

Walkerdine, V. and Lucey, H. (1989) *Democracy In The Kitchen*, London: Virago.

Wenger, E. (1998) *Communities of Practice: Learning, Meaning and Identity*, Cambridge: Cambridge University Press.

Wiltshire Education Authority (1987) *Mathematics For All*, Salisbury, Wiltshire Education Authority

Wittgenstein, L. (1953). *Philosophical Investigations*, Oxford: Basil Blackwell.

Womack, D. (1983) Seeing the Light, *Times Educational Supplement*, 8 April 1983.

Young, R. M. (1979) Why are figures so significant? The role and the critique of quantification, in J. Irvine, I. Miles, I. and J. Evans, Eds., (1979) *Demystifying Social Statistics*, London: Pluto Press, 63–74.

ACKNOWLEDGMENT

This paper is a version of the chapter Ernest, P. (2007) 'Globalization, Ideology and Research in Mathematics Education', that first appeared in Vithal *et al.* (2007).

NOTES

1. Globalization shares a number of characteristics with internationalization and the terms are sometimes used interchangeably, although I prefer to use 'globalization' to emphasize the erosion of the nation state or national boundar-

ies, which is especially important in the growing emergence of a global educational research community.

2. Below, in discussing the concept of progress I shall problematize the terms 'developing' and 'developed' countries. However, for the moment I shall simply use these problematic terms to avoid cumbersome circumlocutions, but put them in 'scare quotes' to show that the terms are used with reservations.

3. This is well known in pharmacology where, for example, teams of ethnobotanists and pharmacognosists visit the Amazon basin, interview indigenous people about pharmacologically active plants and take samples of the plants. They then return to their 'developed' countries of origin where the active substances are extracted, manufactured, marketed and sometimes even patented, by 'Big Pharma, multinational pharmacology corporations whose net profits are in the billions of dollars.

4. The abbreviations used here and elsewhere stand for: Adults Learning Mathematics (ALM), Congress of the European Society for Research in Mathematics Education (CERME), East Asia Regional Conference on Mathematics Education (EARCOME), International Study Group on the Relations between the History and Pedagogy of Mathematics (HPM), International Congress of Mathematical Education (ICME), International Organization of Women and Mathematics Education (IOWME), Mathematics, Education and Society (MES), Political Dimensions of Mathematics Education (PDME), Programme for International Student Assessment (PISA),the International Group for the Psychology of Mathematics Education (PME), Southern African Association for Research in Mathematics, Science and Technology Education (SAARMSTE), Second International Mathematics Study (SIMS), Third International Mathematics and Science Study (TIMSS).

5. Researchers from 'developed' countries do make guest appearances at such regional conferences. However, in addition to the individual's benefits in having more international papers and publications for their CVs, there is also a growing *tourism effect*. As with tourism, travel to exotic locations even if for professional reasons is a consumer good in itself.

6. My basis for these criteria is a combination of the principle of the respect for the value of different cultures (Ernest 1991) and Marx's idea, expressed by Rawls (1972) in his *theory of justice*, that where persons receive less than the value of their contributions they are being exploited.

7. I am stressing the positive features that emerge from the knowledge economy to provide a balanced view, bearing in mind the critique of the domination and ideological effects that follows below.

8. Contrary to the impression it gives, etymologically the country's name does not derive from a hubristic reference to former imperial greatness, but resulted from the amalgamation of separate monarchies and countries (England, Scotland, Wales) into an enlarged, i.e., 'greater' Britain in 1603. However, as the first impression shows, the name is culturally double-coded with residual connotations of colonialisation and imperialism.

9. As a white, heterosexual, middle class origin male I have not experienced the pain of having to overcome stereotyped expectations imposed on me, or the oppression of classism, sexism, racism and homophobia in society. However,

as somebody who did not arrive in the UK until 1951 at the age of 7 with non-British parents (Swedish and Jewish-American) I did experience the feeling of being an outsider for many years, especially in the very conformist Britain of the 1950s.

10. But see Sriraman and Steinthorsdottir (2007) for an interesting critique of the concept of critical thinking.

11. The handbook titles for the modules of these courses are: 1 The Psychology of Mathematics Education, 2 The Mathematics Curriculum, 3 Mathematics and Gender, 4 Mathematics and Special Educational Needs, and 5 Research Methodology in Mathematics Education. These handbooks also treat assessment (in 2), philosophy of mathematics (in 3) and language in mathematics education (in 4). As the students are mid-career education professionals two main themes were deliberately omitted (mathematical content areas, and pedagogy and teaching resources such as ICT) although they do figure incidentally and in some student initiated assignments.

12. All semiotic expressions or signs have a basic representational function whether from a neo-Saussurian or Peircean perspective. In the former case there is the signifier-signified pairing in the sign, in which an expression designates a content. Peirce's theory is similar but introduces a third component (the interpretant) which carries some of the meaning. Some signs purport to model some aspect of reality, which I have termed the modeling function, whereas other only refer to other signs and sign-systems.

13. Although I am an academic with a lifelong commitment to research and publications I must acknowledge that knowledge is not an end in itself but just a means to a greater end, namely that of human happiness. Of course the full achievement of this end necessitates banishing the global causes of unhappiness, such as hunger, poverty, disease, oppression, injustice, etc., which are political problems and issues (to which globalization is contributing). The pursuit of knowledge, which is one of the privileges of the academic life, is for those involved, like myself, a source of happiness. But it must be acknowledged that the pursuit of knowledge (and the happiness it brings) is only possible when its material conditions are supported. In terms of Maslow's (1954) hierarchy of needs, only when our physiological needs, safety and security, and social needs are satisfied, can we seek to develop the self-esteem, prestige and self-fulfillment that come from the pursuit of knowledge and the most beneficial forms that the academic life can take (but which managerialism and performativity are eroding). But this satisfaction in turn presupposes forms of social organisation that produce surpluses to support the lifestyles of academics. Our social critiques depend on the division of labour and the privileging of some classes (including intellectuals) which are their targets. Part of the ingenuity of the global capitalist system that produces such surpluses is that it is able to exploit, incorporate and hence subvert anger, protest and rebellion. Thus the Punk youth movement of 1976, despite its anger and rejection of the bourgeois forms of life, music and fashion ultimately served to inject new blood into the development and marketing of music and fashion. Likewise academic critiques, such as this paper are both something to be marketed as part of the global knowledge economy, as well as performing

a safety-valve function for dissent. They allow writers and readers to channel their outrage in ways that do not destabilize the system, and hence serve to sustain it. Just as mathematics has been able to incorporate and 'tame' the study of uncertainty and chaos throughout its history, so too global capitalism is able to appropriate and market critiques and dissenting voices, from posters of Ché Guevara to the records of Bob Dylan and the films of Michael Moore (Klein 2000).

14. This is the third of three types of interest that drive research. The other two types of interest distinguished by Habermas are the desire to predict and control (technical, scientific interest), the desire to understand (practical, interpretative interest). These form the basis of the scientific and interpretative research paradigms in education, respectively (Bassey 1990–91, Ernest 1994, Schubert 1986). However, these last two paradigms firmly locate the educational research process as types of mode 1 knowledge production. I am also quite clear that I as a researcher also work exclusively in mode 1, writing materials that are aimed at publication, rather than committing my time and energies to mode 2 knowledge production for social reform projects driven by the critical interest.

15. Not surprisingly it is not only social justice and environmentalist groups that use these strategies. There are also right wing think-tanks and commercial pressure groups funded by industry and commercial interests, and even by covert government agencies, which seek to change public opinion and policy. In Ernest (1991) I documented the powerful impact of New Right think-tanks such as the Centre for Policy Studies on British educational and social policy in the 1980s (Gordon 1989, Himmelfarb 1987, Lawlor 1988). Recently it is claimed that American neo-conservatives have similarly manipulated popular opinion to justify the 'War on Terror' and the invasion of Iraq (British Broadcasting Corporation 2005).

16. There were dissenting voices such as Vico (1744/1961). His cyclical theory of historical development does not support the incrementalist features of progressivism that view historical progress as secured, ratchet like, against against forced retreat. Herder also rejects the idea that human history is a linear progression (Sundaram 1998).

17. This rational position was of course the basis for the monstrous 20th century doctrines of 'racial purity' and 'ethnic cleansing'.

18. I am not inverting the received values to claim that tribal knowledge, including ethnomathematics or ethnoscience, is superior to or even comparable to scientific knowledge. Instead I would argue that all such knowledge and belief systems should be viewed in the context of their own cultural spheres. Knowledge and belief systems are much more than rational instruments for achieving material ends. They are the cultural 'glue' that binds peoples into communities and helps to shape identities. However, the commodification of knowledge coupled with rationalism and the ideology of progressivism denies these vital non-instrumental functions of knowledge and belief systems. My plea is instead for mutual understanding and respect between alternative worldviews.

19. Even without the negative connotations of the terms 'developing' or 'underdeveloped', 'development' itself, in terms of what it means and how it is measured, is problematic. Standard measures such as gross national product, equate 'development' with growth in production and consumption of goods and services. Such growth is not automatically good as is assumed by many in 'developed' countries, or those who aspire to 'development' (UNESCO 2002).

20. Not surprisingly technological companies like Casio and Texas Instruments are the biggest sponsors of international conferences in mathematics education where researchers enthused with the latest electronic innovations and the ideals of technological progress act as unpaid marketers and promoters of these goods.

21. I like to believe that I can maintain a critical disengagement from the ideology of the Western society which I inhabit. But I must also acknowledge that this, in the form of consumerism, in turn inhabits and shapes me. I may have internalized academic research ideals and a critical intellectual stance which enables me to distance my ideal self, if not my actual self, from consumerism. But this leaves open the question of the extent to which my judgement is subverted and compromised, and the extent to which my critical stance is a way of dealing with the conflict between my lifestyle and ideals. It also raises the question of whether my academic stance is part of the appropriation effect, i.e., appropriating and making cultural capital out of a critique of the system which in practice I uphold and which sustains my privileges.

22. Traditionally most research in mathematics education drew on psychology as the underlying discipline, and this is reflected in the title of PME. However, over the past two decades research from a broader range of perspectives has been reported at this annual conference, and there have been moves to change the constitution and the focus of the group to reflect this broadening disciplinary base.

CHAPTER 11

WHAT IS THIS THING CALLED SOCIAL JUSTICE AND WHAT DOES IT HAVE TO DO WITH US IN THE CONTEXT OF GLOBALISATION?

Bill Atweh
Curtin University of Technology

This paper has dual aims. It proposes firstly to open the debate about the construction of social justice in mathematics education and secondly to use that construction to reflect on issues affecting the work of mathematics educators in an international arena. It builds on my previous work with many colleagues on issues of globalisation and internationalisation in mathematics education (Atweh & Clarkson, 2001; Atweh, Clarkson & Nebres, 2003) and on social justice in international collaborations in our discipline (Atweh & Ragusa, 2003; Ragusa & Atweh, 2003; Atweh & Keitel, 2008; Atweh, 2007). The theorisation of social justice presented in the previous publications, also adopted here, is informed by the writing of feminist writers such as Iris Marion Young (1990) and Nancy Fraser (1995, 1997; Fraser & Honneth,

Critical Issues in Mathematics Education, pages 111–124
Copyright © 2009 by Information Age Publishing
All rights of reproduction in any form reserved.

2003). In particular, I argue for the importance of engaging with the concept of social justice itself in addition to engaging with practices that promote social justice. Similarly I discuss an approach to understanding social justice practices that goes beyond mere analyses and deconstruction and is capable of providing normative guides for practice. Finally, I raise some problematics in making social justice claims and engaging in practices that promote social justice. While many of the issues raised here might apply for a variety of social justice concerns in the discipline, I will focus the discussion on an issue of increasing importance in mathematic education, namely international collaborations (Atweh & Clarkson, 1991).

WHY ENGAGE WITH THE CONCEPT OF SOCIAL JUSTICE?

Atweh and Keitel (2007) note that social justice concerns are no longer seen at the margins of mathematics education policy, research and practice. Issues relating to gender, multiculturalism, ethnomathematics, and the effects of ethnicity, indigeneity, socio-economic and cultural backgrounds of students on their participation and performance in mathematics are regularly discussed in the literature. Many of these have found their way into policies in educational systems around the world. More recent concerns about access to appropriate mathematics education by students with learning difficulties and special needs, the gifted and talented, and the so called "what about the boys" agendas are increasingly being constructed as social justice issues. Undoubtedly, different writers have different understandings of social justice—at times leading to alternative, if not contradictory conclusions and demands.

Although social justice represents a strong area of educational discourse, the term itself remains under-theorised (Gewirtz, 1998, Rizvi, 1998) and a "contested area of investigation" (Burton, 2003, p. xv). Firstly, at the risk of essentialising the difference between the USA's and Europe's writings on social justice, there seems to be some difference between its conceptualisation in the two contexts–at least in mathematics education. In the US, social justice is often used interchangeably with the constructs of *equity* and *diversity*. For example, Hart (2003) asserts that "Because the terms *equity, equality* and *justice* have been used in different ways in the literature, it is important to briefly consider some of the meanings of these terms" (p. 29). Using Secada's (1989) conceptualisation of equity as "our judgement about whether or not a given state of affairs is just" (p. 29), implies that equity is the measure if social justice has been done. Hart (2003) uses a multidimensional definition of equity as equal opportunity, as equal treatment and as equal outcome and concludes by saying that "I will use equity, as Secada...did, to mean justice" (p. 23). In the same volume, Secada, Cueto and Andrade

(2003) note that "the viewing of group-based inequality as an issue of equity has a long tradition within policy-relevant social science research . . . and in different forms of educational research in particular" (p. 108).

In an attempt to differentiate between equity and social justice, Burton (2003), from the UK, in her introduction to her book "Which Way Social Justice in Mathematics Education," argues that there is a "shift from equity to a more inclusive perspective that embraces social justice" (p. xv). She goes on to say "the concept of social justice seems to me to include equity and not to need it as an addition. Apart from taking a highly legalistic stance, how could one consider something as inequitable as socially just?" (p. xvii)

Further, the equity agenda has been critiqued in the literature in its ability to provide for a normative guide for practice. Wenzel (2001) discuss the difficulties within the traditional equity discourse in determining questions as who is entitled for equity measures and how to avoid the individual selfishness at the expense of the group's benefit. Similarly, he argues that the construction of an individual as a member of a single social group deserving equity measures is problematic. Finally, equity measures tend to deal with a single recipient of the benefits and not as a member of a social group that is systematically excluded.

Similarly, the social justice agenda in mathematics education is at times discussed in relation to *diversity*—also a term that has its origin in the USA literature (Loden & Rosener 1991). While the concept of equity arose from, and is often associated with—though not exclusively–gender concerns, the concept of diversity arose from, and is often associated with—though not exclusively—concerns about cultural and linguistic diversity (Sepheri & Wagner, 2000; Thomas, 1996). Plummer (2003), however, presents an overview of what he calls the "big 8" dimensions of diversity: *race, gender, ethnicity/ nationality, organisational role, age, sexual orientation, mental/physical ability and religion.* In this context, social justice is constructed as "managing diversity" (Cox, 1991; Krell, 2004). Undoubtedly, the increasing diversity of students in most mathematics classrooms and the persistent research evidence that some groups of students are not achieving or participating in mathematics, raise serious social justice issues. However, the diversity agenda is not only concerned with the participation and achievement gaps but also with acknowledging the contribution of the different groups to mathematics and to the consideration of different types of mathematics and different ways on knowing as illustrated in certain feminist writings and the ethnomathematics movement.

However, the diversity discourse might lead to essentialising the differences between the different groups and it may fail to take into consideration the changing constructions of these labels and their contextual understanding in time and place. Similarly, the diversity discourse fails to adequately take into consideration one of the biggest threats to social inequality and exclusion in mathematics education, namely socio-economic background and poverty that are difficult to construct as diversity issues in the same was as cultural differences.

In spite of the overlap in the aims of both agendas of equity and diversity, there is an important difference between them that leads to potentially contradictory outcomes. This relates to their ultimate aims with regard to group status. Equity projects aim at reducing group differences, e.g., in achievement and participation, and hence its ultimate aim is to abolish group differences. Diversity discourse, on the other hand aims at enhancing group differences and status. This is the dilemma that Nancy Fraser (1997) refers to in discussing the multidimensional model of social justice. There are two further limitations of the equity and diversity agendas. On one hand, remediating equity concerns might be vulnerable of a backlash of misrecognition (Fraser, 1995) for the target group by constructing them as victims or as needy of special assistance, while the diversity construction promotes the group status. On the other hand, the diversity agenda might be vulnerable of romanticising difference between groups by treating them as exotic, while the equity agenda highlights the problematic of their exclusion and disadvantage. I will come back to these points later in the paper.

I turn now to question of social justice in international collaborations.

WHY ENGAGE WITH SOCIAL JUSTICE IN INTERNATIONAL COLLABORATIONS?

International contacts in mathematics have a very long history that proceeded the era of globalisation. The transmission of mathematical knowledge from the East (e.g., India) and the South (e.g., Arabia) formed the roots of mathematics as a discipline in Europe (Powell & Frankenstein, 1997). Similarly, in mathematics education, the establishment of the International Commission of Mathematics Instruction (ICMI) in 1908 was both a reflection of the belief that mathematics educational problems can, and need to be solved globally, and at the same time provided promotion of that conviction. With ease of travel and communication and greater awareness of developments and the needs of various countries, contact between mathematics educators has escalated and taken diverse forms. While mathematics educators have always shown an acute awareness of the international status of their profession (as reflected in numerous publications and conferences with the term "international" in their titles), there has been little problematisation of this phenomenon and research activity about the benefits or problems that might arise.

Within the past three decades, mathematics education has witnessed an increase in cross national comparative studies on curriculum and student achievement, perhaps the best known are the TIMSS (Third International Mathematics and Science Study) and PISA (Programme for International Student Assessment) studies. These studies have received considerable attention within and outside the field. International testing has been widely covered by media and featured in public debates about education. The potential benefits, and problems, with international testing have been addressed elsewhere (Clarke, 2003; Kaiser, Luna, & Huntley, 1999; Robitaille, & Travers, 1992). In particular, Keitel and Kilpatrick (1999) raise several political questions about such international comparative studies. They argue that the outcomes of these studies are perceived as biased towards the host country; that is, of those who do the data collection, the analysis and the funding. These authors question if this is to the detriment of other countries and their concerns about improving education systems. Outcomes of such studies are also perceived as necessarily reductionist, as results cannot do justice to the very complex factors involved. The authors claim that the mathematical tasks do not represent the curricula taught in many schools, teachers' questionnaires do not represent the whole range of teaching practices, and the results do not offer valid comparisons between the various countries' curricula with their divergent cultural and social contexts. "No allowance is made for different aims, issues, history and contexts across the mathematics curricula of the systems being studied" (p. 243). They conclude that comparative testing is not really useful as an educational tool, as it does not produce a clear view of what's really happening in the classroom and why.

Of particular relevance here are the differences in performance between industrialised and less-industrialised countries that these tests show. For example, Glewwe and Kremer (1995) show that school students in most less-industrialized countries achieve less than comparable students in more-affluent countries. Moreover, the gaps are estimated to be 3 years of schooling for comparable age groups. Undoubtedly, these gaps can be explained to a large extent by the amount of available resources devoted to mathematics education in different contexts. Jacobsen (1996) discusses the increasing gap between the rich and poor countries and the curtailing of funds from these international agencies that makes it "more difficult to look for governments for improved international co-operation in mathematics education" (p. 1253). He joins Miguel de Guzman, the past President of ICMI, in calling for an increasing role of co-operation between professional mathematics educators and their associations to work to improve mathematics education worldwide. Arguably, solving the problems of inequitable achievement and available resources in less-industrialised countries are beyond the capabilities of single academics or even the profession as a whole working in isolation.

However, such a call presents a challenge for academics who believe that concerns about social justice do not know any boundaries.

Views expressed by mathematics educators about international contacts and activities vary. For some, international interactions lead to greater awareness and understanding of difference which, leads to assisting the less able, to tolerance and conflict resolution. Often these educators achieve greater conscious understanding of their own assumptions and salient aspects of their own practices. To others, such contacts may lead to homogenisation, colonisation and to the marginalisation of the 'have nots'. In any case, there are several social justice issues in international contacts.

First, there are different, and at times conflicting, motivations behind international collaboration. In a globalised world dominated by economic rationality, many of these international collaborations have their roots in financial benefits to the participants. For example, as many universities around the world are facing a reality of reduced government funding, they are turning to international students and projects as a significant source of income. Similarly, many less industrialised countries that depend on international loans to develop their education systems and infrastructure often face additional requirements for specific types of 'reforms' that necessitate contacts with overseas educators and systems. Other international collaborations are based on more altruistic motivations such as the provision of assistance for countries with limited resources to develop their capacity to build their infrastructure and educational reform. Perhaps such collaborations are based on the premise that mathematics is associated with economic development and prosperity, hence assisting poorer countries through establishing a solid mathematics education system may contribute to the reduction of overall poverty. Further, certain types of collaboration may yield direct benefit to individual academics seeking new research sites and markets for their publications.

Second, international collaborations face many factors that limit participation in them by many educators around the world. Not the least of these limitations is financial. The cost of attending international gatherings or subscribing to international journals is a prohibiting factor for many international mathematics educators from less industrialised nations. Similarly, educators from non-English speaking countries often feel excluded from many international activities that are in English. The final report of a recent Discussion Group on International Cooperation at the International Congress of Mathematics Education (Atweh, Boero, Jurdak, Nebres & Valero, 2008) identified further problems arising from language:

> In addition to the dominance of English in many international cooperative activities, the problem of language is also a matter of particular professional jargon used in different national communities to refer to the objects of

their practices. Problems of understanding emerge due to differences in the meanings of commonly used terms. For example, the phrase "didactics of mathematics" carries almost opposite meanings for a native English speaker and speakers of other European languages. Further, care must be given not to exclude some participants from having access to that technical language by oversimplifying it. Hence, genuine cooperation must include a process of communication in which, through languages (natural and specialized), the parties involved negotiate their meanings and intentions for action. (p. 446)

Third, international collaboration may have serious negative effects on some participating countries. Without due care, collaboration between educators with varying backgrounds, interests and resources may lead to domination of the voice of the more able and marginalization of the less powerful. Further, uncritical collaboration may confuse aid to the less resourced countries with a 'missionary' attitude that leads

> to a patronizing relationship, which does not respect and value the diversity of the parties involved. Instead, an attitude of humility and openness to learn from each others should be the basis of international co-operation. (p. 446).

Atweh and Keitel (2008) showed that uncritical contacts between countries can be exploitative, lead into marginalisation and powerlessness, and be considered as a form of symbolic imperialism and violence.

HOW CAN WE UNDERSTAND SOCIAL JUSTICE IN A COMPLEX GLOBALISED WORLD?

Marion Young (1990) argues that principles of social justice are not theorems. Rather, they are claims of some people over others. They are not based in abstract general principles that can be applied to specific practices and situations in all localities and societies. According to Young, "they are [arguments] addressed to others and await their response, in a situated political dialogue" (p. 5).

Traditionally, the conception of a social justice model was based on the redistribution of resources and goods, whether material or symbolic. *Distributive models* of social justice focus more on unequal opportunities in society rather than mere outcomes. McInerney (2004) argues that a society cannot be called just unless "it is characterized by a fair distribution of material and non material resources" (p. 50). Rawls (1973, in McInerney, 2004) claims "the primary subject of social justice must be the basic structure of society, or, more precisely, 'the way in which the major social institutions distribute fundamental rights and responsibilities and determine the division of advantages from social cooperation'" (p. 50). At the same time as he is affirming the individual rights to pursue goods, he is insisting that distribution of wealth, income, power and authority are justifiable if they work to maximize the benefit of the least advantaged in society. Gewirtz (1998) identifies two forms of distributive justice: a weak form, equality of opportunity, and a strong form, equality of outcome.

In education, distributive models of social justice are reflected in compensatory programs allocating designated resources for the disadvantaged. However, this model does not question the curriculum itself, the pedagogy or the regimes of testing used in the classroom and their role in creating educational inequality. Further, it constructs the disadvantaged as individuals and not as parts of a collective. Finally, it does not take into account the reasons for the inequality that have historical roots and are socially and politically determined. Arguably the majority of compensatory programs to increase the achievement of target groups in education follow this construction of social justice.

Several poststructuralist feminist writers have critiqued distributive models. Gewirtz (1998) argues that *relational* understandings of social justice are needed in order to "theorize about issues of power and how we treat each other, both in the micro face-to-face interactions and in the sense of macro social and economic relations which are mediated by institutions such as the state and the market" (p. 471). Relational models of social justice deal with "the *nature* and *ordering* of social relations" (p. 471, italics in original). Gewirtz goes on to indicate that "the relational dimension is holistic and non-atomistic, being essentially concerned with the nature of inter-connections between individuals in society, rather than with how much individuals get" (p. 471). Marion Young present a critique of traditional conceptions of social justice in that they are based on "having" rather than "doing." Grounding social justice in individual solutions that allow little room for the consideration of multiple social groups is inadequate. Furthermore, extending such models, developed on the distribution of material goods to other goods such as self-respect, honour opportunity, and power, is problematic. To understand the struggles for social justice by a variety of groups, such as women, African Americans, and gay and lesbian people, feminist theorists posited a discourse of social justice based on the principle of *recognition*. Nancy Fraser (1995) expounds:

> Demands for "recognition of difference" fuel struggles of groups mobilised under the banners of nationality, ethnicity, 'race', gender and sexuality. . . . And cultural recognition replaces socioeconomic redistribution as the remedy of social injustice and the goal of political struggle. (p. 68)

In response to the critique that giving attention to cultural recognition might have devalued economic inequality that is best alleviated through a distribution model, Fraser (2001) argues that social justice today requires *both* redistribution and recognition measures. She presents a model of "parity of participation" as a guiding principle that incorporates both models. In later publications (Fraser, 1997; Fraser & Honneth, 2003) she presents what she calls a "critical theory or recognition" that avoids reducing one dimension to the other and avoids falling into postmodern non-normative deconstruction. Importantly, Fraser argues that redistribution and recognition remedies analytic tools that are not mutually exclusive and in practice most social justice action contains elements of both.

The two constructions of social justice as *distribution* and as *recognition* correspond roughly to the construction of equity and diversity respectively, terms more familiar in mathematics education. By using the bi-dimensional model to understand both agendas can provide for a better understanding on the relationship between the two discourses. However, this does not yet contribute to a resolution of the difficulties identified above. The conflict between equity and diversity agendas has been translated into the dilemma that Fraser (1997) calls the distribution-recognition dilemma. To deal with this dilemma, the author introduces two further analytic tools to describe remedial action for social injustice. Fraser differentiates between affirmative and transformative remedies for injustice and argues that they cut across the redistribution-recognition divide. *Affirmative* remedies include those "aimed at correcting inequitable outcomes of social arrangements without disturbing the underlying framework that generates them" (p. 82), whilst *transformative* remedies are "aimed at correcting inequitable outcomes precisely by restructuring the underlying generative framework" (2001, p. 82). It remains to be shown how these theoretical tools assist in a resolution of the dilemma discussed above. I will turn to this in the concluding section after a reconsideration of social justice remedies in international collaborations.

HOW TO ENGAGE IN SOCIAL JUSTICE ACTION IN INTERNATIONAL COLLABORATIONS?

Based on this discussion and on a similar model suggested by Fraser, in another context (Atweh, 2007) I have put forward a model comprised of four modes characterising possible cross national collaborations towards achieving social justice in international collaborations (see Table 11.1). In the previous publication I have discussed these modes of collaboration in some details based on results of a study about internationalisation and globalisation conducted with leading mathematics educators from Latin America (Brazil, Colombia and Mexico) and Asia (South Korea, The

TABLE 11.1

	Affirmative	Transformative
Redistribution	Mode 1: *Aid* *Attributes:* Sharing of information and resources among countries. Represents cultural classification based upon access to knowledge. Can generate misrecognition.	Mode 2: *Development* *Attributes:* Restructuring of relations of knowledge production. Blurs group identification. Can help remedy misrecognition.
Recognition	Mode 3: *Multiculturalism* *Attributes:* Acknowledging cultural differences, such as cross cultural research. Supports group identification.	Mode 4: *Critical Collaboration* *Attributes:* Deep restructuring of relations of recognition. Blurs group differentiation.

Philippines, and Vietnam) that I conducted between 2001 and 2002.[1] I will only provide a summary of these points here.

Affirmative-redistribution for social injustice remedies target the lack of resources and absence of traditions of internationally acceptable research and theorising of mathematics education and the exclusion of academics from less affluent countries in many international collaborations, gatherings and publications. Remedies that are often provided to overcome these indicators of social injustice take the form of sharing of programs and curriculum or financial assistance to academics from less affluent countries to enable a few of them at least to participate in such international gatherings. This aid mode of collaboration is based on the transmission of goods, (either material or symbolic) from one culture to another can give rise to serious concern. To start with they often lack reciprocity among the players leading to a form of colonialisation of mathematics education from the North to the South and from West to East. Further, they give rise to problems of misrecognition of the aid recipients as not having something worthwhile and original of their own to contribute to the international status of the discipline. Finally, they lead into a condition of dependency on donor countries since their contribution to capacity building is minimal.

The affirmative-recognition mode of remedies targets the lack of recognition to the mathematical knowledge that different cultural groups have developed and use in everyday transaction as valid mathematical knowledge and the contribution of the different groups to the mainstream mathematics education often identified as Anglo-European (Powell & Frankenstein, 1997). The contribution of the ethnomathematics movement to increasing awareness of complex mathematics in the daily practices of many social and cultural groups has undoubtedly lead to the abolishing of the myth of underdevelopment and primitiveness from these societies. However, its contribution to the emancipation of members of these societies is not as clear (Dowling, 1998; Vithal & Skovsmose, 1997). Hence, it contributes to the recognition of the other without necessarily contributing to alter or change access to, or production of, material and/or symbolic goods.

The transformative-redistribution mode of remedies targets the enabling of the marginalised academics and cultures to develop their own capacity to generate their own knowledge, research and theory about mathematics education. Hence it effects a change of pre-existing patterns and norms of knowledge production and may have short or long-term effects. International interactions under this model include international postgraduate students from less industrialised countries and programs that contribute to the professional development of educators. However, it is usually unidirectional with clear demarcation between the providers and recipients of development. Similarly, this mode of interactions does not necessarily problematise differences in interests and needs of the different participants; hence it leads into blurring of cultural differences.

Finally, the transformative-recognition mode of remedies targets the deconstructions of the binaries that construct academics from affluent/developed/industrialised and those from poor/underdeveloped/less industrialised countries and attempts to develop critical collaborations that are mutual and lead into reciprocal learning. Like multiculturalism, critical collaboration aims to give recognition and respect to the knowledges different cultural groups and countries provide. However, in this category effort is made to challenge the structures that give rise to inequality in status, as well as the knowledge shared, among nations. Critically collaborative activities are necessarily based on participation from educators in different countries as all work to develop local knowledge and simultaneously contribute to collective international knowledge albeit it is not universal but always contextualised.

WHAT CAN WE CONCLUDE ABOUT SOCIAL JUSTICE AND ABOUT INTERNATIONAL COLLABORATIONS?

This paper has set to promote the debate on the construction of social justice in mathematics education and on social justice in international collaborations. It argued that even though social justice concerns and actions are prevalent in mathematics education research and practice, the term remains undertheorised. The two related agendas of equity and diversity, more familiar in mathematics education, are often used interchangeably with each other and with the construct of social justice. Undoubtedly, the three agendas overlap; however, their basic assumptions and differences often remain unexamined. In this context I discussed a particular construction of social justice that is more general than equity and diversity and can provide the means to examine essential issues that arise as a result of action designed to achieve social justice. In this construction, social justice can aim at redistribution of goods (material or symbolic) to alleviate inequality and disadvantage as well as increase the recognition to the marginalised and misrecognised. I followed Nancy Fraser's arguments that neither of these dimensions is reducible to the other. However, there are potential sources of conflict between them. The critical theory of social justice summarised here may lead into a more informed and self reflective action that attempts to identify and address achievements and limitations of social justice action—hence it is built on a transformative agenda that avoids falling into the two extremes of unreflective action and actionless reflection.

Secondly, in this paper I have turned a critical gaze on the practice of mathematics education in an international context. With the rising evidence about inequality in achievement in mathematics from students in different cultures and the lack of participation of academics from less industrialised countries in international efforts to solve problems of mathematics education, the social justice implications of international collaborations can and should not be overlooked. However, not all action towards achieving equity, no matter how well intentioned it is, is sufficient to achieve social justice. Here I argued that modes of interactions that are based on concerted efforts to increase the agency of all mathematics educators to collaborate across country lines and access to resources and at the same time challenge the binaries in terms of their backgrounds is a worthwhile endeavour.

Mathematics educators who are committed to issues of equity, diversity and social justice in any of the manifestations of the injustice in the discipline can not be consistent in their commitment without paying attention to issues of social justice in international collaborations. Social justice knows no national and cultural boundaries. More generally, in an increasing internationalised and globalised world, the work of the majority of mathematics educators has international dimension. Such components can either be part of the problem or part of the solution of social injustice in the discipline.

REFERENCES

Atweh, B. (2007). International interactions in mathematical education. In U. Gellert & E. Jablonka, (Eds.), *Mathematisation and Demathematisation: Social, Philosophical and Educational Ramifications* (pp. 171–186). Rotterdam/Taipei: Sense Publishers.

Atweh, B., Boero, P., Jurdak, M., Nebres, B. & Valero, P. (2008) International cooperation in mathematics education: Promises and challenges. In M. Niss (Ed.), *Porceedings of ICME 10*. Rosskilde, Denmark: Rosskilde University.

Atweh, B. & Clarkson, P. (2001). Internationalisation and globalisation of mathematics education: Towards an agenda for research/action. In B. Atweh, H. Forgasz, & B. Nebres (Eds.), *Sociocultural Research on Mathematics Education: An International Perspective* (pp. 77–94). New York: Erlbaum.

Atweh, B. Clarkson, P. & Nebres, B. (2003). Mathematics education in international and global context. In A. Bishop, M. A. Clements, C. Keitel, J. Kilpartick, & F. Leung (Eds.), *The Second International Handbook of Mathematics Education*, (pp. 185–229). Dordrecht: Kluwer Academic Publishers.

Atweh, B. & Keitel, C. (2008). Social justice in international collaborations. In B. Atweh; M. Borba; A. Calabrese Barton; N. Gough; C. Keitel; C. Vistro-Yu; & R. Vithal (Eds.), *Internationalisation and Globalisation in Mathematics and Science Education* (pp. 95–112). Dordrecht, Netherlands: Springer.

Atweh, B. & Ragusa, A. (November, 2003). *Issues in Social Justice in International Collaborations: Views of educators from around the world.* Conference paper presented at The Australian Sociological Association annual conference. Armidale: University of New England.

Burton, L. (Ed) (2003). *Which Way Social Justice in Mathematics Education?* London: Praeger.

Clarke, D.J. (2003). International comparative studies in mathematics education. In A.J. Bishop, M.A. Clements, C. Keitel, J. Kilpatrick, & F.K.S. Leung (Eds.), *Second International Handbook of Mathematics Education* (pp. 145–186). Dordrecht: Kluwer Academic Publishers.

Cox, T.H. (1991) *The multicultural organization.* Academy of Management Executive, 5 (2) 34–47.

Dowling, P. (1998). *The Sociology of Mathematics Education: Mathematical Myths/Pedagogic Texts.* London: The Falmer Press.

Fraser, N. (1995). From redistribution to recognition: Dilemmas of justice in a post-socialist society. *New Left Review,* July-August, 68–93.

Fraser, N. (1997) *Justice Interruptus: Critical Reflections on "Postsocialist" Condition.* New York: Routledge.

Fraser, N. (2001). *Social justice in the knowledge society.* Invited keynote lecture at conference on the "Knowledge Society," Heinrich Böll Stiftung, Berlin.

Fraser, N. & Honneth, A. (2003). *Redistribution or Recognition? A Political-Philosophical Exchange.* London: Verso:

Gewirtz, S. (1998). Conceptualizing social justice in education: Mapping the territory. *Journal of Educational Policy, 13*(4), 469–484.

Glewwe, P. & Kremer. M. (1995). *Schools, Teachers, and Education Outcomes in Developing Countries. Centre for International Development Working Paper No. 122.* Cambridge: Harvard University.

Hart, L. (2003). Some directions for research on equity and justice in mathematics education. In L. Burton, (Ed.), *Which Way Social Justice in Mathematics Education?* (pp. 25–49). London: Praeger.

Jacobsen, E. (1996). International co-operation in mathematics education. In A. Bishop, et al. (Eds.), *International Handbook of Mathematics Education* (pp. 1235–1256). Dordrecht: Kluwer.

Kaiser, G., Luna, E. & Huntley, I. (Eds.). (1999). *International Comparisons in Mathematics Education.* London: Falmer Press.

Keitel, C. & Kilpatrick, J. (1999). Rationality and irrationality of international comparative studies. In G. Kaiser, I. Huntley, & E. Luna (Eds.) *International Comparisons in Mathematics Education* (pp. 241–257). Falmer Press.

Krell, G. (2004). Managing diversity: Chancengleichheit als Wettbewerbsfaktor. (Managing diversity: Equity of chances as a factor of competitiveness). In G. Krell (Ed.) *Chancengleichheit durch Personalpolitik.* Wiesbaden: Gabler, 41–56.

Loden, M., & Rosener, J. (1991) *Workforce America: Managing Employee Diversity as a Vital Resource.* Homewood IL: Irvin Inc.

McInerney, P. (2004). Making *Hope Practical: School Reform for Social Justice.* Queensland: Post Pressed.

Plummer, D.L. (2003) (Ed.). *Handboook of Diversity Management.* Lanham, MD: University Press of America.

Powell, A., & Frankenstein, M. (Eds.). (1997). *Ethnomathematics: Challenging Eurocentrism in Mathematics Education.* Albany: Sunny Press.

Ragusa, A. & Atweh, B. (2003). *Analysing global collaborations using a post-structuralist model of social justice and social change.* Paper presented at the Social Change in the 21st Century conference. Carseldine: Queensland University of Technology: Centre for Social Change Research.

Rizvi, F. (1998). Some thoughts in contemporary theories on social justice. In B. Atweh, S. Kemmis & P. Weeks (Eds.), *Action Research in Practice: Partnerships for Social Justice in Education (pp. 47–56).* London: Routledge.

Robitaille, D. F., & Travers, K. J. (1992). International studies of achievement in mathematics. In D. Grouws (Ed.), *Handbook of Research on Mathematics Education.* New York: Macmillan, 687–709.

Secada, W. (1989). *Equity in Education.* Philadelphia: Falmer.

Secada, W., Cueto, S., & Andrade, F. (2003). Opportunity to learn mathematics among Aymara-, Quechua-, and Spanish-speaking rural and urban fourth- and fifth-graders in Puno, Peru. In L. Burton, (Ed.), *Which Way Social Justice in Mathematics Education?* London: Praeger, 103–132.

Sepheri, P. & Wagner, D. (2000) Managing Diversity—Wahrnehmung und Verständnis im Internationalen Management. Personal, *Zeitschrift für Human Resource Management,* 52(9) 456–462.

Thomas, R. R. (1996). *Redefining Diversity.* New York: Amacom.

Vithal, R. & Skovsmose, O. (1997). The end of innocence: A critique of "ethnomathematics." *Educational Studies in Mathematics, 34,* 131–157.

Wenzel, M. (2001). A social categorization approach to distributive justice: Social identity as the link between relevance of inputs and need for justice. *British Journal for Social Psychology, 40,* pp. 315–335.

Young, I. M. (1990). *Justice and the Politics of Difference.* NJ: Princeton University.

NOTE

1. The interviews were part of a project was supported by a grant from the Australian Research Council conducted in conjunction with Philip Clarkson.

CHAPTER 12

DEFROSTING AND RE-FROSTING THE IDEOLOGY OF PURE MATHEMATICS

An Infusion of Eastern–Western Perspectives on Conceptualising a Socially Just Mathematics Education

Bal Chandra Luitel and Peter Charles Taylor
Curtin University of Technology

ABSTRACT

Adopting a method of *writing as inquiry*, the paper deconstructs the overriding image of *mathematics as a body of pure knowledge*, thereby constructing an integral perspective of a socially just mathematics education in Nepal, a south Asian nation that is spiritually and historically rich and culturally and linguistically diverse. Combining a bricolage of storied, interpretive, reflective and poetic genres and an Integral philosophy, we envision a culturally contextualized mathematics education that is inclusive of Nepalese cultural, linguistic and spiritual diversities. This socially just mathematics education would enable Nepalese learners to: (a) co-generate mathematics from their

Critical Issues in Mathematics Education, pages 125–152
125

cultural contexts; (b) connect their lived cultural experiences with formal mathematics and *vice versa*; (c) take up social, cultural and situated inquiry approaches to learning mathematics; and (d) solve real world problems by using different forms of mathematics.

INTRODUCTION

How can the notion of social justice be incorporated into mathematics education in Nepal? This question comes to my mind whilst I (Bal Chandra Luitel) begin to reflect upon my recent professional activities as a mathematics teacher educator, thereby generating a number of nodal moments that demonstrate how the dualist nature of *mathematics as a body of pure knowledge* together with an arid teacher-centred pedagogy causes prospective teachers to undergo painful learning experiences. Such events resonate with my experience as an undergraduate student who could neither find a meaningful link between mathematics and his lived experiences nor enjoy his mathematics classes (Luitel, 2003). Embedded in the school mathematics curriculum of Nepal, the image of 'pure' mathematics is likely to have contributed to the rampant underachievement of Nepali students in mathematics, as reported by recent national studies (Koirala & Acharya, 2005; Mathema & Bista, 2006). The major consequence of such a phenomenon is to disadvantage students from gaining better opportunities in their present and future lives. This case of gross social injustice has prompted us to write this paper as a means of unpacking social justice perspectives in mathematics education in Nepal, a rapidly modernizing southern Asian nation with a largely agrarian based economy, and a nation that is spiritually and historically rich and culturally and linguistically diverse (with over 92 distinct languages).

The paper emanates from my ongoing doctoral research that employs an arts-based auto/ethnographic method of inquiry so as to construct a culture-sensitive transformative philosophy of mathematics teacher education in Nepal. Auto/ethnography is characterised by the method of *writing as inquiry* which affords a performativity of self-culture dialectics and critical reflexivity, an approach that recognises the development of the researcher's subjectivity during the process of inquiry (Richardson & St. Pierre, 2005; Roth, 2005). In this method, writing is constitutive of the process of inquiry, rather than being an add-on activity performed on completion of the inquiry, and gives rise to an emergent research design not dissimilar to investigative journalism or novel writing. In this arts-based approach we employ the notion of data generation and perspectival visioning (Clough, 2002) via a bricolage of storied, interpretive and poetic genres and reflective 'interludes' located strategically throughout the paper.

In the writing as inquiry process, Peter and I have performed varying roles as co-constructors of this paper. Elsewhere we have used the metaphors of 'architect' and 'builder' to portray our co-generative writing roles (Taylor, Luitel, Tobin, & Desautels, 2007). My primary role as builder-architect is to construct coherent texts whereas, as architect-builder, Peter reads critically and refashions my text, engages me in co-generative dialogue, and at times adds another brick to the wall. Of course, this dichotomy is somewhat simplistic inasmuch as the roles of builder and architect overlap and merge as we engage in the complex tasks of co-generative inquiry.

The paper begins with three semi-fictive cameos constructed on the basis of my experience as a mathematics teacher educator at the University of Himalaya[1] where I have been involved in developing and implementing a mathematics teacher education program that aims to produce secondary schoolteachers for Nepali schools. These cameos, which depict the image of *mathematics as a body of pure knowledge,* provide a basis for generating a hypercritical commentary incorporating three dimensions of social justice: recognition, inclusion and meaningfulness. In the first section, the commentary embodies an antithesis of the image of *mathematics as a body of pure knowledge,* a hitherto established view of mathematics that promotes a universalist agenda of mathematics as neutral in relation to cultural and political values. In the second section, a fictive dialogue ensues between me and three characters of the cameos as a means of generating synthesised perspectives about the nature of mathematics, inclusive pedagogy, meaningfulness of mathematics learning, and recognising non-Western knowledge traditions in mathematics education. This dialogue serves as an example of how we can rescue mathematics education from unhelpful social injustice promoting dualisms such as East versus West, content versus pedagogy, theory versus practice and knowledge versus activity.

Taking Integralism on board, the final section of the paper makes use of recent philosophical and political perspectives of education, historical-contextual information related to mathematics education in Nepal, and 'boxed poems' and dialectical reasoning as sources of integral vision making. Integralism derives mainly from Eastern wisdom traditions, such as Buddhism and Hinduism, and considers the process of knowing as organic, evolutionary and wisdom-oriented (Sri Aurobindo, 1952; Wilber, 2004) and dialectical (Wong, 2006). One of the many tenets of this philosophy is to emphasize the transformative synergy of inner self (Spirit) and exterior realities (Maya), thereby harnessing alternative logics of knowing, such as dialectical thinking, nondualism, metaphor and poetizing. Integral Philosophy (Wilber, 2004) is a referent for generating 'vision logic' to develop a socially justifiable mathematics education for Nepal.

DECONSTRUCTING THE SOCIAL INJUSTICE-LADEN MYTH
OF PURE MATHEMATICS: AN ANTITHESIS

Cameo I

After completing postgraduate studies at an Australian university in 2003, I continue to work at the University of Himalaya where I am responsible for mathematics education programs in the Institute of Curriculum and Teaching. I have a strong desire to upgrade the one-year diploma program into a fully-fledged masters program specialising in mathematics education. Pondering several possibilities, I quickly write an application to the director attaching a proposal that explains the needs of a master's course in our institute. Next day, I am summoned by the director and find myself discussing several issues related to the proposed program. One of his questions puts me in a difficult situation. The question is similar to this: "Will the new program incorporate enough 'pure mathematics'?"

Cameo II

Now, the Subject Committee is formed. I am in a meeting with members of the committee. I present a structure of the proposed two year masters course. After the completion of my presentation, four members start making comments. "There is no Advanced Pure Math," says Member One. Immediately Member Two comments, "There should be a unit on scientific decision-making process in the course." Member Three's concern is on the proposed credit hour of Pure Math II, which according to him is not enough to teach its content. Whilst I am thinking about how to respond to these questions, Member Four's blunt comment, "What has sociology to do with a mathematics education course?," situates me in yet another dilemma.

Cameo III

The program is launched with 22 students. The students soon start feeling under the weather with two units, Pure Math I and Pure Math II. I start hearing that students are not satisfied with these units. Then, I meet with the unit tutors, and soon find them blaming students for being lazy, disrespectful, incompetent and unmathematical. Tutor One laughingly blames me for teaching them 'unmathematical stuff', such as philosophy, pedagogy and ethnomathematics. Tutor Two prefers his class to be mathematically oriented in which, perhaps, he does not entertain questions and interactions. What should I do? I start one-to-one consultations with students. Many of them point to the tutors' didactic pedagogy and the highly abstract nature of the subject matter as contributing factors to the dilemma situation. In the midst of this dilemma situation, one student raises a serious question. He asks me: "Why have you prescribed the units of *Pure Math I* and *Pure Math II*, which have no direct connection with our professional practice?." He further indicates that these units are not helping him to be a good mathematics teacher; rather they are contributing to his pain and suffering.

INTERLUDE I

Now, what should I do with these cameos? My plan is to unpack the hegemonic nature of pure mathematics. Wait a minute! Am I going to be impressionistically critical? Yes, because I want to use a hypercritical genre (Van Maanen, 1988) so as to construct an antithesis to the thesis of pure mathematics being all-powerful and all-pervasive. Perhaps, this genre also partly shares the notion of a resistant reading which helps me (Faust, 1992) to interpret the cameos from the vantage point of my lived reality in which I experienced an unhelpful social hierarchy associated with the dominance of pure mathematics in Nepali mathematics education. Perhaps, my unfolding critique of pure mathematics can also be read from a subaltern perspective in order to compel readers to listen to the prevailing social injustice (Beverley, 2005).

Arriving at this juncture, I realise that Adorno's negative dialectic (Wong, 2006) is going to help me to 'discharge' a deconstructionist standpoint about pure mathematics. My hypercritical standpoint also garners support from Chinese dialectic that regards opposition as the precondition of changes (xiang-fan-xiang-yin; in Wong, 2006). And I gain insight from Shad-darshan (six Hindu schools of thought) (Radhakrishnan, 1927) that debates help me to generate understanding of the eternal.[2] Therefore, this critical view of mine has privileged Yang over Ying, *bibaad* over *baad*,[3] and antithesis over synthesis. For now, please read it that way.

I shall navigate my journey of interpreting the three cameos by means of three dimensions of social justice: recognition, inclusion and meaningfulness. The concept of 'recognition' helps me to uncover the perpetual ideology of non-recognition of difference in the field of mathematics education. Indeed, the notion of recognition 'could involve upwardly revaluing disrespected identities and the cultural products of maligned groups. It could also involve recognising and positively valorising cultural diversity *within the field of mathematics education*' (Fraser, 1997, p. 5, emphasis added). The notion of 'inclusion' (Young, 2000) refers to the extent to which participation is ensured for all those who are affected by the process of discussion and decision-making in mathematics education. The idea of 'meaningfulness' (D'Ambrosio, 2006a; Luitel & Taylor, 2007) is useful for considering the relevance and applicability of mathematics education in relation to the cultural lifeworlds of learners. In what follows, my unfolding interpretation of the three cameos aims to clarify pertinent issues of social justice in Nepali mathematics education.

Recognition

There seems to be a dissonance between the metaphor of *mathematics as a pure body of knowledge* and the idea of recognising differences in mathematics education. The term 'purity', from both literal and metaphorical perspectives, appears to entail a notion of superiority, thereby involving students in following a rigid dogmatism. Can superiority and recognising others go together? In what follows, I argue that the othering discourse of pure mathematics seems to create an entanglement with other knowledge systems which entitles them as inferior, powerless and non-mathematical. I see the transmission of the message of *pure-mathematics-is-all-powerful* as undermining the inventiveness and emergence of cultural activities.

The problem deepens further as pure mathematics recognises only a particular knowledge system based in Westocentric ontology, epistemology and axiology (D'Ambrosio, 2006b; Taylor & Wallace, 2007). A question arises: Whose interest is being served by pure mathematics? It seems to me that *mathematics as a body of pure knowledge* promotes the twin myths of hard control and cold reason (Taylor, 1996) so as to camouflage the authentic image of mathematics as uncertain and unfolding human activity. By subscribing to uncertainty as an epistemic metaphor, we can facilitate learners becoming constructors of mathematics from their *own* lifeworlds. Where Skovsmose and Valero (2001) use the notion of 'internalism' to criticise the self-satisfying nature of mathematics education research, I use this concept to critique the dominant nature of pure mathematics that imposes a circular 'self-justificatory system' (Lerman, 1990) in an attempt to misrecognise the local, implicit and cultural nature of mathematics.

Pure mathematics seems to subscribe to a Platonist standpoint that regards mathematical knowledge as independent of the knower, leading to the notion that mathematics is an ideology-, culture-, and worldview-free subject. In an era of democracy, this perspective has major implications for the education of young men and women, amongst which is the concern that they may develop a narrow seemingly ideology-free view of the nature of mathematics. However, mathematics has never been free from ideologies (Gutstein, 2003); rather, it has developed from certain interpretive, linguistic and observational standpoints. It seems to me that depicting mathematics as an ideology-free subject has helped to colonise non-Western cultures through scientific, technological and educational interventions by materially rich Western countries (D' Ambrosio, 2006b). In a (World-Bank-defined) 'developing' country such as Nepal, importing pure mathematics from materially rich Western countries and then 'stuffing' it into students without due recognition of their cultural worldviews creates a chain of social injustices within the landscape of mathematics education (Luitel & Taylor, 2007).

Historians of mathematics (Boyer, 1968; Eves, 1983) point out that mathematical knowledge has not been developed overnight; rather it has been brought forward by human endeavours and then shaped by contemporary social, cultural and political factors. If pure mathematics used this historical insight to enhance its pedagogy there could be the possibility of recognising the different mathematical worlds of learners. However, by becoming the bastion of epistemic certainty, pure mathematics seems to ignore the historical contingency of mathematical knowledge. This ignorance of its own history further creates an illusion in which to see mathematical purity as extra-human, extra-cultural and extra-social. There is a high chance that the image formed by the many 'extras' will continue to steer the discourse of mathematics education, thereby harbouring a pedagogy of non-recognition.

Inclusion

Let me start this section by referring to Cameo 3 in which a character, Tutor One, indicates that *his* pure mathematics is all-powerful, and that the students are not capable of *doing* this. The students are all mathematics graduates, and so I wonder from where and when on Earth pure mathematics might recruit its ideal students, those who are teachable according to the archaic philosophy of cultural reproduction (Bourdieu & Passeron, 1977) by which one reproduces an hierarchical, elitist, meritocratic and competitive culture of pure mathematics? Do the underlying notions of cultural reproduction and inclusion match? As 'inclusion allows for maximum expression of interests, opinions, and perspectives relevant to the problems and issues' (Young, 2000, p.23), the intention of instilling a dominating pedagogy of reproduction may cause inclusion not to flourish within the pedagogical landscape of pure mathematics.

Historically, the term 'inclusion' seems to have a close link with 'special education' in that inclusiveness refers to a non-segregated educational practice for differently able learners (Pickles, 2004). For me, the notion of inclusion refers to increasing the participation of all learners in mathematics, thus reducing the possibility of their exclusion from classroom activities. If I ask Tutor One and Tutor Two about how they might ensure that all students participate in *their* mathematics and how they might avoid the possibility of exclusion, they would likely answer that they need to find a way to make students listen attentively to their lectures, to rote memorise (oxymoron) theorems, to be present on their test days, and to submit their tutor-imposed assignments. Having taught according to the Logicist-Formalist school, they may apply the 'Excluded Middle Principle' (Kline, 1980) so as to claim that there is no possibility of exclusion as they have already claimed students' participation.

In defence of pure, formal and academic mathematics Rowlands and Carson (2002; 2004) argue that mathematics represents an artificial cultural system that is independent of human culture. They seem to renew the defence of the canon of pure mathematics by shifting the age-old argument from *mathematics as a culture-free enterprise* to *mathematics as an artificial culture*. This new argument, however, seems to be a renewed interest in claiming pure mathematics to constitute a non-human culture. Does the metaphor of *mathematics as an artificial culture* permit more than technical participation of learners in the discourse of mathematics? Perhaps, the idea of artificiality seems to fit well with the idea of *technical* or *artificial participation* through lectures, assignments, narrowly-focused exams and so forth. Thus, there are few possibilities of moving forward to *practical participation* in which to emphasise the meaning-making enterprise of pure mathematical content. How about critical participation in the discourse of pure mathematics (Taylor & Campbell-Williams, 1993)? *Critical participation* entails situations in which learners can question the relevance of pure mathematics and its pedagogy to their teaching profession. However, it may be hard to marry the self-proclaimed purity of pure mathematics and critical participation of learners because criticality may be difficult to sustain within a closed and self-referential system.

In the quest for an inclusive mathematics education, I believe we need to recognize two types of exclusion, 'external' and 'internal' (Young, 2000). External exclusions are generally attributed to official decisions about curriculum, learning style, assessment and other issues that are directly concerned with students' well being. Internal exclusions may be subtle, and are generally caused by hidden curricula that are prevalent in sites of pedagogical enactment such as the mathematics classroom. For instance, although students are told that they can participate in the discourse of mathematics, only a select group may be able to fully participate because of a number of inhibitive factors, such as the low language proficiency of learners and nonrecognition of participation. Can mathematics as a body of pure knowledge be an appropriate metaphor for developing an inclusive pedagogy of mathematics education? I have strong doubts. Perhaps by embracing alternative metaphors, such as *mathematics as contingent, corrigible, fallible* and *an ever-developing knowledge system* (Ernest, 2006), we can pave the way for developing an inclusive pedagogy.

Meaningfulness

My recent reading of ethnomathematics, critical mathematics education and various forms of constructivism (Cobb, 1994; D'Ambrosio, 2006a; Ernest, 1997, 1994a; Glasersfeld, 1995; Skovsmose, 2005) gives me a sense that one of the many common aspirations is to enhance students' experience of meaningfulness in mathematics education. Ethnomathematicians argue that mathematics is a cultural and historical construct, and therefore it should be linked with the cultural practices of learners in order to make it meaningful according to their lifeworlds. Similarly, different constructivist perspectives emphasise a meaning-making pedagogy of mathematics. These perspectives, to varying degrees, allow learners to develop multiple standpoints on mathematical concepts, thereby internalising viable concepts through an active construction process. In particular, critical perspectives regard learners as ends-in-themselves rather than as objects to be manipulated as an instructional means for achieving something else.

The metaphor of *mathematics as a body of pure knowledge* tenders a vexing image representing the told and untold stories of meaningless mathematics (Luitel, 2003). Viewed from an ethnomathematical perspective, such *a* mathematics seems to impose on learners a fixed and culturally dislocated image. As the image of pure mathematics seems to subscribe to instructivist pedagogy, there may be few opportunities to afford learners opportunities to develop multiple sensibilities of mathematics or to engage actively in making meaning out of their daily lifeworld activities. Considering the idea of meaningfulness from the perspective of liberatory pedagogy (Freire, 1996) seems to lay emphasis on bolstering the individual learner's agency, thereby awakening them to understand their own socio-cultural valuing, becoming and being. Does pure mathematics allow learners to act as ends-in-themselves? Does it help widen learners' sense of being and valuing through their participation in mathematics learning?

The notion of the meaningfulness of mathematics seems to have links with its direct applicability to the cultural lifeworld of the learner. In saying so, I am aware that there are layered interpretations of key terms such as *applicability, lifeworld* and *learner.* The idea of applicability can be described as the extent to which mathematics is translatable to particular cultural contexts. The cultural context of application that I am referring to is education, and thus the lifeworlds of learners constitute primarily schools and educational institutions. Now a couple of questions arise: Is pure mathematics applicable to the immediate lifeworlds of learners? In what follows, I briefly address this question.

Perhaps, it is not an over interpretation to argue that pure mathematics has little to say about the cultural realities of many people. Restivo (1994) argues that one of the main qualities of a good mathematician is the extent to which s/he is able to represent his/her mathematics abstractly. This can be put simply as: the more abstract the mathematics the higher the recognition as a mathematician. The emphasis on abstractness, the hidden rule of being incomprehensible by common people, and the convention of using symbolic language that has little relevance to day-to-day realities are key characteristics of pure mathematics (Restivo & Bauchspies, 2006). In totality, these characteristics convey the image that pure mathematics tries to dispel the cultural (Luitel & Taylor, 2007) and embodied (Nuñez, 2006) nature of knowing, thus endorsing a culturally dislocated mathematics education.

INTERLUDE II

Peter's question, "Do you want to deconstruct pure mathematics or the hegemony of pure mathematics?" causes me to think for a while. At this stage I think that I may be deconstructing both. Does this mean that I am *rejecting* pure mathematics? Can I maintain my personal sustainability by rejecting pure mathematics?

While writing the above commentary I felt that I was constantly agitating against pure mathematics. Does this mean that I am also writing against myself as some years back I was teaching a similar type of mathematics? Did I recognise students' ideas? Was I inclusive to all students? What strategies did I use to make pure mathematics meaningful? So what should I plan for? Perhaps I cannot live forever in a hypercritical world. This is as dangerous as the reclusive world of pure mathematics. I need alternative logics that can help me reconceptualise a balanced view of mathematics education in Nepal.

In the upcoming section, I use 'hyphen-logic' to defuse many unhelpful dichotomies. Perhaps, this is the stage of my attempting to know 'tatva', a Sanskrit word that means 'eternal truth'. You may notice that the following section helps me to synthesise *Iam-notIam, Ying-Yang, Bad-Bibad,* and *thesis-antithesis* in order to paint a holistic picture of a socially just mathematics education in Nepal.

UNPACKING SOCIAL JUSTICE PERSPECTIVES THROUGH SEMI-FICTIVE DIALOGUES: A SYNTHESIS

Nurturing an Inclusive Mathematics Education

Director: Throughout my career as an educational administrator, I held the belief that pure mathematics is universal and objective. Therefore, it has nothing to do with politics, culture, day-to-day affairs or social justice. To me, there is a danger of your critical standpoints being interpreted as idiosyncratic rather than social justice-oriented.

Bal Chandra: One can hold the view that mathematics is an 'objective' and 'universal' knowledge system, however, my concern is that this is not recognised as *a view*, rather it has served as *the only view* of mathematics in the educational landscape of Nepal. Although I agree that earlier in this paper I was too critical of pure mathematics, some aspects of my critique mirror, perhaps, prevailing unhelpful dichotomies of *pure* versus *impure* mathematics, *hard* versus *soft* mathematics, *Western* versus *local mathematics*, *content* versus *pedagogy* mathematics. Perhaps holding an historical view that mathematics evolves through different historical, cultural and political junctures can help us to achieve a balance between such dichotomies. A developmental synthesis can be foreseen by providing learners with an epistemic context in which different natures of mathematics co-exist and co-interact.

Tutor One: First of all, I am surprised that you choose the issue of social justice in mathematics education. This issue is overly political, thus it has nothing to do with the mathematics I teach. Do you think that your overly critical approach helps teachers to be aware of social justice?

Bal Chandra: At some stage in my thinking, the vantage point of a critical perspective helped me to understand social phenomena as constitutive of power relationships. This can be useful for viewing pedagogical activities as an enactment of power relationships amongst different stakeholders. For instance, the existing power relationships between mathematics teacher and students can be oppressive to students because of the prevailing teacher-centred pedagogy in Nepal. However, I acknowledge that this perspective often promotes an unhelpful oppressor-oppressed dichotomy. Perhaps, it is worth exploring an alternative perspective that regards teaching as a desire-less, selfless and compassionate (Sri Aurobindo, 1952) act in which to devote one's professional life to the wellbeing of the other. And more so, perhaps an approach that combines both views, *teaching as a self-less act* and *teaching as a political act*, can help promote a socially just mathematics education in Nepal. As a teacher I can act selflessly and compassionately whilst, at the same time, being aware of the possibility of power misuse by me during my pedagogical enactment.

Tutor Two: If you acknowledge that there are disciplines then you need to understand that they have agreed upon standards of what counts as knowledge. Pure mathematics is not an exception. While taking social justice on board, you could argue that teaching powerful ideas of pure mathematics could empower prospective teachers and then their future students. Because if the prospective teachers understand theorems, definitions and mathematical problems, they can teach mathematics to their students more effectively. In this way social justice would prosper as a result of mathematics education.

Bal Chandra: I accept the fact that different branches of mathematics have been unified according to their respective standards. At times, however, these standards are contested, questioned, and reformulated (Ernest, 1994a, 1994b). This does not mean that *all* such standards are changed to the same degree; rather the varying degrees of change, from minimal modification to total replacement, is a possibility.

I agree with your view that provisions for exploring many powerful ideas of mathematics are desirable in order to enhance rich and sustainable understanding of the world lived in by learners. Although I acknowledge the importance of pure mathematics in providing us with rules, patterns and structures, I beg to offer an inclusive view that powerful mathematical ideas emanate from our day-to-day lifeworlds and cultural activities. Perhaps, this view also gives room to pure mathematics to provide learners with discursive opportunities to search for meaningful and sustainable mathematical ideas.

What can be the salient features of such powerful mathematical ideas? I envision them to be: (a) adaptable to the cultural realities of learners, (b) useful for enhancing rich understandings of social and cultural phenomenon, (c) meaningful for each learner according to his/her needs and interests, and (d) emancipatory for learners' development of critical understandings of themselves and the world around them. Perhaps an inclusive view of (mathematics) intelligence that takes into account different modes of thinking and acting practised by the linguistically and culturally diverse Nepali population would help to formulate a compatible framework for assessment of such powerful mathematical ideas (Sternberg, 2007).

I find that *mathematics as cultural activity* is a more inclusive image than foundationalist metaphors such as *mathematics as a body of knowledge* and *mathematics as a culture-free enterprise.* Perhaps by enacting a cultural activity metaphor we can offer an alternative view that mathematical theorems, definitions and problems are human constructions in which to celebrate differing natures of mathematics.

Developing a Recognition-Oriented Pedagogy

Director: While interpreting your stories from social justice perspectives, you have identified an interesting dimension of recognition. In what follows, you argue that pure mathematics is very weak in recognising others, thereby creating an socially unjust pedagogy of mathematics education. This idea seems to be very philosophical. What is your vision for bringing the idea of recognition into pedagogical practice?

Bal Chandra: I have benefited greatly from following an ongoing philosophical debate about the nature of mathematics (Ernest, 1991; Hersh,

1997; Lakoff & Núñez, 2000). Such debates have unpacked different views of mathematics such as *mathematics as metaphorical embodiment, human activity,* and *invented knowledge.* As an educator, I believe that philosophical debates offer ways to develop a comprehensive vision for developing a social justice oriented mathematics education in Nepal. Perhaps the use of different philosophical referents is indicative of an imminent global recognition of different viewpoints about mathematics. There may be a critical view that pure mathematics is elitist and non-recognisant whereas an opposing view asserts that pure mathematics is useful for developing analytic-speculative reasoning. But can we afford to live by a single viewpoint? Perhaps we need to be cognisant of these differing views in order to instil a recognition-oriented pedagogy of mathematics.

What might the mathematics education landscape of Nepal look like when a recognition-oriented pedagogy is incorporated? Perhaps we would see: (a) mathematics teachers developing the sensibility that they need to respect students' ideas; (b) members of curriculum committees recognising that there are alternative knowledge systems that should be included in mathematics curricula; and (c) mathematics teacher educators providing prospective mathematics teachers with an authentic experience of this pedagogical tool. Beside this, I envisage that the notion of recognition as a pedagogical referent can be helpful for conceptualising a multicentric pedagogy in which each student can develop the feeling of being valued and acknowledged.

Tutor One: I also have a problem with the term *recognition.* I think you are trying to create a utopia in which all students have different mathematics and will enjoy a 'whatever goes' mathematics. This kind of social justice may dumb down the students.

Bal Chandra: I see two antithetical worlds evolving in the hypercritical genre that I used earlier to interpret my cameos. The world of pure mathematics appears to be the promoter of social injustice in the mathematics education of Nepal while the hypercritical world seems to reject the presence of pure mathematics thereby promoting a utopic view of mathematics. Now, it seems to me that both views, *mathematics as a body of knowledge* and its alternative critique constitute a dualistic tendency of avoiding the other. How is it possible to develop an encompassing alternative that recognises both views?

Perhaps, a synergy that amends the current monological mathematics pedagogy by recognising alternative natures of mathematics is inevitable. It also would help to be open to the emergence of different mathematical knowledge systems that are prevalent in the diverse linguistic and cultural landscapes of Nepal. Instilling a recognition-oriented pedagogy can help learners to become aware of their agency. I further envisage that a helpful synthesis of *mathematics as a pure fixed knowledge system* and *mathematics as cultural activity* can be represented by *mathematics as mediated social discourse* which gives rise to multiple forms of mathematics. Perhaps, enacting multiple mathematical discourses will eventually reinforce *mathematical literacy*, a key component of students becoming a responsible citizenry.

Tutor Two: You state that pure mathematics enforces its own way of validating a single form of knowledge through enculturation. I am particularly anxious about your framework of social justice which seems to demean mathematical enculturation by means of which we develop mathematical thinking among students. To get rid of mathematical enculturation we need to stop teaching pure mathematics, which I think is suicidal to the future of mathematics education. The more effective the mathematical enculturation the better the mathematical thinking. I will subscribe to this type of social justice rather than demeaning the power of pure mathematics.

Bal Chandra: I cannot deny some degree of enculturation in mathematics education. However, we also need to be aware of pure mathematics unhelpful emphasis on enculturation as one-way cultural border crossing. The prevailing practice of enculturation seems to have more demerits than merits, particularly when the aim is to change students' beliefs rather than promote multiple (sometimes contradictory) conceptual understandings. Perhaps an inclusive alternative to enculturation is *acculturation* (Taylor & Cobern, 1998) which 'comprehends those phenomena which result when groups of individuals having different cultures come into continuous first-hand contact, with subsequent changes in the original culture patterns of either or both groups' (Redfield, Linton, & Herskovits, 1936, p. 49).

One of the benefits of acculturation is being inclusive of some aspects of enculturation. For instance, when two students share their differing views of a mathematical concept they need a language that is shared by both. The substantial focus of acculturation, which is on crossing multiple cultural borders, seems to promote a progressive form of enculturation which raises awareness of mainstream interpretations of mathematical concepts as well as their limitations. I think that subscribing to acculturation as a pedagogical referent would enrich mathematical thinking among learners. Would this not help to enhance the rigour of mathematics?

Enacting Pedagogical Inclusion through Integralism

Tutor One: While talking about the social justice dimension of inclusion, you have criticised conventional educational activities such as lecturing, presentations, exams and assignments. I think we need a framework with which we can help students to develop correct understandings of pure mathematics. Does your social justice framework regard lecturing, tutor-made assignments and exams as sources of social injustice?

Bal Chandra: Perhaps multiple pedagogical referents are appropriate for addressing the diverse needs of students. By using metonymical representations, such as *teaching as lecturing, assessment as probing,* and *students as object,* we have narrowly reduced the purpose of education to passing an exam. Indeed, there is nothing wrong with lecturing as long as it is a pedagogical means to a socially just end, rather than serving as an end in itself. Perhaps by using integralism as a pedagogical referent for promoting inclusivity we can unite the spiritual (Inner) and worldly (Outer) realities of the learner.

The antithesis that tutor-made assignments are a source of social injustice has some merit inasmuch as tutors may be habituated to using a one-size-fits-all approach to assessing students. However, offering only an ideological critique may not solve the problem. Perhaps a developmental assessment approach that also considers 'contextual intelligence' (Sternberg, 2007) can promote participatory, teacher-student mediated, explanatory and learner-owned pedagogies in mathematics education.

Tutor Two: I have a problem with making sense of your different modes of participation. It is clear that you do not like technical participation in which students gain mastery of mathematical concepts, mathematical proofs and mathematical problem solving. Your second preferred mode of practical participation does not help without students firstly knowing the basic knowledge and skills of pure mathematics because the making of meaning cannot be accomplished without fundamental ideas of the discipline. In what follows from your writing, critical participation seems to provide a space for being critical about the subject they study. I think this is a very awkward formulation because students are not capable of critiquing the subject matter until they develop a high level of expertise in mathematics.

Bal Chandra: It may appear in my discussion that the three modes of participation—technical, practical, critical—are hierarchical and thus separate from each other. However, my intention is instead to explain relationships between the three modes of participation by means of the concept of 'holarchy' which helps reformulate mistakenly conceptualised hierarchies by using the notion of holism (Wilber, 1996). In what follows, technical participation is the basic element, with other modes of participation being embraced successively (see Figure 12.1). I am critical of *mere* technical participation of learners because it seems to focus solely on one-way cultural border crossing. Given the nature of holarchy, practical participation is inclusive of technical participation. Similarly critical participation is inclusive of technical and practical modes of participation. I do not think that critical participation is only the business of experts in mathematics; rather it can help students to be aware of their *own* false consciousness about the nature of mathematics being studied. Perhaps employing the notion of holarchy would help change our perception of these three modes of participation from ordered and hierarchical to

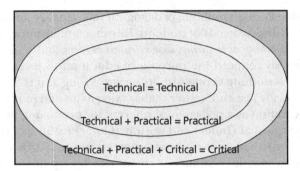

Figure 12.1 A *holarchy* of three modes of *participation*.

developmental and interactive. Employing technical participation alone can be imposing while combining its use inclusive with other modes of participation can be empowering.

Meaningfulness for Social Justice

Director: I have not seen that much of a direct relationship between meaningfulness and social justice. Do you intend to say that *every mathematics* we teach in school should be applicable to the immediate lifeworlds of the learners? If so, you have narrowly defined the notion of meaningfulness. I do not think that whatever we teach today should be totally applicable to their immediate world. Delimiting everything to learners' day-to-day realities may not address the future aims of education. To me the notion of meaningfulness needs to be interpreted from a broad vision which is capable of prescribing 'future mathematics' that prospective teachers will eventually use in their teaching.

Bal Chandra: Although meaningful mathematics can empower learners by helping them to make sense of their lifeworlds through mathematical concepts this, however, is not to subscribe to the view that we should stop thinking about the future. It would be helpful to consider the present-future dialectic; what is in the present can also be part of the future and what is not today can be a basis of what will constitute the future. I find it helpful to consider the present and future not as separate entities but as always being connected. Given the dialectical relationship between present and future, it is important that we consider *education as life* (Dewey, 1943) rather than preparation for life. By employing the *education as life* metaphor we start to think about the type of mathematics that is useful for learners' unfolding lifeworlds. Does this not solve the present-future dichotomy embedded in our conventional conceptualisation of mathematics?

Tutor One: I like the notion of meaningfulness which tends to imply the applicability of mathematics to one's immediate lifeworld. However, I would argue it differently. Mathematical meanings are very powerful and learners can get them only when they are richly acquainted with enough of the content knowledge of mathematics. How can a learner realise, for example, that Roll's theorem[4] can be applied to find the maximum height of an undulating mountain without actually being exposed sufficiently to Roll's theorem? In order to be able to apply any such theorem students need to understand correctly the fundamentals of pure mathematics.

Bal Chandra: You seem to subscribe to the view that meaningfulness is achieved only after being adequately exposed to the subject matter of mathematics. An antithesis of this view is that the subject matter unfolds after a relevant activity grounded in the learners' lifeworlds. The conventional idea of privileging subject matter has some merits, such as (a) the subject matter can be a guiding tool for learning, and (b) learners can benefit from having an advance organiser before starting a mathematical application. On the other hand, starting mathematics through an activity grounded in the learner's context can be empowering for them as they start mathematising contextual problems, thereby generating meaningful mathematical knowledge. Which pedagogical approach can enhance such a notion of meaningfulness in mathematics? Perhaps an approach that promotes interaction between algorithmic-deductive (subject matter first) and contextual-inductive (activity first) in which learners use both approaches thereby complementing the inefficiency of one by the effectiveness of the other. For instance, perhaps starting with a contextual problem, then making links with Roll's and other theorems, and then making several backwards and forwards movements between theorems (subject matter) and contextual problems, can help make Roll's theorem more fully meaningful to learners.

Tutor Two: While discussing the notion of meaningfulness we need to remind ourselves that pure mathematics has the power to differentiate mathematical and non-mathematical objects, concepts and problems. Mathematically, applicability can be interpreted as the power of recognising and applying appropriate mathematical ideas. What do you think about this?

Bal Chandra: An antithesis of *some-phenomena-can-be-more-mathematical-than-others* is that categories such as mathematics and non-mathematics constitute our contingent forestructures. Some aspects of this antithesis help us to be open about the emerging and evolving nature of mathematical knowledge. When some aspects of emergence are incorporated into mathematics pedagogy learners can benefit by the underlying metaphor of *mathematics as somewhat open inquiry.*

How and when is the power of mathematics demonstrated? One view is that mathematical power is demonstrated by solving algorithmic problems of mathematics. In this model students need to gain a mastery of the technical knowledge of pure mathematics before applying it. Alternatively, students can be provided with rich contexts in which to explore (pure) mathematical ideas through contextualised activities such as social inquiry, collaborative projects, and other culturally-situated actions. Perhaps these contextualised activities can help students develop multiple sensibilities of mathematical concepts thereby helping them to realise the power of mathematics-in-the-making.

INTERLUDE III

By using dialectical thinking I have attempted thus far to develop an inclusive social epistemology and pedagogy of mathematics, particularly for teacher education. This "hyphen logic" now leads me to incorporating "vision logic" and "poetic logic" with which to see into the present-future in a way that regards both categories as One, just like a full day comprises day and night. This juncture is crucial for me because I need to think about how the upcoming section will emerge.

My recent readings of Aurobindo, Shantideva and the I Ching encourage me to use the power of poetic thinking to explain the unexplainable. As the popular saying goes, "a poem reaches beyond the sunrays,"[5] I also find Sama Veda, one of the Vedic sacred epics, promoting poetic thinking as important as other types of thinking.[6] These poems can be read as my layer of thinking which is hard to unpack by means of a straight sentence writing style.

ENACTING SOCIAL JUSTICE VIA CULTURALLY CONTEXTUALISED MATHEMATICS EDUCATION

The popular Nepali adage, *'Don't forget your soil'*, is worth mentioning here as we attempt to make connections between Nepali spirituality and the notion of a culturally contextualised mathematics education. The soil metaphor gives a sense of identity, worldview and cultural activities of people situated in a particular context. Elsewhere (Luitel & Taylor, 2007; Luitel & Taylor, 2005; Taylor et al., 2007) we have discussed ways in which mathematics education in Nepal can start to *remember its soil* and begin to address the outstanding problem of gross underachievement and exclusion in mathematics education. In this paper we renew our vision of culturally contextualised mathematics education in Nepal, and extend it in relation to the notion of social justice. In the process we have anticipated a number of questions being directed to us: Which perspectives and theories are behind the notion of contextualisation? How does contextualised mathematics education reconcile with conventional mathematics education? In what ways does the idea of contextualisation promote social justice in mathematics education? In what follows, we try to explore possible answers to these and other evolving questions, thereby further crystallising our idea of contextualised mathematics education.

...Philosophically Speaking

In the last three decades, we have witnessed an upsurge of educational philosophies giving rise to new visions of how to improve the outstanding problems of education. In this process, radical constructivism, social constructivism and social constructionism have emerged to redefine the conventional knowledge system and propose new sets of knowledge standards. In so doing they have critiqued the long-standing Platonist view of mathematics and offered exciting possibilities for mending injuries caused by earlier foundational philosophies. These alternative philosophies tend to offer an inclusive pedagogical practice with which to enrich learning through continuous, developmental and authentic assessment approaches.

> _Poetically speaking_
>
> _By the help of 'that'_
> _I explain 'this'_
> _Fusing this and that_
> _I generate a vision of_ thisthat.
>
> Thisthat _is not_ this
> _Because it has that within it._
> Thisthat _is not_ that
> _Because it has_ this _within it._
>
> Thisthat _is also this_
> _Because it is inclusive of that_
> Thisthat _is also that_
> _Because it is inclusive of this._
>
> _I start with two: this versus that_
> _I make three: this, thisthat and that_
> _Finally I come to one: the_ spirit
> _which helps me to be inclusive._

More than that, the post-epistemological nature of radical constructivism has opened up a powerful new framework of 'viability' by means of which the mathematical knowledge system can be explained as temporal, situated and experiential. Social constructivism has challenged the foundational philosophies, refining further Lakatosian quasi-empiricism and offering alternative metaphors to represent the nature of mathematics: _mathematics as a dialogic, fallible, corrigible_ and _socially constructed knowledge system_ (Ernest, 1994). And, in recent years, social constructionism has challenged the basic conventional assumption of knowing as a dualistic enterprise and posited the notion of knowing as constructing the world through social artifacts, 'products of historically situated interchanges among people' (Gergen,

1985, p. 267). To construct mathematical knowledge according to this perspective requires a cooperative enterprise in which to consider the multiple historical and cultural bases of the knower in relationship with 'the other'. In such a situation different mathematical knowledge systems are validated through the vicissitudes of social process that are based largely on the cultural practices that are shared by people-in-relationships.

...Politically Speaking

In critiquing the overriding conventional view of culture-free mathematics D'Ambrosio (2007) argues that modern mathematics has been a means of oppression. His rescue mission, *program ethnomathematics*, is influenced by the work of Paulo Freire (1996) who has been regarded as the champion of liberatory pedagogy. Whereas much of Freire's work has been situated in the field of adult literacy, D'Ambrosio (1999) makes a case against the widespread cultural deprivation perpetrated by Westernised mathematics. D'Ambrosio's (2007, p. 179) *program ethnomathematics* contributes to "restoring cultural dignity and offers the intellectual tools for the exercise of citizenship." We feel a solidarity with D'Ambrosio to the extent that his view of ethnomathematics offers an inclusive vision of mathematics education in Nepal.

Critical mathematics education is also useful for examining the empowering and disempowering roles of mathematics education (Skovsmose, 2005) in Nepali society. It offers a critical view of how formal/academic mathematics is responsible for perpetuating through conventional mathematics education a hierarchical social structure. The *apartheid* education system of South Africa can be considered to be a visible example of how academic mathematics, together with other political influences, became a tool of segregation and oppression (Khuzwayo, 1998). In a similar vein, concerns about the introduction of ethnomathematically inspired pedagogical approaches in mathematics education have been raised. It is argued that if ethnomathematics is used as the sole pedagogical referent for mathematics education, learners' would become alienated from conventional mathematics and lose access to social choices which arise from developing this knowledge (Vithal & Skovsmose, 1997). We are sympathetic to this concern because we believe that history demonstrates clearly that the imposition of a single nature/referent/mode of mathematics education has not catered well to the needs of a culturally diverse Nepali population. A narrowly conceived mathematics education, whether it be conventional or radical, is likely to fail to enable learners to cross multiple mathematical (and cultural) borders, thereby restricting development of their lifeworlds.

...Contextually Speaking

The widespread underachievement in mathematics in Nepal is evident as more than 60% of the students fail each year in mathematics in the national exam of School Leaving Certificate (SLC) (Mathema & Bista, 2006). Similarly, a number of national studies have shown general underachievement and poor participation rates of pupils in mathematics at the primary level ((EDSC, 1997, 2003). The status of female students is worse: a recent UNESCO-initiated study (Koirala & Acharya, 2005) found that Nepalese girls' achievement in school mathematics is consistently below that of boys.

A brief historical perspective can help interpret the phenomenon. It is interesting to note that formal education in Nepal started initially with the blind importation of a British model in 1853. Although there was only one school and it catered to the ruling class, the Rana and Shaha families, present day education in Nepal seems to have followed in the same footsteps by continuing to import blindly education and curriculum models from materially rich counties of the North and the West. However, the methods and modes of importation have changed in different historical periods. Until the 1950s, Nepal relied on foreign textbooks, teachers, examination boards and affiliating bodies. After the 1970s, Nepali education entered into a new phase of importing educational 'software' and 'hardware', by swinging back and forth as per the interest of politicians, bureaucrats and donors. In

> *Poetically speaking*
>
> *What is history?*
> *Perhaps a ~~h~~istory*
> *That has been ~~h~~istor^{ed}~~y~~*
> *In our lifeworlds*
>
> *Multiple ways of telling-*
> *verse, prose, song and dance*
> *Different voices of reading-*
> *Responsive, resistant and dialogical*
>
> *History flows as though a river*
> *Changing its pathways with floods and*
>   *streams*
> *Some worship the river*
> *for bringing hope to them*
> *Some term it a devil for causing all sorts*
>   *of pain*
> *Some see themselves as the source of the*
>   *problem*
> *Not the river.*

this process, a culture deficit theory seems to have prevailed in justifying the importation of education and curriculum models. This is not to say that we should stop learning from others' experiences. Intercontextual and intercultural transferability can be an alternative to this blind importation.

Another reading of Nepali history generates a positive image of growth in which the literacy rate has risen to about 55%, almost every Village

Development Committee has at least one school, the small country has five functional universities, a significant number of professionals are produced within the country, and a realisation that education should be grounded in day-to-day realities (mainly language) of the learner is becoming pertinent. However, these questions arise: Is this educational growth distributed evenly amongst the population? To what extent has rural Nepal benefited from this growth?

Given these historical realities, Nepali mathematics education remains foreign to most Nepali students even for those who are performing well (Luitel, 2003). Perhaps, the influence of blindly imported mathematics curriculum and textbooks continues to bolster an image of *mathematics as a body of knowledge*, thereby undermining otherwise progressive images of mathematics such as *mathematics as cultural activity*. Perhaps, the overriding influence of pure mathematics in mathematics pedagogy, together with a narrowly conceptualised notion of *assessment as an add-on activity to teaching*, has fuelled the existing problem of underachievement.

...Dialectically Speaking

In what ways will a culturally contextualised mathematics education be different from the existing conventional mathematics education in Nepal? Perhaps first it is important to discuss the notion of culture. Metaphorically, culture can be understood as artefacts, schema, refinements, cultivation, human activities, or a production site of meaning (Schech & Haggis, 2000). Owing to the limitation of any particular metaphorical image, we conceptualise culture in terms of a range of dialectical pairs: consciousness|representation, artefacts|meanings, schema|activities, and reification|enactment.[7] Metaphorically, the symbol for *Nand* (|) represents a dynamic synthesis of opposing pairs as a means of developing holism. Each dialectical pair represents a recursive relationship between the two adversaries, at times producing internal contradictions and complementation whilst maintaining an evolving coherent system. This dialectical perspective helps us to generate a view that culture never remains unchanged, rather it is always in flux, offering a forestructure for viewing and interpreting the complexity of the world.

> *Poetically speaking*
>
> () *is neither this.*
> *Therefore* () *is not that.*
> () *is here.*
> *Therefore* () *is there.*
> () *is not here.*
> *Therefore* () *is not there.*
> () *is this and that.*
> () *is here and there.*
>
> *This co-exists with that.*
> *Here makes sense of there.*
> *Yes exists because of no.*
> () *makes sense of* ~ ().
> *'I am' because 'I am not'*

Similarly, drawing on the Sanskrit equivalent of culture, *sanskriti*, gives us a meaning for culture as involving a constant refinement of our actions.

In employing this dialectical perspective a mosaic of multiple natures of mathematics forms in our discursive landscape. In this process, we realise that the notion of dualism does not fit well as a theoretical referent because of its inability to explain the enactment of a complex mosaic of multiple mathematical natures. By contrast, our nondualist-integral perspective (Sri Aurobindo, 1998; Wilber, 2004) helps to dispel many unhelpful dichotomies, such as mathematics versus culture, knowledge versus activity, and meaning versus interpretation. In the dialectical process of reconceptualising mathematics, alternative natures of mathematics, such as *mathematics as activity, mathematics as informal problem solving, mathematics as socio-cultural enactment, and mathematics as cultural meaning-making*, combine to build a synergy with the conventional nature of *mathematics as a body of decontextualised pure knowledge*. By imagining mathematics as both a noun and a verb the possibility of mathematics education being dominated by the conventional image of *mathematics as a decontextualised body of pure knowledge* greatly reduces. With such a complex image of contextualised mathematics education the cultural lifeworlds of learners can more readily interact with mathematics, thereby bringing their cultural capital to the forefront.

How might Nepali teachers and students act in a culturally contextualised mathematics education? The teacher would strive to recognise students' cultural and individual differences, promote inclusive participation of students, and create caring and collaborative learning environments; thereby promoting meaningful mathematical acculturation according to which students would often engage in crossing and recrossing the borders of local and formal mathematics. Engaged in exploring connections between formal and local (village or street) mathematics, Nepali students would be seen (a) co-generating mathematics from their cultural contexts; (b) linking their cultural experiences with formal mathematics; (c) taking up social inquiry, project, and cultural inquiry approaches to learning mathematics; (d) developing local classifications of mathematical ideas, based on their uses in local cultural contexts; and (e) solving real world problems by using different forms of mathematics.

INTERLUDE IV

I have arrived at the end of this journey. The journey with many detours and different stopping points has proved to be adventurous. Knowingly or unknowingly I seem to have been guided by the idea of a proscriptive logic which says: "This is what's forbidden; everything else is allowed". (Davis & Simmt, 2003, p. 147). This act of depicting complexity seems to be as old as the early Vedic and Buddhist poets who wrote about the Shunya, the Pancha Tatva, the Mokshya and the Nirvana. I (and ~I) am in a crisis of representation. Sounds, alphabet, words and sentences seem to be ceasing to the One that is characterised by emptiness. What is that which is not empty? The beginning is empty so is the end. Perhaps, I have arrived at a new beginning.

CONCLUDING *FOR NOW*

Given the seemingly adversarial metaphors of mathematics *as a body of pure knowledge* and mathematics as *cultural activity*, mathematics education in Nepal requires a social justice oriented transformation that produces an inclusive image of mathematics education. Given the cultural, lingual and spiritual diversities of Nepal, a credible approach to incorporating social justice in mathematics education is to implement a culturally contextualised mathematics education which integrates multiple natures of mathematics, thereby opening room for celebrating not just deductive-analytic logic but also others, such as dialectical and poetic logics, that have been preserved by wisdom traditions of the East and West. The transformative image of culturally contextualised mathematics education can be a healing metaphor for Nepali mathematics education because it promotes a recognition-oriented epistemology, an inclusive view of mathematics, and a meaningful pedagogy. By enacting such a transformative vision of mathematics education, we would likely witness: (a) Nepali teachers and students working collaboratively to construct mathematics from their cultural sites, (b) Nepali teacher educators and teachers embarking on a journey of exploring culturally contextualised pedagogical tools in mathematics, (c) Nepali mathematicians and teacher educators setting up a new venture to explore different mathematical practices (and logics) of diverse Nepali communities, (d) other stakeholders of the Nepali education system being committed to adhering to the notion of social justice in mathematics education.

REFERENCES

Beverley, J. (2005). *Tenstimonio,* subalternity and narrative authority. In N. K. Denzin & Y. S. Lincoln (Eds.), *The SAGE handbook of qualitative research* (3rd ed., pp. 547–557). Thousand Oaks: Sage Publications.

Bourdieu, P., & Passeron, J.-C. (1977). *Reproduction in education, society, and culture.* Beverly Hills, CA: Sage.

Boyer, C. B. (1968). *A history of mathematics.* New York: Wiley.

Clough, P. (2002). *Narratives and fictions in educational research.* Buckingham: Open University Press.

Cobb, P. (1994). Where is the mind? Constructivist and sociocultural perspectives on mathematical development. *Educational Researcher, 23*(7), 13–20.

D'Ambrosio, U. (1999). Literacy, matheracy, and technocracy: a trivium for today. *Mathematical Thinking and Learning, 1*(2), 131–153.

D'Ambrosio, U. (2006a). *Ethnomathematics: Link between traditions and modernity.* Rotterdam: Sense Pub.

D'Ambrosio, U. (2006b). The Program Ethnomathematics: A theoretical basis of the dynamics of intra-cultural encounters. *Journal of Mathematics and Culture, 1*(1), 1–7.

D'Ambrosio, U. (2007). The role of mathematics in educational systems. *ZDM, 39*(1), 173–181.

Davis, B., & Simmt, E. (2003). Understanding learning systems: Mathematics education and complexity science. *Journal for Research in Mathematics Education, 34*(2), 137–167.

Dewey, J. (1943). *The school and society.* Chicago: University of Chicago Press.

EDSC. (1997). *National achievement level of grade three students.* Kathmandu: Educational Development Service Centre.

EDSC. (2003). *National achievement level of grade five students.* Kathmandu: Educational Development Service Centre.

Ernest, P. (1991). *The philosophy of mathematics education.* London: Falmer Press.

Ernest, P. (Ed.). (1994a). *Constructing mathematical knowledge: Epistemology and mathematics education.* London.: Falmer Press.

Ernest, P. (Ed.). (1994b). *Mathematics, education, and philosophy: An international perspective.* London: Falmer Press.

Ernest, P. (2006). Nominalism and conventionalism in social constructivism *Philosophy of Mathematics Education Journal, 19,* Downloadable from: http://www.people.ex.ac.uk/PErnest/pome19/index.htm.

Eves, H. (1983). *An introduction to the history of mathematics.* New York: CBS College.

Faust, M. A. (1992). Ways of reading and "The Use of Force." *The English Journal, 81*(7), 44–49.

Fraser, N. (1997). *Justice interrupts: critical reflections on the "postsocialist" condition.* New York: Routledge.

Freire, P. (1996). *Pedagogy of the oppressed* (New rev. ed.). London: Penguin.

Gergen, K. J. (1985). The social constructionist movement in modern psychology. *American Psychologist, 40*(3), 266–275.

Glasersfeld, E. v. (1995). *Radical constructivism: A way of knowing and learning.* London: Falmer Press.

Gutstein, E. (2003). Teaching and learning mathematics for social justice in an urban, Latino school. *Journal for Research in Mathematics Education, 34*(1), 37–73.

Hersh, R. (1997). *What is mathematics, really?* New York: Oxford University Press.

Khuzwayo, H. (1998). "Occupation of our Minds": A dominant feature in mathematics education in South Africa. In P. Gates (Ed.), *Conference proceedings of the First*

International Mathematics Education and Society (pp. 219–231). Nottingham: Centre for the Study of Mathematics Education.

Kline, M. (1980). *Mathematics : The loss of certainty.* New York: Oxford University Press.

Koirala, B. N., & Acharya, S. (2005). *Girls in science and technology education: A study on access and performance of girls in Nepal.* Kathmandu: UNESCO.

Lakoff, G., & Núñez, R. E. (2000). *Where mathematics comes from: How the embodied mind brings mathematics into being* (1st ed.). New York, NY: Basic Books.

Lerman, S. (1990). Alternative perspectives of the nature of mathematics and their influence on the teaching of mathematics. *British Educational Research Journal, 16*(1), 53–61.

Luitel, B. C. (2003). *Narrative explorations of Nepali mathematics curriculum landscapes: An epic journey* Unpublished Master's Project, Curtin University of Technology, Perth: Downloadable from http://pctaylor.com

Luitel, B. C., & Taylor, P. C. (2005, Apr). *Overcoming culturally dislocated curricula in a transitional society: An autoethnographic journey towards pragmatic wisdom.* Paper presented at the annual meeting of the American Educational Research Association (AERA), Montreal, Quebec.

Luitel, B. C., & Taylor, P. C. (2007). The shanai, the pseudosphere and other imaginings: Envisioning culturally contextualised mathematics education *Cultural Studies of Science Education 2,* 621–635.

Mathema, K. B., & Bista, M. B. (2006). *A study on student performance in SLC.* Kathmandu: Ministry of Education and Sports

Nuñez, R. (2006). Do real numbers really move? : Language, thought, and gesture: The embodied cognitive foundations of mathematics In R. Hersh (Ed.), *18 unconventional essays on the nature of mathematics* (pp. 160–181). New York: Springer.

Pickles, P. A. C. (2004). *Inclusive teaching, inclusive learning: Managing the curriculum for children with severe motor difficulties.* London: David Fulton.

Radhakrishnan, S. (1927). *Indian philosophy* (Vol. II). London: George Allen & Unwin Ltd.

Redfield, R., Linton, R., & Herskovits, M. J. (1936). Memorandum for the study of acculturation. *American Anthropologist* (38), 149- 152.

Restivo, S. (1994). The social life of mathematics. In P. Ernest (Ed.), *Mathematics, education, and philosophy: An international perspective* (pp. 209–220). London: Falmer Press.

Restivo, S., & Bauchspies, W. (2006). The will to mathematics: Minds, morals, and numbers. *Foundations of Science, 11*(1), 197–215.

Richardson, L., & St Pierre, E. (2005). Writing: a method of inquiry. In N. Denzin & Y. Lincoln (Eds.), *The SAGE handbook of qualitative research* (3rd ed., pp. 959–578). Thousand Oaks: Sage.

Roth, W. -M. (2005). Auto/biography and auto/ethnography: Finding the generalized other in the self. (In Roth, W. -M. (Ed.), *Auto/biography and auto/ethnography: Praxis of research method* (pp. 3–21). Rotterdam: Sense Publishers.)

Rowlands, S., & Carson, R. (2002). Where would formal, academic mathematics stand in a curriculum informed by ethnomathematics? A critical review of ethnomathematics. *Educational Studies in Mathematics, 50*(1), 79–102.

Rowlands, S., & Carson, R. (2004). Our response to Adam, Alangui and Barton's ``A Comment on Rowlands & Carson `Where would Formal, Academic Mathematics stand in a Curriculum informed by Ethnomathematics? A Critical Review'''. *Educational Studies in Mathematics, 56*(2), 329–342.

Schech, S., & Haggis, J. (2000). *Culture and development: A critical introduction.* Oxford: Blackwell Publishers.

Sharma, G. (1986). *History of education in Nepal (in Nepali).* Kathmandu: Makalu Publication.

Skovsmose, O. (2005). *Travelling through education: Uncertainty, mathematics, responsibilities.* Rotterdam: Sense.

Skovsmose, O., & Valero, P. (2001). Breaking political neutrality: The critical engagement of mathematics education with democracy. In B. Atweh, H. Forgasz & B. Nebres (Eds.), *Sociocultural research on mathematics education: An international perspective* (pp. 37–55). Mahwah N.J.: Lawrence Erlbaum Associates.

Sri Aurobindo. (1952). *Integral education.* Pondicherry: Sri Aurobindo International University Centre.

Sternberg, R. J. (2007). Who are the bright children? The cultural context of being and acting intelligent. *Educational Researcher, 36*(3), 148–155.

Taylor, P. (1996). Mythmaking and Mythbreaking in the Mathematics Classroom. *Educational Studies in Mathematics, 31*(1/2), 151–173.

Taylor, P., & Campbell-Williams, M. (1993). Discourse toward balanced rationality in the high school mathematics classroom: Ideas from Habermas's critical theory. In J. A. Malone & P. C. S. Taylor (Eds.), *Constructivist interpretations of teaching and learning mathematics* (pp. 135–148). Perth, Australia: Curtin University of Technology.

Taylor, P. C., & Cobern, W. W. (1998). Towards a critical science education. In W. W. Cobern (ed.), *Socio-cultural perspectives on science education: An international dialogue* (pp. 203–207). Dordrecht, Netherlands: Kluwer Academic Publishers.

Taylor, P. C., & Wallace, J. (Eds.). (2007). *Contemporary qualitative research: Exemplars for science and mathematics educators.* Dordrecht, The Netherlands: Springer.

Taylor, P. C., Luitel, B. C., Tobin, K. G., & Desautels, J. (2007). Forum: Contextualism and/or decontextualism, painting rich cultural pictures, and ethics of co-authorship. *Cultural Studies of Science Education, 2,* 639–655.

Van Maanen, J. (1988). *Tales of the field: On writing ethnography.* Chicago: University of Chicago Press.

Vithal, R., & Skovsmose, O. (1997). The end of innocence: A critique of 'Ethnomathematics'. *Educational Studies in Mathematics, 34*(2), 131–157.

Wilber, K. (1996). *A brief history of everything* (1st ed.). Boston: Shambhala.

Wilber, K. (2004). *The simple feeling of being: Visionary, spiritual and poetic writings.* Boston, MA: Shambhala.

Wong, W.-c. (2006). Understanding dialectical thinking from a cultural-historical perspective. *Philosophical Psychology, 19*(2), 239–260.

Young, I. M. (2000). *Inclusion and democracy.* London: Oxford University Press.

NOTES

1. I have used this pseudonym to protect the identity of my research participants associated with this institution.
2. वादे वादे जायते तत्त्वबोध: (*Baade baade jaayate tatvabodha:*)
3. *Baad* and *bibaad* are Sanskrit words and their English equivalent terms can be proponent and opponent respectively (see, http://sanskritdocuments.org/dict/).
4. Roll's theorem can be stated as: If $f(a) = f(b) = 0$ then $f'(x) = 0$ for some x with $a \leq x \leq b$.
5. जहाँ पुग्दैनन रवी त्यहाँ पुग्छन कवी!
6. A translation of Sama Veda is available at http://www.sacred-texts.com/hin/sv.htm.
7. '|' represents a logical operator that consists of a logical AND followed by a logical NOT and returns a false value only if both operands are true.

CHAPTER 13

MATHEMATICS EDUCATION AND THE BRAZILIAN LANDLESS MOVEMENT

Three Different Mathematics in the Context of the Struggle for Social Justice

Gelsa Knijnik
Unisinos, Brazil

ABSTRACT

This paper aims to discuss issues related to mathematics education and social justice in our Empire times taking as an empirical base for the discussion the work developed by the author in the last 16 years with the Brazilian Landless Movement. The paper analyzes this peasant social movement, focusing on the political role it is assuming in what Hardt and Negri (2001) called *Empire*, and more specifically, the educational work it is improving in the country. It also presents the theoretical background that informs the author's ethnomathematics thinking based on a Post-Modern perspective in its connections with Post-Structuralist theorizations, more specifically, those associated with the work of Michel Foucault. Moreover, using the work of the "Second Wittgen-

Critical Issues in Mathematics Education, pages 153–169

153

stein" (which corresponds to his book "Philosophical Investigations") three different mathematics: are shown: a mathematics produced by a *form of life* associated to MST peasants, another one produced by a *form of life* of the urban sawmill men and a third, produced by a *form of life* found in the Western Eurocentric school, even considering that all of them have *family resemblances*.

INTRODUCTION

The well-known Brazilian educator Paulo Freire, in the language of his time, stated, in his book *Pedagogy of the Oppressed*, that education had a political dimension. Later, he himself reformulated that first statement he had made, saying that education is political. Freirian thinking, in particular its emphasis on the politicity of education and the central role given to culture in the constitution of educational processes, in its time had a major impact on the peripheral countries and also on the central ones, an impact that, possibly, gradually also reached the area of Mathematics Education.

Decades after Freire's initial ideas, we are examining the politicity of mathematics education from other theoretical perspectives, opening the possibility of assigning new meanings to this key issue. Such theories were aligned with the new world economic, social and political designs that shape these times of Empire, marked by large numbers of the poor crossing borders, producing culturally plural and socially even more unequal scenarios. Today it seems that the aims of Enlightenment thinkers have failed. In fact, as pointed out by twentieth-century authors like Sarup (1996:94), the ideas of linear progress, absolute truths, the rational planning of ideal social orders and the standardizations of knowledge and production embraced by Modernity, the extraordinary intellectual effort produced by its project in developing "objective science, universal morality and autonomous art" and its beliefs "in justice and possibility of happiness of human have been cruelly shattered." More than ever, the majority of the world's population is living in subhuman conditions, wars are being waged everywhere and nature is being destroyed all over the planet. Even so, or precisely because of this, our hopes of a more just and egalitarian world still remain. The "old" project of Modernity is gone but new ones are being built. It can be considered that at least punctually educators can contribute to their implementation. In fact, our curricular practices as well as our work as researchers are not neutral, since they have the potentialities to favor or not the inclusion of those who are "the other," the "different": unequally different, since their cultural differences—such as gender, social class, race/ethnicity, sexuality and ageness—are those which have less value in our society.

There are many questions that we have to ask ourselves as mathematics educators committed to understanding—and resisting—the social injustice of these Empire times. This critical exercise cannot avoid problematizing the position assigned to mathematics as well as to mathematics education by Western society. Following Dunne and Johnston, our analytic exercise must consider

> (and deconstruct) the privilege that a access to mathematics confers on its 'chosen few', to understand its 'gate-keeping' role in relation to further education and future careers and consider this in the production and reproduction of hierarchical gender (and class and race) relations. What constitutes mathematics, what counts as valued mathematical knowledge, how things came to be this way and how they are sustained are critical questions. (Dunne and Johnston, 1994: 227)

This paper aims to problematize some of these questions, taking as an empirical base for the discussion the work I have been doing in the last 16 years with the Brazilian Landless Movement. An analysis of this peasant social movement, focused on the political role it is assuming in what Hardt and Negri (2001) called "Empire," and more specifically, the educational work it is improving in the country are the subject of the next section of the paper. The following one will outline the theoretical background that informs the Ethnomathematics thought which I have been developing with the Landless peasants. The third section presents empirical data that show three different mathematics: a mathematics produced by a *form of life* associated to MST peasants, another one produced by a *form of life* of the urban sawmill men and a third, produced by a *form of life* found in the Western Eurocentric school, even considering that all of them have *family resemblances*.

BRAZILIAN LANDLESS MOVEMENT IN THE TIME OF THE EMPIRE

Michel Hardt and Antonio Negri (2001) begin their well-known book "Empire," saying that "it is materializing before our very eyes (...) [since] we have witnessed an irresistible and irreversible globalization of economic and cultural exchanges" (Ibidem:11) which instituted "a global order, a new logic and structure of rule—in short, a new form of sovereignty. Empire is the political object that effectively regulates these global exchanges, the sovereign power that governs the world." (Ibidem:11). Sovereign power, according to the authors, ultimately scrambled the spatial divisions between the First, Second and Third World, since

> (w)e continually find the First World in the Third, the Third in the First, and the Second almost nowhere at all(...). In the postmodernization of the global economy, the creation of wealth tends even more toward what we will call biopolitical production, the production of social life itself, in which the

economic, the political and the cultural increasingly overlap and invest one another" (Ibidem:13).

This new imperial order is taken as a background to this paper, considering the importance of attempting to understand adult education as a field of knowledge as well as the contemporary social movements and their educational processes within this new world configuration characterized by the "absence of boundaries," in which "the rule of the Empire operates on all registers of the social order, extending down to the depths of the social world" (Ibidem:15).

Hardt and Negri, on examining the potentials for constructing alternatives that counter the imperial power, highlight the role taken on by the struggles of the proletariat—a social subject that they consider beyond the industrial working class—constituted, in fact, by "all those exploited and subject to capitalist domination" (Ibidem:72) but who do not make up a "homogeneous or undifferentiated unit:" They further consider that there are new forms of struggle through which this new proletariat "expresses its desires and need" (Ibidem:72), struggles which must be identified, not as "the appearance of a new cycle of internationalist struggles, but rather [as] the emergence of a new quality of social movements" (Ibidem:74).[1]

This "new quality" is expressed by "fundamentally new" characteristics of the struggles of the social movements:

> first, each struggle, though firmly rooted local conditions, leaps immediately to the global level and attacks the imperial constitution in its generality. Second, all the struggles destroy the traditional distinction between economic and political struggles. The struggles are at once economic, political and cultural—and hence they are biopolitical struggles, struggles over the form of life. They are constituent struggles, creating new public spaces and new forms of community." (Ibidem: 56).

Among the many struggles of social movements that could be analyzed in their relationship with education, especially mathematics education, we can consider the struggles for land reform carried out by the Brazilian Landless Movement, which is well known on the international scene, mainly but not only because of "new" aspects that have been instituted in the sphere of education. In fact, as written in the official MST's website (http://www.mstbrazil.org):

> Landless Movement, in Portuguese, Movimento Sem Terra (MST) is the largest social movement in Latin America with an estimated 1.5 million landless members organized in 23 out of 27 states. The Landless movement carries out long-overdue land reform in a country where less than 3% of the population owns two-thirds of the land on which crops could be grown. Since 1985,

the MST has occupied unused land where they have established cooperative farms, constructed houses, schools for children and adults and clinics, promoted indigenous cultures and a healthy and sustainable environment and gender equality. The MST has won land titles for more than 250,000 families in 1,600 settlements as a result of MST actions, and 200,000 encamped families currently await government recognition. Land occupations are rooted in the Brazilian Constitution, which says land that remains unproductive should be used for a larger social function.

The paths followed by the Brazilian Landless Movement in the 22 years of its history, and especially the educational processes that are being produced there, have been the subject of discussion in many academic forums and in publications involving this theme. Thus, in this paper, I am interested in highlighting a few more recent strategies that the MST has implemented, and which may be creating fissures in the *smooth* space of imperial sovereignty, in which "there is no place of power—it is both everywhere and nowhere" (Hardt & Negri, 2001:210) and "there is progressively less distinction between inside and outside" (ibidem:206). It is in this space "crisscrossed by so many fault lines that it only appears as a continuous, uniform space" (ibidem: 210) that one can conjecture on the potentials of social movement struggles to subvert the imperial order.

The MST strategy of occupying large unproductive rural properties as a way of pressuring the State to carry out Land Reform, which marked the initial years of struggle, was gradually expanded to include the occupation of other spaces, such as public buildings, organization of regional and national marches, and setting up camps at the side of main highways. Thus, MST has aimed at occupying many and "all" possible territories. Moving continuously along the roads, staying for short periods in small towns and large cities constitute a strategy which counters the Empire's need to "restrict and isolate the spatial movements of the multitude to stop them from gaining political legitimacy" (ibidem: 422).

While the Empire seeks to "isolate, divide and segregate," MST defined fighting strategies that, in a sense, undermine this segregationist operation. Its slogan: "Land Reform, everybody's struggle" indicates that if it is "everybody's struggle," divisions must be smoothed down, more and more joint actions are needed. It is in this sense that one may understand two of the strategies developed by MST in recent years. The first concerns the implementation of MST actions organized with other popular social movements which are attuned with their position of repudiating the different social repercussions of neoliberal policies, as well as its participation in large national and international demonstrations of opposition to the imperial order. The second strategy that has been acted upon by MST, with a less occasional character, refers to its integration to the Via Campesina, "an organization that brings together the main rural social movements in the world, to fight against neoliberalism and defend the peasant life and culture."(JST, 2004).

In brief, what is presented here points at new strategies that have been more recently implemented by the Movement, strategies that can be considered as having the potential to undermine the Imperial order, contributing to "the constitution of a society in which the basis of power is defined by the expression of the needs of all" (Hardt & Negri, 2001:434).

But there are other kinds of strategies which were implemented since the beginning of Landless Movement struggle and can be seen as reinforcing those above mentioned. I am referring to the work coordinated by its Educational Sector, in providing education to their members. First of all it is important to highlight that the educational process, which has been developed by the MST over its 22-year history must be understood beyond schooling, since each Landless subject educates her/himself through her/his participation in the everyday life of their communities and also through the wide range of political activities developed by the Movement. This means that the children, youth and adult peasants are educated by the multiple facets of the struggle for land, which produce very specific social identities. Nevertheless, these social identities do not form something compact, uniform, in which hundreds of family from different social strata would ultimately become a unified whole, homogenized by the struggle for land.

To look at this social movement with such lenses implies considering that if there is some kind of intention of establishing a "Landless identity," this intention is never completely fulfilled. Summing up, the Landless educate themselves in the struggle—in the occupations, the marches, in their ways of organizing the settlements, through their cultural artefacts—learning the many possible meanings of "being landless." But in this educational process there is a sort of rebellion against fixing *one* social identity. There are many axes—such as those of gender, sexuality, race/ethnicity, ageness—which in their crossovers ultimately shape multiple Landless identities, multiple ways of giving meaning to the struggle for land. This position allows us to say that the peasant culture of the Brazilian Landless Movement—in Wittgenstein's words, its *form of life*—is marked by difference.

The schooling activities developed by the Landless Movement cover Child Education, Elementary and High School Education, Teacher Training Courses and projects of Education of Youths and Adults. As shown on the MST website, the Landless Movement Schooling project involves 1800 schools in camps or settlements (grade 1 to 8), with 160 thousand students and 3900 teachers; 250 educators who work with children up to 6 years; 3000 educators working with 30 thousand peasants of literacy and numeracy projects of Adult Education; and Teacher Training Courses implemented in partnership with public and private universities around the country.

This schooling project, according to one of the Landless Movement official documents, sees the need for "two articulated struggles: to extend the right to education and schooling in the rural area; and to construct a school that is *in* the rural area, but that also *belongs* to the rural area: a school that is politically and pedagogically connected to the history, culture, social and human causes of the subjects of the rural area (...)" (Kolling et al., 2002:19). The movement has dedicated itself to conceiving the schooling of its children, youths and adults paying attention to these two struggles. In particular, such struggles are providing the guidelines for its adult mathematics education. This means that landless educators considered peasant culture a key issue also for those teaching and learning processes related to mathematics. But they explicitly mention that this valorization cannot deny the relevance of acquiring mathematical tools connected to academic mathematics that can improve the use of new technologies for managing the production in rural areas and can allow the learners to go further in their schooling trajectory. As will be shown in the next section these ideas are strongly connected to the field of Ethnomathematics, more specifically to the ethnomathematics thinking I have been developing, which is rooted in the Landless culture and in my experience in working with the peasants living in southern Brazil.

THE THEORETICAL BASIS OF AN ETHNOMATHEMATICS THOUGHT

The basis of the ethnomathematics thought I have been elaborating is based on a Post-Modern perspective in its connections with Post-Structuralist theorizations, more specifically, those associated with the work of Michel Foucault. According to such theorizations, I have considered that Ethnomathematics may consist of a toolbox which allows analyzing: a) the Eurocentric discourses that institute academic mathematics and school mathematics; b) the effects of truth produced by the discourses of academic mathematics and school mathematics; c) issues of difference in mathematics education, considering the centrality of culture and the power relations that constitute it (Knijnik, 2006).

Operating with a toolbox that has this configuration is attuned with the positions of authors like Santos (1995), who argue about the need to cast suspicion on the education practiced today in the Western world, which he described as centrally Eurocentric:

> The cultural map underlying the modern educational systems is, cartographically speaking, a map with a Mercator projection. The central characteristic of this projection is that it places the European continent at the center of the

map, inflating its size to the detriment of the other continents. In symbolical terms, the modern educational map is a Mercator map. The Eurocentric culture occupies almost all of the size of the map and, only marginally and always taking the central space into account, are the other cultures drawn (...) This is the map of the Imperial culturalism of the West. In this map the conflict between cultures either does not appear completely, or it appears as a conflict solved by the superiority of Western culture in relation to the other cultures. (ibidem:26).

We are facing an issue that concerns the politics of knowledge, in the dispute around the definition of which knowledges are included and which excluded in the schooling processes. This dispute is marked by power-knowledge relations, which ultimately legitimate and are the legitimizer of some discourses, which interdict others, precisely those that are about the knowledges, the rationalities, the values, the beliefs of cultural groups we place in the position of "the others."

One should then ask how a single rationality among other rationalities—the rules by which individuals and cultures deal with space, time and quantification processes—all that which Western civilization associates with the notion of mathematics—became a "truth," the only "truth" that could be accepted as mathematics in the school curriculum. What is at stake here is to problematize the sovereignty of the Modern rationality, which scorns all other rationalities associated to "other" *forms of life,* the existence of a single mathematics—"the official one"—with its Eurocentric bias and its rules marked by abstraction and formalism. To be more precise, we must say that this "official" mathematics—the academic one—is composed by a set of branches, including all those associated with so-called "pure mathematics" and "applied mathematics." The so-called school mathematics—the traditional set of knowledges taught at school—inherits at least part of the formal and abstract *grammar* that constitutes academic mathematics, through pedagogical recontextualized processes, in Bernstein's words. In summary, it can be said that all these different mathematics offers a "dream of order, regularity, repeatability and control (...) and with it the idea of a "pure," disembodied reason" (Rotman, 1993:194).

These issues lead us to Ludwig Wittgenstein's ideas presented in his book "Philosophical Investigations" (2004) in which he criticized not only his earlier work (presented in *Tractatus*) but also "the whole tradition to which it belongs" (Glock, 1996:25), "the foundationist schools, and dwells at length upon knowing as a process in mathematics" (Ernest, 1991:31). In his remarkable book "The Philosophy of Mathematics Education" Paul Ernest (1991) examines philosophical schools and their contributions to the conceptualization of (academic) mathematics. In his analysis about the conventionalist view of mathematics, which considered that "mathematical knowledge and truth are based on linguistic convention" (ibidem: 30) the author refers to the work of Wittgenstein, saying that the philosopher "proposes that the logical necessity of mathematical (and logical) knowledge rests on linguistic conventions, embedded in our social linguistic practices" (ibidem: 32).

In shaping a new philosophy of mathematics—social constructivism—Ernest refers to Wittgenstein as one of the philosophers who considers knowledge not only as a product, giving "great weight to knowing and the development of knowledge" (ibidem: 90). He considers that "social constructivism employs a conventionalist justification for mathematical knowledge" (ibidem: 64), assuming that "the basis of mathematical knowledge is linguistic knowledge, conventions and rules, and language is a social construction" (ibidem:42). Ernest argues that his philosophical perspective "assumes a unique natural language" showing that "an alternative (i.e., different) mathematics could result" (ibidem:64) as a consequence of this position. Mentioning the work of Alan Bishop as an evidence of different mathematics (ibidem: 67), Ernest will say that "such evidence of cultural relativism strengthens rather than weakens the case in favour of social constructivism" (ibidem: 64).

These ideas are strongly connected to the ethnomathematics thinking presented in this paper based on the work of the "Second Wittgenstein." In fact, viewing mathematics "not as a body of truths about abstract entities, but as part of human practice" (Glock, 1996: 24), the philosopher's work gives us tools for thinking about rationality as forged from social practices of a *form of life*, which implies to consider it as "invention," as "construction" (Condé, 2004: 29). Moreover, with the support of the philosopher's ideas—and using the expressions that he coined—one can admit the existence of distinct mathematics—distinct ethnomathematics, in D'Ambrosio's words.[2] The basis of this statement can be found in the argument that these different mathematics—in Wittgenstein's words, different *language games*—are produced by different *forms of life*, a term conceived by the "Second Wittgenstein" as "stress[ing] the intertwining of culture, world-view and language" (Glock, 1996:124), as "patterns in the weave of our life" Glock (1996: 129)).

In Wittgenstein's late work, especially in the new conception of language presented by the philosopher, Condé (1998, 2004) argues about the crucial role of the notion of *use*:

> In such work, *use* is directly connected to the concept of meaning (...) the meaning is determined by the *use* we make of the words in our ordinary language. (...) The meaning of a word is given based on the *use* we make of it in different situations and contexts. (...) the meaning is determined by the *use*. (ibidem: 47)

It is in this sense that this notion of use, according to Wittgenstein, is considered pragmatic, not "essentialist." Meaning is determined by the use of words and such a use respects rules, which are themselves produced in social practices, constituting *language games*. As pointed out by Condé (1998:91) "the notion of *language games* involves not only expressions, but also the activities with which these expressions are linked." *Language games* are produced based on sets of rules (that are rooted in social practices), each of them constituting a specific grammar. So, the grammar that marks each *language game* is itself a social institution. Moreover, authors like Spaniol (see Condé, 1998:110) argue that "the grammar constitutes the logic itself, the grammar is the logic. (...) It is impossible to analyze the logic without considering the language."

From what was briefly explained here based on the work of the "Second Wittgenstein" and some of his interpreters (like Condé and Glock before mentioned), it follows that different *forms of life* produce different *language games*, each of them marked by a specific grammar and such grammar, as a set of rules, constitutes the specific logic. This rationale drives us to admit that there is more than a single *language game*: there are different *language games*. Is there some kind of relationship between them? If the answer is positive, how does it operate? The response to these questions is given by the "Second Wittgenstein" through the notion of *family resemblances*. The philosopher would say (as shown in aphorisms 66 and 67 of Philosophical Investigations) that *language games* form "a complicated network of similarities overlapping and criss-crossing: sometimes overall similarities, sometimes similarities of detail" (Wittgenstein, 2004: 320) and adds:

> I can think of no better expression to characterize these similarities than family resemblances; for the various resemblances between member of a family. Build, features, color of eyes, gait, temperament, etc. etc overlap and criss-cross in the same way—and I shall say: 'games' form a family.

Operating with the ideas of the "Second Wittgenstein" in the context of the struggle for land in the south of Brazil, leads us to assume the existence of three different mathematics: a mathematics produced by a *form of life* associated to MST peasants, another one produced by a *form of life* of the urban sawmill men and a third, produced by a *form of life* found in the Western Eurocentric school, even considering that all of them have *family resemblances*.

LANDLESS' LANGUAGE GAME OF *CUBAGEM OF WOOD*
AND TWO OTHERS MATHEMATICS

Cubagem of wood (in Portuguese *Cubagem da madeira*)—to calculate "how many cubics[3] there are in a truck load"—is a common practice in the Landless' culture. The peasants perform it when it is necessary to build houses or animal shelters in camps and settlements and to purchase or sell planks, i.e., "in our negotiations with the sawmill men," as said one MST member. Throughout my work with MST groups I have realized the importance they give to such language games, produced by their form of life. In teacher education courses and at settlement schools, particularly, I have found great interest in discussing that practice, constituted by a specific grammar, a specific set of rules.

On one of the occasions when I was working with *cubagem of wood* in a Teacher Education Course class with lay teachers, Edinei, a student who was then living in a camp, explained about how important it was for his community to learn about *cubagem of wood*:

> My sponsor took an interest in this question of "cubagem of wood." So I also didn't know about it. So, I said that I'd be coming here and bring him the answer. Because he was working like this, as a workman, they [he and his fellows] logged and sold the wood, and sold it to the sawmill, in this case to the sawmill owner (...) When one took those meters there to sell, one got so much. When the fellow who worked as a workman himself bought the wood that had been sawed, there were many more cubic meters. So, how could this be?

During my work with that group of students I found that Edinei's sponsor's question was shared by many in the class. They were expecting that I help them to go further in learning about the grammar that marks *cubagem of wood*' language game. But it was expected that I also assume another role. In consonance with the Sector of Education pedagogical guidelines (as discussed in the previous section) they aimed to acquire the school mathematics knowledge—the one called by them "book mathematics." Avoiding a naïve perspective, they were aware of the social importance of such a language game and the need to learn its specific grammar as part of their struggle to undermine the Empire sovereignty.

The starting point of the pedagogical work was a student's narrative. Roseli, a municipal teacher at the time, still without a degree like her colleagues, told what she had learned from her father about *cubagem of wood*. Going to a place where tree trunks lay the group of students took up position around one of them, helped by Helena (who had an electronic calculator), Antonia (who took notes on a sheet of paper) and Nelci and Cleci, (who contributed in issues concerning measurement). With their assistance, Roseli described what would henceforth be called by the group *Roseli's method of cubagem of wood.*[4]

> **Roseli:** First one takes hold of this by the middle [of the log] because there it is thicker and here it is thinner [pointing to the ends of the trunk]. Then, around the middle, one has more or less the average, it is the average. Now I take this string and pass it around it. Done. Now I fold it in four, then after I fold it in four I will measure it to see how many centimeters one will have..
>
> **Cleci:** 37.
>
> **Roseli:** There, the result is 37 centimeters. Now I take these 37 and multiply by itself, multiply by 37.
>
> **Helena:** [using a calculator to perform the multiplication] 37 by 37 gives 1369.
>
> **Roseli:** [talking to her colleague] Write it down, Antonia, so what we will not forget it. Now I'm going to measure the length. After this, now, I know that there is 37, so now is when I measure the length. The result was 1 meter and 46. Now then, I multiply the length by the number I had before, which came from the small piece of string, which had given 37 times 37:1369.
>
> **Helena:** I make this 1369 by 1 and 46.
>
> **Roseli:** It is all centimeters. One has to do 1369 by 146.
>
> **Helena:** [with the calculator] 1360 times 146 gives . . . 199874.
>
> **Roseli:** That is the number one gets.
>
> **Rosane:** 199874 what?
>
> **Roseli:** 199874 cubic [centimeters] of wood.
>
> **Rosane:** It is the same as doing side time side times length.
>
> **Juarez:** She did almost the same thing that Jorge did. She went and measured the trunk diameter,[5] then she made a square and multiplied by the length.

Even considering the theoretical difficulties involved in translating language games, I found it important to express *Roseli's method* using words and the syntax of the school mathematics' language game, the one we are more familiar with. I am aware that in doing so some (or maybe most of the) specificities that constitute the Landless' form of life which produced the *cubagem of wood* language game are suppressed. So, it can be said that *Roseli's method* became a "hostage" of school mathematics language game when it is said that "her" method basically involves two steps: the first, to identify, by modeling, a tree trunk with a cylinder whose circumference coincides with that of the middle part of the trunk, and the second, identification, also by modeling the cylinder in a quadrangular prism, whose measure on the side is one fourth of the perimeter of the cylinder base. Thus, Roseli's Method for "cubagem" of wood finds, as trunk volume, the volume of the quadrangular prism whose side of the base was obtained by determining the fourth part of a circumference. This, in turn, corresponds to the cylinder base, obtained by modeling from the initially given tree trunk. Roseli explained "her" method[6] step by step, as she pointed to the different parts of the trunk involved in the process. This narrative triggered the study on *cubagem of wood* which we developed from then on.

During the discussion of *Roseli's method* there were students who immediately related its grammar to the one that constitutes the *land cubação* language game called by the group *Jorge's method*, which was studied before (Knijnik, 1997). In effect, both grammars have one rule in common: the identification process which associates a cylinder base (in the case of *cubagem* of wood) or a quadrilateral (in the case of *land cubação*) with a square. The relationship established by the group between both language games was an interesting pedagogical issue linked to what Wittgenstein called *family resemblances*.

This notion of Wittgenstein can be helpful in understanding another language game which emerged in the pedagogical process. At some point in the discussions, a two-student dialogue produced a shift in the debate about *Roseli's method*.

> **Jorge:** The measurement process that I know is almost the same [as Roseli's method], except that we measure at the narrow end of the wood.
>
> **Ildemar:** The point is that the right thing would be to do it in the middle. But the purchasers do not want to buy a piece that will fall away, if they want it for square wood[7] or things like that. They will want a piece that goes from here to there [which goes from one end to the other of the log]. Those chips that are produced will only be for burning.

According to these students, there were urban sawmill men who did not use the "middle of the log" as reference, considering only its narrower end, since they were interested in obtaining whole planks.[8] For this purpose a different rule of calculation was introduced, conforming a specific grammar, which leads to a new *language game*, different from *Roseli's*. The *sawmill men's method* was mentioned by most of the group as being practiced at sawmills in the urban areas close to their communities. We found that we were dealing with a language game which is produced by a specific *form of life*, different from that of the Landless' peasant.

But the pedagogical process was not circumscribed to *Roseli' method* and to the sawmill men one. The "book mathematics" *language game* with the specific rules that shape its grammar was also analyzed. Moreover, the *family resemblances* of these three language games were emphasized. The work involved studying the modeling process of Roseli's Method and learning mathematical tools such as relations between a cubic meter and its multiples. In different situations the results of calculating the "amount of wood" obtained by *Roseli's method* were compared empirically to the volume of the cylinder produced by "her" method, which would correspond to a better approach to the total quantity of wood of the trunk, reckoning not only the part useful to obtain "whole planks." The group also found that the results of *Roseli's method* minimize those obtained using the cylinder volume. The group showed particular interest in learning "the formulas of book mathematics" connected to the discussion we were holding. In learning how to calculate volumes of the cylinder and rectangular prisms (which also requires knowing how to determine the length of the circumference, the area of a circle) the group was dealing with the specific grammar which constitutes the *language game* of Western Eurocentric school mathematics.[9]

Bringing those three *language games* into the mathematics class enabled the group to go further in the appropriation of rules that shape the *grammars* produced by each form of life. They learned more about the Landless peasant *cubagem of wood* practiced in their communities. When the *family resemblances* of those *language games* were analyzed, the students were able to identify the "remnants" of wood that were produced by *Roseli's method*, which were even greater when the initial measure of the log circumference was determined at the "narrow end," as considered by the *sawmill men's method.* So, in this case, the wood not used for making planks could be useful for other purposes and therefore, in given situations, it should also be included in the accountancy of their calculations.

Summing up, it can be said that learning about different mathematics and their *family resemblances* allowed the peasant students to broaden not only their mathematical world, but also their ways of seeing the complex social relations involved in different *forms of life* that produce such different *language games.*

SOME CLOSING WORDS

I would like to end saying that the issues I attempted to discuss here are no more than provisional, unmarked by hopes for certainty, in the sense given by Stronach and Maclure (1997). I follow them when they say that we must recognize and try to work within the necessary *failure* of methodology's hope for certainty, and its dream of finding an innocent language in which to represent, without exploiting or distorting, the voices and ways of knowing of its subaltern 'subjects'" (ibidem: 4). The ideas I brought to this paper are inspired by this position. In fact, throughout my trajectory as a researcher I have always tried to mobilize all my efforts in order to never forget to problematize my own discourse, since it is necessarily marked by my 'privileged' voice as an intellectual working with "subaltern 'subjects'" like the Brazilian landless people. As all discourses, it is marked by power-knowledge relations. In my attempts I have been favoured by this social movement, which is very much aware of the risk of exposing themselves to academic research and of being narrated by "the others," of being represented by us.

The Brazilian Landless Movement peasants chose to take this risk not only because of my "good intentions" of being vigilant about my role in the work I have been developing with them for all these years. It is also because they see education as one of the central issues of their struggle to undermine the "Empire" and consider the importance—at least at this point of their trajectory as a social movement—to have academics contributing to the construction of their educational schooling processes, which, for them, is a key element for the social justice project they are attempting to build.

REFERENCES

Condé, M. (1998). *Wittgenstein: linguagem e mundo.* São Paulo: Annablume.

Condé, M. (2004). *As teias da Razão:* Wittgenstein e a crise da racionalidade moderna. Belo Horizonte: Argvmentvm.

D'Ambrosio, U. (2001). *Etnomatemática:* elo entre a tradição e a modernidade. Belo Horizonte: Autêntica.

Dunne, M. & Johnston, J. (1994) *Research in Gender and Mathematics Education: The Production of Difference.* In: Ernest, P. (ed.) *Mathematics Education and Philosophy:* An International Perspective. London: The Falmer Press.

Ernest, P. (1991). *The Philosophy of Mathematics Education.* London: The Falmer Press.

Glock, H. (1996). *A Wittgenstein Dictionary.* Oxford: Blackwell Publishers.

Hardt, M. & Negri, A. (2001). *Empire.* London: Harvard University Press.

JST—*Jornal Sem Terra* (2004). Landless Movement Newspaper.

Knijnik, G. (1997). Politics of Knowledge, Mathematics Education and the peasants' struggle for land. Educational Action Research Journal, v. 5, n. 3.

Knijnik, G. (2006). *Educação matemática, culturas e conhecimento na luta pela terra.* Santa Cruz do Sul: EDUNISC.

Kolling, E.; Cerioli, P.; Caldart, R. (org.). (2002). *Educação do Campo:* Identidade e Políticas Públicas. Brasília, DF: Articulação Nacional por uma Educação do Campo. Coleção Por Uma Educação do Campo, n. 4.

Klusener, R. & Knijnik, G. (1986). *A prática de cubação de madeira no meio rural: uma pesquisa na pespectiva da Etnomatemática.* (Porto Alegre, UFRGS, Instituto de Matemática). Texto datilografado

Mattos, M. Nepstad, D. & Vieira, I. (1992). *Cartilha sobre mapeamento de área, cubagem de madeira e inventário florestal.* (Belém do Pará, EMBRAPA/ Woods Hole Research Center).

Rotman, B. (1993). *Ad Infinitum: The ghost in Turing's Machine.* Stanford: Stanford University Press.

Santos, B. (1995). Para uma pedagogia do conflito. In: Silva, L. (org). *Novos mapas culturais, novas perspectivas educacionais.* Porto Alegre: Sulina. p. 15–34.

Sarup, M. (1996). *Identity, culture and the postmodern world.* Edinburgh: Edinburgh University Press.

Stronach, I. & Maclure, M. (1997). *Educational research undone.* The Postmodern Embrace. Buckingham, Philadelphia: Open University Press.

Wittgenstein, L. (2004). *Philosophical Investigations.* Oxford: Publishers.

NOTES

1. Here a point should be highlighted. In their formulations about social movements, which are considered central by Hardt and Negri, as I mentioned previously, it is the "new proletariat", defined based on the domination of work by capital and by the exploitation processes associated with it. In this sense, possibly one could say that social movements that articulate around other axes of submission (such as that of ethnicity, gender, sexuality), remain outside the discussions by the authors.

2. In fact, D'Ambrosio (2001) considers that each branch of academic mathematics shapes an ethnomathematics; school mathematics is an ethnomathematics and also the ways in which specific cultural groups—like the Brazilian Landless peasant—deal with numbers, space, measurement, etc are considered different ethnomathematics.

3. The terms "*cúbicos*" and "*cúbicos de madeira*" are used in the Brazilian rural areas to mean cubic meters of wood. The term "*metros de madeira*", in English, "meters of wood", is also used.

4. It is interesting to observe that on this occasion—different from other pedagogical situations involving Landless peasant practices—the girl students were conducting the explanation. At that time I considered that this gender issue could be connected to the following fact: The whole group was (previously) asked to interview members of their communities about the *cubagem of wood*. As I had observed, it was a male bias practice in the Brazilian countryside but the "disciplined" women did their "homework" in such a detailed way that they felt more confident than the male students—who "know it only by practice"—about it.

5. The student used the expression "trunk diameter" to refer to what is considered in the language game of school mathematics the trunk circumference.

6. Several students referred to the use of Roseli's Method in their communities. The so called *Roseli's method* had already been identified in fieldwork previously performed in the south of Brazil (Klüsener & Knijnik, 1986) and it was also practiced in the state of Acre, in the north of the country (Mattos, Nepstad & Vieira, 1992).

7. At that time, some students used the expression "square wood" to refer to a wooden plank.

8. Taking into account the remarks made before, concerning the "translation" issues from one language game to another, it could be said that the *sawmill men's method* consists of calculating the volume of a quadrangular prism whose height is given by the tree trunk. The quadrangular base, however, different from *Roseli's method*, is obtained by the inscription of a square with a maximum side at the log base, considered as a circle.

9. The group questioned the possibility of applying what they had studied in the context of both *Roseli's method* and the *sawmill men's method* to other peasant practices. One of the students mentioned that the rules he learned in those mathematics classes could be used in planning the construction of silos for crop storage, at the time one of the main goals of his comrades to render the settlement economically feasible.

CHAPTER 14

KERALA MATHEMATICS AND ITS POSSIBLE TRANSMISSION TO EUROPE

Dennis Francis Almeida
University of Exeter

George Gheverghese Joseph
University of Manchester

ABSTRACT

Mathematical techniques of great importance, involving elements of the calculus, were developed between the 14th and 16th centuries in Kerala, India. In this period Kerala was in continuous contact with the outside world, with China to the East and with Arabia to the West. Also after the pioneering voyage of Vasco da Gama in 1499, there was a direct conduit to Europe. The current state of the literature implies that, despite these communication routes, the Keralese calculus lay confined to Kerala. The paper is based on the findings of an ongoing research project, which examines the epistemology of the calculus of the Kerala school and its conjectured transmission to Europe.

Critical Issues in Mathematics Education, pages 171–188

171

INTRODUCTION

According to the literature the general methods of the calculus were invented independently by Newton and Leibniz in the late 17th century[1] after exploiting the works of European pioneers such as Fermat, Roberval, Taylor, Gregory, Pascal, and Bernoulli[2] in the preceding half century. However, what appears to be less well known is that the fundamental elements of the calculus including numerical integration methods and infinite series derivations for π and for trigonometric functions such as sin x, cos x and tan^{-1} x (the so-called Gregory series) had already been discovered over 250 years earlier in Kerala. These developments first occurred in the works of the Kerala mathematician Madhava and were subsequently elaborated on by his followers Nilakantha Somayaji, Jyesthadeva, Sankara Variyar and others between the 14th and 16th centuries.[3] In the latter half of the 20th century there has been some acknowledgement of these facts outside India. There are several modern European histories of mathematics[4] which acknowledge the work of the Kerala school. However it needs to be pointed out that this acknowledgement is not necessarily universal. For example, in the recent past a paper by Fiegenbaum on the history of the calculus makes no acknowledgement of the work of the Kerala school.[5] However, prior to the publication of Fiegenbaum's paper, several renowned publications detailing the Keralese calculus had already appeared in the West.[6] Such a viewpoint may have its origins in the Eurocentrism that was formulated during the period of colonisation by some European nations.

EUROPEAN PERSPECTIVES ON INDIAN AND KERALA MATHEMATICS

In the early part of the second millennium evaluations of Indian mathematics or, to be precise, astronomy were generally from Arab commentators. They tended to indicate that Indian science and mathematics was independently derived. Some, like Said Al-Andalusi, claimed it to be of a high order:

> "[The Indians] have acquired immense information and reached the zenith in their knowledge of the movements of the stars [astronomy] and the secrets of the skies [astrology] as well as other mathematical studies. After all that, they have surpassed all the other peoples in their knowledge of medical science and the strengths of various drugs, the characteristics of compounds, and the peculiarities of substances."[7]

Others like Al-Biruni were more critical. He asserted that Indian mathematics and astronomy was much like the vast mathematical literature of the 21st century—uneven with a few good quality research papers and a majority of error strewn publications: "I can only compare their mathematical and astronomical literature, as far as I know it, to a mixture of pear shells and sour dates, or of pearls and dung, or of costly crystals and common pebbles."[8]

Nevertheless a common element in these early evaluations is the uniqueness of the development of Indian mathematics. However by the 19th century and contemporaneous with the establishment of European colonies in the East, the views of European scholars about the supposed superiority of European knowledge was developing racist overtones. Sedillot[9] asserted that not only was Indian science indebted to Europe but also that the Indian numbers are an 'abbreviated form' of Roman numbers, that Sanskrit is 'muddled' Greek , and that India had no chronology. Although Sedillot's assertions were based on imperfect knowledge and understanding of the nature and scope of Indian mathematics, this did not deter him from concluding: "On one side, there is a perfect language, the language of Homer, approved by many centuries, by all branches of human cultural knowledge, by arts brought to high levels of perfection. On the other side, there is [in India] Tamil with innumerable dialects and that Brahmanic filth which survived to our day in the environment of the most crude superstitions."

In a similar vein Bentley[10] also cast doubt on the chronology of India by locating Aryabhata and other Indian mathematicians several centuries later than was actually the case. He was of the opinion that Brahmins had actively fabricated evidence to locate Indian mathematicians earlier than they existed:

> We come now to notice another forgery, the *Brahma Siddhanta Sphuta*, the author of which I know. The object of this forgery was to throw Varaha Mihira, who lived about the time of Akber, back into antiquity...Thus we see how Brahma Gupta, a person who lived long before Aryabhata and Varaha Mihira, is made to quote them, for the purpose of throwing them back into antiquity....It proves most certainly that the *Braham Siddhanta* cited, or at least a part of it, is a complete forgery, probably framed, among many other books, during the last century by a junta of Brahmins, for the purpose of carrying on a regular systematic imposition.

For the record, the actual dates are Aryabhata b 476 AD, Varamihira existed c 505, Brahmagupta c 598, and Akbar c 1550.[11] So it is justifiable to suggest that Bentley's hypothesis was an indication of either ignorance or a fabrication based on a Eurocentric history of science. Nevertheless Bentley's altered chronology had the effect not only of lessening the achievements of the Indian mathematics but also of making redundant any conjecture of transmission to Europe.

Inadequate understanding of Indian mathematics was not confined to run of the mill scholars. More recently Smith,[12] an eminent historian of mathematics, claimed that, without the introduction of western civilization in the 18th and 29th centuries, India would have stagnated mathematically. He went on to say that: "Not since Bhaskara (i.e., Bhaskara II, b. 1114) has she produced a single native genius in this field."

This inclination for ignoring advances in and priority of discovery by non-European mathematicians persisted until even very recent times. For example there is no mention of the work of the Kerala School in Edwards' text[13] on the history of the calculus nor in articles on the history of infinite series by historians of mathematics such as Abeles[14] and Fiegenbaum.[15] A possible reason for such puzzling standards in scholarship may have been the rising Eurocentrism that accompanied European colonisation.[16] With this phenomenon, the assumption of white superiority became dominant over a wide range of activities, including the writing of the history of mathematics. The rise of nationalism in 19th-century Europe and the consequent search for the roots of European civilisation, led to an obsession with Greece and the myth of Greek culture as the cradle of all knowledge and values and Europe becoming heir to Greek learning and values.[17] Rare exceptions to this skewed version of history were provided by Ebenezer Burgess and George Peacock. They, respectively, wrote:

> Prof. Whitney seems to hold the opinion, that the Hindus derived their as-
> tronomy and astrology almost bodily from the Greeks. . . . I think he does not
> give the Hindus the credit due to them, and awards to the Greeks more credit
> than they are justly entitled to.[18]

> . . . (I)t is unnecessary to quote more examples of the names even of distin-
> guished men who have written in favour of a hypothesis [of the Greek origin of
> numbers and of their transmission to India] so entirely unsupported by facts.[19]

However, by the latter half of the twentieth century European scholars, perhaps released from the powerful influences induced by colonisation, had started to analyse the mathematics of the Kerala School using largely secondary sources such as Rajagopal and his associates.[20] The achievements of the Kerala School and their chronological priority over similar develop-ments in Europe were now being aired in several Western publications.[21] However these evaluations are accompanied by a strong defence of the Eu-ropean claim for the invention of the generalised calculus .

For example, Baron[22] states that:

> The fact that the Leibniz-Newton controversy hinged as much on priority in
> the development of certain infinite series as on the generalisation of the op-
> erational processes of integration and differentiation and their expression in
> terms of a specialised notation does not justify the belief that the [Keralese]
> development and use for numerical integration establishes a claim to the in-
> vention of the infinitesimal calculus.

Calinger[23] writes:

> Kerala mathematicians lacked a facile notation, a concept of function in trigonometry....Did they nonetheless recognise the importance of inverse trigonometric half chords beyond computing astronomical tables and detect connections that Newton and Leibniz saw in creating two early versions of calculus? Apparently not.

These comparisons appear to be defending the roles of Leibniz and Newton as inventors of the generalised infinitesimal calculus. While we understand the strength of nationalist pride in the evaluation of the achievements of scientists, we do find difficulty in the qualitative comparison between two developments founded on different epistemological bases. It is worthwhile stating here that the initial development of the calculus in 17th century Europe followed the paradigm of Euclidean geometry in which generalisation was important and in which the infinite was a difficult issue.[24] On the other hand, from the 15th century onwards the Kerala mathematicians employed computational mathematics with floating point numbers to understand the notion of the infinitesimal and derive infinite series for certain targeted functions[25] (Whish, 1835). In our view it is clear that qualitatively different intellectual tools and in different eras to investigate the similar problems are likely to produce qualitatively different outcomes. Thus the sensible way to understand Kerala mathematics is to understand it within the epistemology in which it was developed. To do otherwise is akin to trying to gain a full appreciation of the literature of Shakespeare by literally translating it into Urdu—the semantic and cultural connotations would undoubtedly be lost.

A DISCUSSION ON TRANSMISSION

The basis for establishing the transmission of science may be taken to be *direct* evidence of translations of the relevant manuscripts. The transmission of Indian mathematics and astronomy since the early centuries AD via Islamic scholars to Europe has been established by direct evidence. The transmission of Indian computational techniques was in place by at least the early 7th century for by 662 AD it had reached the Euphrates region.[26] A general treatise on the transmission of Indian computational techniques to Europe is given by Benedict.[27] Indian Astronomy was transmitted Westwards to Iraq, by a translation into Arabic of the *Siddhantas* around 760 AD[28] and into Spain. This transmission was not just westwards for there is documentary evidence of Indian mathematical manuscripts being found and translated in China, Thailand, Indonesia and other south-east Asian regions from the 7th century onwards.[29]

In the absence of such direct evidence the following is considered by some to be sufficient[30] to establish transmission:

i. the identification of methodological similarities
ii. the existence of communication routes
iii. a suitable chronology for the transmission.

Further there is van der Waerden's 'hypothesis of a common origin' to establish the transmission of (mainly Greek) knowledge.[31] Neugebauer uses his paradigm to establish his conjecture about the Greek origins of the astronomy contained in the *Siddhantas*. Similarly van der Warden uses the 'hypothesis of a common origin' to claim that Aryabhata's trigonometry[32] was borrowed from the Greeks. Van der Waerden makes a similar claim about Bhaskara's work on Diophantine equations and, whilst offering an argument based on methodological similarities, he is sufficiently convinced about the existence of an unknown Greek manuscript which was available to Bhaskara and his students.[33] Van der Waerden concludes his work on the Greek origins of these works of Aryabhata and Bhaskara work by stating that scientific discoveries are, in general, dependent on earlier prototypical works.[34]

What we see from these paradigms is that a case for claiming the transmission of knowledge from one region to another does not necessarily rest on documentary evidence. This is a consequence of the fact that many documents from ancient and medieval times do not now exist, having perished due to variety of reasons. In these circumstances priority, communication routes, and methodological similarities appear to establish a socially acceptable case for transmission from West to East. Despite these elements being in place, the case for the transmission of Keralese mathematics to Europe seems to require stronger evidence. One has merely to survey the literature of the history of mathematics to date to see hardly any credible mention about the possibility of this transmission.[35]

So how can our conjecture of transmission of Kerala mathematics possibly be established? The tradition in renaissance Europe was that mathematicians did not always reveal their sources or give credit to the original source of their ideas. However the activities of the monk Marin Mersenne between the early 1620s to 1648 suggest some attempt at gathering scientific information from the Orient. Mersenne was akin to being "the secretary of the early republic of science."[36] Mersenne corresponded with the leading renaissance mathematicians such as Descartes, Pascal, Hobbes, Fermat, and Roberval. Though a minim monk, Mersenne had had a Jesuit education and maintained ties with the Collegio Romano. Mersenne's correspondence reveals that he was aware of the importance of Goa and Cochin (in a letter from the astronomer Ismael Boulliaud to Mersenne in Rome[37]), he also wrote of the knowledge of Brahmins and "Indicos"[38] and took an active interest in the work of orientalists such as Erpen—regarding Erpen he mentions his "les livres manuscrits Arabics, Syriaques, Persiens, Turcs, Indiens en langue Malaye"[39]

It is possible that between 1560 to 1650, knowledge of Indian mathematical, astronomical and calendrical techniques accumulated in Rome, and diffused to neighbouring Italian universities like Padua and Pisa, and to wider regions through Cavalieri and Galileo, and through visitors to Padua, like James Gregory. Mersenne may have also had access to knowledge from Kerala acquired by the Jesuits in Rome and, via his well-known correspondence, could have helped this knowledge diffuse throughout Europe. Certainly the way James Gregory acquired his Geometry after his four year sojourn in Padua where Galileo taught suggests this possibility.

All this is circumstantial–to make our case for the transmission of Keralese mathematics we will use stronger criteria. In addition to the Neugebauer criteria of *priority, communication routes, and methodological similarities*, we propose to test the hypothesis of transmission on the grounds of *motivation* and *evidence of transmission activity by Jesuits missionaries*. In the next section all these aspects will be discussed.

THE CASE FOR THE TRANSMISSION OF KERALESE CALCULUS TO EUROPE: PRIORITY, COMMUNICATION ROUTES, AND METHODOLOGICAL SIMILARITIES

The priority of Keralese developments in the calculus over that of Newton and Leibniz is now beyond doubt.[40] Madhava (1340–1425) is credited with the original ideas in the Keralese mathematics. These ideas led to derivation for the infinite series for π which we have illustrated earlier and to infinite series for a range of trigonometric functions. These developments in the calculus, therefore, precede the late 17th century calculus of Newton and Leibniz by at least 250 years.

A communication route between the South of India and the Arabian Gulf (via the port of Basrah) had been in existence for centuries.[41] The arrival of the Portuguese Vasco da Gama to the Malabar coast in 1499 heralded a direct route between Kerala and Europe via Lisbon. Thus, after 1499, despite its geographical location, which prevented easy communication routes with the rest of India, Kerala was linked with the rest of the world and, in particular, directly to Europe.

Whilst the two aspects of priority and communication routes are readily established, the existence of methodological similarities requires further discussion. Firstly there is a similarity in the approach to calculus in the

Yuktibhasa and the approach to calculus adopted by Fermat, Pascal, Wallis, and others. In the *Yuktibhasa* the following key result is proved:[42]

$$\lim_{x \to \infty} \frac{1}{n^{k+1}} \sum_{i=1}^{n} i^k = \frac{1}{k+1}, \quad k = 1, 2, 3, \dots \quad (14.1)$$

This exact result was adopted by Fermat, Pascal, Wallis, and others in the 17th century to evaluate the area under the parabolas $y = x^k$, or, equivalently, calculate $\int x^k \, dx$. At this point we remark that our conjecture of the transmission of Kerala mathematics is given credence by the fact that Wallis used reasoning similar to the ones given in the *Yuktibhasa*. That is, Wallis replaces the term n^2 by $n(n+1)$ implicitly implying that, as n tends to ∞, $(n+1)$ can be replaced by n so that he can use the approximation $n(n+1) \approx n^2$.[43] Methodological similarities between the mathematics of Aryabhata school, upon which Kerala mathematics is based, and the works of the renaissance mathematicians are not infrequent. We give two further examples.

In 1655, John Wallis stated in his *Arithmetica Infinitorum* that the convergence of the continued fraction satisfied certain relations. Exactly these recurrence relations were discovered by Bhaskara II some 500 years earlier in his *BeejGanita*.[44] By itself this fact may not be much more than a coincidence except that Wallis gives exactly Bhaskara's proof of *Pythagoras'* theorem in his treatise on angular sections.[45] Brezinski posits that Wallis may have derived his results independently,[46] but does objectively suggest a scenario for the possible transmission of Bhaskara's work.[47]

Another similarity with Bhaskara works is the challenge problem that Fermat issued in 1657 "What is for example the smallest square which, multiplied by 61 with unity added, makes a square?" As Struik[48] notes, Indian mathematicians already had a solution to this problem—the case of $A = 61$ is given as a solved example in the *BeejGanita* text of Bhaskara II. This coincidence is not trivial when we consider that the solution [$x = 1,766,319,049$, $y = 226,153,980$] involves rather large numbers. A similar problem had earlier been suggested by the 7th century Brahmagupta, and Bhaskara II provides the general solution with his *chakravala* method. Thus, this challenge problem suggests a connection of Fermat with Indian mathematics.

THE CASE FOR THE TRANSMISSION OF KERALESE MATHEMATICS TO EUROPE: MOTIVATION AND EVIDENCE OF TRANSMISSION ACTIVITY BY JESUITS MISSIONARIES

Motivation

The motivation for the import of knowledge from India to Europe arose from the need of greater accuracy in arithmetical computation [as is well attested by historian of mathematics], the calendar, and in astronomy.

For example, by the middle of the 16th century there was an error in the calculations that formed the basis of the existing Julian calendar. The true solar year was around 11.25 minutes less than the assumed 365.25 days thus causing a cumulative error which was offsetting the date of Easter appreciably. For example, the vernal equinox was scheduled by the calendar to take pace on 21 March but it actually took place on March 11—thus, without correction Easter would eventually take place in Summer rather than in Spring. The evidence for this may be found in the use of the Indian calendrical term 'tithi'[49] in Viete's critique of the Gregorian calendar reform.

Is astronomy we point to the attested remarkable similarities between the planetary model by the Kerala mathematician Nilakantha and the later one by Tycho Brahe and the adoption by Kepler of the 10th century Indian lunar model of the astronomer Munjala.[50]

Evidence of Transmission Activity by Jesuits Missionaries

The arrival of Francis Xavier in Goa in 1540 heralded a continuous presence of the Jesuits in the Malabar till 1670. While the early Jesuits were interested in learning the vernacular languages and conversion work, the latter Jesuits who arrived after 1578 were of a different mould. The famous Matteo Ricci was in the first batch of Jesuits trained in the new mathematics curriculum introduced in the Collegio Romano by Clavius. Ricci was an accomplished mathematician.[51] Ricci also studied cosmography and nautical science in Lisbon prior to his arrival in India in 1578. Ricci's arrival in Goa was significant in respect of Jesuit acquisition of local knowledge. His specialist knowledge of mathematics, cosmography, astronomy and navigation made him a candidate for discovering the knowledge of the colonies and he had specific instructions to investigate the science of India.[52]

Subsequently several other scientist Jesuits trained both by Clavius or Grienberger (Clavius' successor as Mathematics Professor at the Collegio Romano) were sent to India. Most notable of these, in terms of their scientific activity in India, were Johann Schreck and Antonio Rubino. The former had studied with the French mathematician Viete, well known for his work in algebra and geometry. At some point in their stay in India these Jesuits went to the Malabar region including the city of Cochin, the epicentre of developments in the infinitesimal calculus.[53] Later we shall see that all these scholarly Jesuits attempted to acquire local knowledge. For the moment we point out that the Jesuits had an interest in the calendar that stemmed from the Church's desire to reform the erroneous dating of Easter and other festivals—Clavius headed the commission that ultimately reformed the Gregorian calendar in 1582. Others have conjectured that these Jesuits were part of an interchange of scientific ideas between Europe on the one hand and India and China on the other.[54]

We have mentioned above that there was a batch of mathematically able Jesuits who arrived in India in the late 15th and early 16th centuries and who had specific objectives to study local knowledge. We should point out that the mathematics of the Kerala school was essentially astronomy which had application to astrological prediction. Consequently it is our conjecture that the Jesuits had the opportunity and the motivation to transmit the calculus back to Europe bundled in the knowledge of the astronomy and calendrical science.

With regard to the earlier Jesuits in the Malabar Coast we observe that several references in the historical works of Wicki[55] indicate that they were interested in arithmetic, astronomy and timekeeping of the region. They were able to appreciate this knowledge by their learning of the vernacular languages such as Malayalam and Tamil.[56] The rationale for learning the vernacular languages was to aid their work in converting the local populace to Jesuit Catholicism by understanding their science, culture and customs. A prominent early Jesuit de Nobili, for instance, in 1615, wrote a critical paper on the Varamihira's *Vedanga Jyotisa*.[57] Indeed, it appears that the Jesuits tried to formalise this policy by including local sciences such as astrology or *jyotisa* in the curriculum of the Jesuit colleges in the Malabar Coast.[58] The early Jesuits were also active in the transmission of local knowledge back to Europe. Evidence of this knowledge acquisition is contained in the collections *Goa* **38**, **46** and **58** to be found in the Jesuit historical library in Rome (ARSI). The last collection contains the work of Father Diogo Gonsalves on the judicial system, the sciences and the mechanical arts of the Malabar region. This work started from the very outset.[59] The translation of the local science into European languages prior to transmission to Europe was epitomised by Garcia da Orta's popular *Colloquios dos simples e drogas he cousas mediçinas da India* published in Goa in 1563—there may have been other publications of this type which remain obscured because possibly due to linguistic and nationalistic reasons.[60]

If the early and late Jesuits were involved in learning the local sciences then, given the academic credentials of the Jesuits such as Ricci, Schreck and Rubino of the middle period, it is a plausible conjecture that this work continued and with greater intensity. There is some evidence that this did happen. It is also known that Ricci made enquiries about Indian calendrical science—in a letter to Maffei he states that he requires the assistance of an "intelligent Brahmin or an honest Moor" to help him understand the local ways of recording and measuring time or *jyotisa*.[61] Then there is de Menses who, writing from Kollam in 1580, reports that, on the basis of local knowledge, the European maps have inaccuracies.[62] There were other later Jesuits who report of scientific findings on such diverse things as calendrical sciences and inaccuracies in the European maps and mathematical tables. Antonio Rubino wrote, in 1610, similarly about inaccuracies in European mathematical tables for determining time.[63] Then there is the letter from Schreck, in 1618, of astronomical observations intended for the benefit of Kepler[64]—the latter had requested the eminent Jesuit mathematician Paul Guldin to help him to acquire these observations from India to support his theories.[65]

Whilst this does not establish the fact that these Jesuits obtained manuscripts containing the Kerala mathematics it does establish that their scientific investigations about the local astronomy and calendrical sciences would have lead them to an awareness of this knowledge. There are some reports[66] that that the Brahmins were secretive and unwilling to share their knowledge. However this was not an experience shared by many others. For example in the mid-seventeenth century Fr. Diogo Gonsalves, who learnt the local language Malayalam well, was able to write a book about the administration of justice, sciences and mechanical arts of the Malabar. This book is to be found in the MS Goa 58 in the Jesuit Historical library (ARSI), Rome. And a Brahmin spent eight years translating Sanskrit works for Fr. Frois during the same time.[67]

The information gathering and transmission activities of the Jesuit missionaries are thus not in doubt. In addition after the 1580 annexation of Portugal by Spain and subsequent loss of funding from Lisbon, the rationale for transmission acquired another dimension, that of profit.[68] Whatever the nature of the profit, intellectual or material, the motivation may have been sufficient for the learned Jesuits to have acquired the relevant manuscripts containing the Keralese mathematics.

A CONJECTURE FOR THE MODE OF ACQUISITION BY THE JESUITS OF THE MANUSCRIPTS CONTAINING THE KERALA CALCULUS

Even if the conjecture of the transmission of the Kerala calculus by the Jesuit missionaries is accepted it leaves open the question as to how might the Jesuits have obtained key manuscripts of Indian astronomy such as the *Tantrasangraha* and the *Yuktibhasa*? Such manuscripts would require the Jesuits being in close contact with scholarly Brahmins or Kshatriyas. We have already mentioned above that at least one scholarly Brahmin was working for the Jesuits. In addition, as we shall now show, the Jesuits were in communication with the Kshatriya kings of Cochin whose scholarship enabled them to be in very close proximity to the Kerala mathematics.

The kings of Cochin came from the scholarly Kshatriya Varma 'Tampuran' family who were knowledgeable about the mathematical and astronomical works of medieval Kerala. This is attested by Whish[69] and Srinivasiengar[70] who state that the author of the *Sadratnamala* is Sankara Varma, the younger brother of Raja of Cadattanada near Tellicherry and further states that the Raja is a very acute mathematician. Srinivasiengar[71] further refers to the Malayalam *History of Sanskrit Literature in Kerala* which identifies the King of Cochin, Raja Varma, as being aware of the chronology of the *Karana-Paddhati.* Rama Varma Tampuran who, in 1948 (together with A.R. Akhileswara Iyer) had published an exposition in Malayalam on the *Yuktibhasa* was one of the princes of Cochin. Ragagopal and Rangachari[72] state that Rama Varma Tampuran supplied them with the manuscript material relating to Kerala mathematics. Sarma[73] identifies the valuable contribution to the analysis of Kerala astronomy by Rama Varma Maru Tampuran. Mukunda Marar (who is the eldest son of the last king of Cochin) stated in a personal communication to the Aryabhata Forum that the kings of Cochin would have, at the least, been aware both of the astronomical methods for astrological prediction and of the manuscripts that contained these methods. Several were scholarly enough to publish commentaries of the mathematical and astronomical works. Mukunda Marar, himself, worked with Rajagopal and published a work.[74] Moreover, various authors, from Charles Whish in the 19th century to Rajagopal and Rangachari in the 20th have acknowledged that members of the royal household were helpful in supplying these manuscripts in their possession.

This suggests that the former royal family in Cochin, which was in possession of a large number of MSS, had not only a scholarly tradition, but also a tradition of helping other scholars. Thus, the royal family could itself have been a possible source of knowledge for the Jesuits. Indeed the Jesuits working on the Malabar Coast had close relations with the kings of Cochin.[75] Furthermore, around 1670, they were granted special privileges by King Rama Varma[76] who, despite his misgivings about the Jesuit work in conversion, permitted members of his household to be converted to Christianity.[77] The close relationship between the King of Cochin and the foreigners from Portugal was cemented by King Rama Varma's appointment of a Portuguese as his tax collector.[78] Given this close relationship with the Kings of Cochin, the Jesuit desire to know about local knowledge, and the royal family's contiguity to the works on Indian astronomy, it is quite possible that the Jesuits may have acquired the key manuscripts via the royal household.

CONCLUSION

In conclusion, it is worth noting that we have focused so far on evidence of direct transmissions of Kerala ideas to Europe. But, as pointed out by Balasubramaniam[79] the transmission of the discoveries of Kerala mathematics could have been as 'know-how' and computation techniques through the channel of craftsmen and technicians. This may explain the absence of documentary evidence in Jesuit communications. Even if there is documentary evidence in 16th century European manuals used for navigation, map-making and calendar construction of the use of approximate series derived from the discoveries of the Kerala School, it would hardly have been directly communicated to European mathematicians. After all craftsmen oriented to practical rather than theoretical concerns are not likely to write to leading mathematical figures in Europe or be taken seriously if they do. For more a detailed discussion of this interesting conjecture, see the aforementioned article by Balasubramaniam.

ACKNOWLEDGMENTS

The authors would like to acknowledge the Arts and Humanities Research Board (AHRB) support for a project on the investigation of Kerala mathematics during the period 2002–2005. Also acknowledged is the input from C. K. Raju during the period preceding the AHRB project. Parts of this paper were included in a presentation at the XXI International conference on the history of science in Sept 2001, Mexico City and in a published article by the authors: 'Eurocentrism in the History of Mathematics: The case of the Kerala School', *Race and Class*, 2004, 45(4), 45–59.

NOTES

1. See, for example, Margaret Baron, *The Origins of the Infinitesimal Calculus*, Oxford, Pergamon, 1969, p 65.
2. See, for example, Charles Edwards, *The Historical Development of the Calculus*, New York, Springer-Verlag, 1979, p189, and Victor Katz, "Ideas of calculus in Islam and India", *Mathematics Magazine*, Washington, 68 (1995), 3: 163–174, p 163 and p 164.
3. See the work of K Venkateswara Sarma, *A History of the Kerala School of Hindu Astronomy*, Hoshiarpur, Vishveshvaranand Vedic Research Institute, 1972, p 21 and p 22 and the paper by Charles Whish, "On the Hindu quadrature of the circle and the infinite series of the proportion of the circumference to the diameter exhibited in the four Shastras, the Tantrasamgraham, Yukti-Bhasa, Carana Padhati, and Sadratnamala", *Transactions of the Royal Asiatic Society of Great Britain and Ireland*, London, 3 (1835): 509–523, p 522 and p 523.
4. For example, Margaret Baron, *Origins of Calculus*, op cit, p 62 and p 63; Ronald Calinger, *A Contextual History of Mathematics to Euler*, New Jersey, Prentice Hall, 1999, p 284
5. Leone Fiegenbaum, "Brook Taylor and the Method of Increments", *Archive for History of Exact Sciences*, Baltimore, 34 (1986): 1–140, p 72

6. For example, Charles Whish, "On the Hindu quadrature of the circle and the infinite series of the proportion of the circumference to the diameter exhibited in the four Shastras, the Tantrasamgraham, Yukti-Bhasa, Carana Padhati, and Sadratnamala", *Transactions*, 3 (1835): 509–523; C T Rajagopal and M S Rangachari, "On an Untapped Source of Medieval Keralese Mathematics", *Archive for History of Exact Sciences*, Baltimore, 18 (1978): 89–102; C T Rajagopal and T V Vedamurthi, "On the Hindu proof of Gregory's series", *Scripta Mathematica*, New York, 18(1951): 91–99

7. S. Al-Andalusi, c 1068, *Science in the medieval world*, translated by S. I. Salem and A. Kumar, University of Texas Press, 1991, p 11–12

8. Al-Biruni, 1030, *India*, translated by Qeyamuddin Ahmad, New Delhi, National-al Book Trust, 1999. p 70

9. L. A. Sedillot, 1873, 'The Great Autumnal Execution', in the *Bulletin Of The Bibliography And History Of Mathematical & Physical Sciences* published by B. Boncompagni, member of Pontific Academy, Reprinted in *Sources of Science*, no. 10 (1964), New York & London

10. J. Bentley, 1823, *A Historical View of the Hindu Astronomy*, Baptist Mission Press, Calcutta. See esp. p 151.

11. G. G. Joseph, 2000, *The Crest of the Peacock: non-European Roots of Mathematics*, Princeton University Press

12. D. E. Smith, 1923/5, *History of Mathematics*, 2 volumes, Boston, MA. Ginn & Co. (Reprinted by Dover, New York, 1958). See Vol. 1, p 435.

13. C. H. Edwards, 1979, *The Historical Development of the Calculus*, Springer-Verlag, New York

14. F.F. Abeles , 1993, 'Charles Dodgson's geometric approach to arctangent relations for π, *Historia Mathematica*, 20 (2), 151–159

15. L. Fiegenbaum, 1986, 'Brook Taylor and the Method of Increments, *Arch Hist Ex Sci*, 34(1), 1–140

16. G. G. Joseph, 1995 'Cognitive Encounters in India during the Age of Imperialism' *Race and Class*, 36 (3), 39–56

17. G. G. Joseph, 2000, *The Crest of the Peacock: non-European Roots of Mathematics*, Princeton University Press. See p 215.

18. E. Burgess, 1860, *The Surya Siddhanta: A Text-Book Of Hindu Astronomy*, Reprinted, 1997, Motilal Banarsidass Publishers Private Limited, New Delhi

19. G. Peacock, 1849, 'Arithmetic—including a history of the science' in *Encyclopedia Metropolitana Or Universal Dictionary Of Knowledge;* Part 6 First Division, J. J. Griffin and Co, London. P 420

20. For example, C. T. Rajagopal and T. V. Vedamurthi, 1952,'On the Hindu proof of Gregory's series', *Scripta Mathematica*, **18**, 65–74.C. T. Rajagopal and M. S. Rangachari, 1978,'On an Untapped Source of Medieval Keralese Mathematics', *Archive for the History of Exact Sciences*, **18**, 89–102

21. See, for example, V. J. Katz, 1992, *A History of Mathematics: An Introduction*, Harper Collins, New York. V. J. Katz, 1995, "Ideas of calculus in Islam and India", *Mathematics Magazine*, Washington, **68**, 3: 163–174. R. Calinger, 1999, *A Contextual History of Mathematics to Euler*, Prentice Hall, New Jersey. M. E. Baron, 1969, *The Origins of the Infinitesimal Calculus*, Pergamom, Oxford

22. M. E. Baron, 1969, *The Origins of the Infinitesimal Calculus*, Pergamon, Oxford. p 65.

23. R. Calinger, 1999, *A Contextual History of Mathematics to Euler*, Prentice Hall, New Jersey. p 28

24. See, for example, V. J. Katz, 1992, *A History of Mathematics: An Introduction*, Harper Collins, New York

25. A. M. Whish, 1835, "On the Hindu quadrature of the circle and the infinite series of the proportion of the circumference to the diameter exhibited in the four Shastras, the Tantrasamgraham, Yukti-Bhasa, Carana Padhati, and Sadratnamala." *Tr. Royal Asiatic Society of Gr. Britain and Ireland*, **3**, 509–523

26. John Berggren, *Episodes in the Mathematics of Medieval Islam*, New York, Springer Verlag, 1986, p 30

27. Susan Benedict, *A Comparative Study of the Early Treatises Introducing into Europe the Hindu Art of Reckoning*, Ph.D. Thesis, University of Michigan, Rumford Press, 1914

28. B V Subbarayappa and K Venkateswara Sarma, *Indian Astronomy-A Source-Book*, Bombay, Nehru Centre Publications, 1985, p XXXVIII

29. B V Subbarayappa and K Venkateswara Sarma, *Indian Astronomy*, op cit, 1985, p XXXVIII

30. Otto Neugebauer, *The Exact Sciences in Antiquity*, New York, Harper, 1962, p 166 and p 167

31. Bartel van der Waerden, *Geometry And Algebra In Ancient Civilizations*, Berlin, Springer-Verlag, 1983, p 211

32. Bartel van der Waerden, *Geometry And Algebra*, op cit, p 133

33. Bartel van der Waerden, "Pell's equation in Greek and Hindu mathematics", *Russian Mathematical Surveys*, 31 (1976): 210–225, p 210. Here he states that "the original common source of the Hindu authors was a Greek treatise in which the whole method was explained."

34. Bartel van der Waerden, "Pell's equation in Greek and Hindu mathematics", *Russ Math Surveys*, 31 (1976): 210–225, p 221: "...in the history of science independent inventions are exceptions: the general rule is dependence"

35. A few notable exceptions are: Victor Katz, "Ideas of calculus in Islam and India", *Math Magazine*, 68 (1995), 3: 163–174, p 173 and p 174; Ronald Calinger, *Contextual History of Mathematics*, op cit, p 282; George Joseph, *The Crest of the Peacock: non-European Roots of Mathematics*, Princeton, Princeton University Press, 2000, pp 354–356

36. Ronald Calinger, *Contextual History of Mathematics*, op cit, p 475

37. Marin Mersenne, *Correspondance du P. Marin Mersenne*, 18 volumes, Paris, Presses Universitaires de France, 1945-, Vol XIII, p 267

38. Marin Mersenne, *Correspondance*, op cit, Vol XIII, p 518–521

39. Marin Mersenne, *Correspondance*, op cit, Vol II, p 103–115

40. Margaret Baron, *Origins of Calculus*, op cit, p 63; Victor Katz, "Ideas of calculus in Islam and India", *Math Magazine*, 68 (1995), 3: 163–174, p 173 and p 174

41. B V Subbarayappa and K Venkateswara Sarma, *Indian Astronomy*, op cit, 1985, p XXXVIII; Victor Katz, "Ideas of calculus in Islam and India", *Math Magazine*, 68 (1995), 3: 163–174, p 174

42. Victor Katz, "Ideas of calculus in Islam and India", *Math Magazine*, 68 (1995), 3: 163–174, p 169

43. Joseph Scott, *The Mathematical work of John Wallis*, New York, Chelsea, 1981, p 30

44. Claude Brezinski, *History of Continued Fractions*, op cit, p 43

45. Henry Colebrooke, *Miscellaneous essays*, op cit, vol 2, p 395

46. Claude Brezinski, *History of Continued Fractions*, op cit, p 32. He states that "...the recurrence relationship of continued fractions (of Bhaskara II) will only be discovered in Europe by John Wallis in 1655, which is 500 years later! Thus it is important to notice that the first English translation of Bhascara II's work appears in 1816."

47. Claude Brezinski, *History of Continued Fractions*, op cit, p 34. Here he posits that "...it is highly probable that Muslim mathematicians were in close contact with their Indian colleagues during this period, and that they brought back their writings to western countries. Since Arabic mathematicians were translated much earlier into European languages, it is possible that the Indian mathematical writings were known in Europe before their translation."

48. Dirk Struik, *A Source Book in Mathematics, 1200–1800*, Cambridge, Mass., Harvard University Press, 1969, p 29 and p 30

49. R Bein, "Viète's Controversy with Clavius Over the Truly Gregorian Calendar", *Archive for History of Exact Sciences,* 61(1), 2007, 39–66

50. Dennis Duke, 'The second Lunar anomaly in ancient Indian astronomy, *Archive for the History of the Exact Sciences,* 61(2007), 147–157

51. Vincent Cronin, *The Wise Man From The West*, London, Collins, 1984, p 22

52. Henri Bernard, *Matteo Ricci's Scientific Contribution to China*, Westport, Conn., Hyperion Press, 1973, p 38: "Ricci had resided in the cities of Goa and of Cochin for more than three years and a half (September 13, 1578-April 15, 1582): he had been requested to apply himself to the scientific study of this new and imperfectly know country, in order to document his illustrious contemporary, Father Maffei, the 'Titus Livius' of Portuguese explorations."

53. See, for example, Isaia Iannaccone, *Johann Schreck Terrentius*, Napoli, Instituto Universitario Orientale, 1998, p 50- p 58 and Ugo Baldini, *Studi su filosofia e scienza dei gesuiti in Italia 1540–1632*, Firenze, Bulzoni Editore, 1992, p 214 and p 215

54. Ugo Baldini, *Studi su filosofia*, op cit, p 70: "It can be recalled that many the best Jesuit students of Clavius and Geienberger (beginning with Ricci and continuing with Spinola, Aleni, Rubino, Ursis, Schreck, and Rho) became missionaries in Oriental Indies. This made them protagonists of an interchange between the European tradition and those Indian and Chinese, particularly in mathematics and astronomy, which was a phenomenon of great historical meaning"

55. Josef Wicki, *Documenta Indica*, 16 volumes, Rome, Monumenta Historica Societate Iesu, 1948-, vol IV, p 293 and vol VIII, p 458

56. Josef Wicki, *Documenta Indica*, op cit, vol XIV p 425 and vol XV p 34*

57. Vincent Cronin, *The Wise Man*, op cit, p 178- p180

58. Josef Wicki, *Documenta Indica*, op cit, Vol III, p 307

59. Domenico Ferroli, *The Jesuits in Malabar,* 2 Vols, Bangalore, Bangalore Press, 1939, Vol 2, p 402: "In Portuguese India, hardly seven years after the death of St. Francis Xavier the fathers obtained the translation of a great part of the 18 Puranas and sent it to Europe. A Brahmin spent eight years in translating the works of Veaso (Vyasa) ... several Hindu books were got from Brahmin houses, and brought to the Library of the Jesuit college. These translations are now preserved in the Roman Archives of the Society of Jesus. (*Goa* **46**)"

60. George Sarton, *The Appreciation of Ancient and Medieval Science During the Renaissance (1450–1600),* Philadelphia, University of Pennsylvania, 1955, p102

61. Josef Wicki, *Documenta Indica,* op cit, vol XII, p 474

62. Josef Wicki, *Documenta Indica,* op cit, vol XI, p185: "I have sent Valignano a description of the whole world by many selected astrologers and pilots, and others in India, which had no errors in the latitudes, for the benefit of the astrologers and pilots that every day come to these lands, because the maps theirs are all wrong in the indicated latitudes, as I clearly saw"

63. Ugo Baldini, *Studi su filosofia,* op cit, p 214: "... comparing the real local times with those inferable from the ephemeridis [tables] of Magini, he [Rubino] found great inaccuracies and, therefore, requested other ephemeridi

64. Isaia Iannaccone, *Johann Schreck,* op cit, p 58

65. Carola Baumgardt, *Johannes Kepler: Life and Letters,* New York, Philosophical Library, 1951, p 153

66. Pascual D'elia, *Galileo In China Relations through the Roman College Between Galileo and the Jesuit Scientist-Missionaries (1610–1640)* :Translated by Rufus Suter and Matthew Sciascia, Cambridge, Mass, Harvard University Press, 1960, p15

67. Domenico Ferroli, *The Jesuits in Malabar,* op cit, p 402

68. Domenico Ferroli, *The Jesuits in Malabar,* op cit, volume 2, p 93: "Most of the Jesuit missionaries set to work to master the vernaculars ... some of their number studied Indian books and Indian philosophy, not merely with the idea of refuting it, but with the desire of profiting by it."

69. Charles Whish, "On the Hindu quadrature of the circle and the infinite series of the proportion of the circumference to the diameter exhibited in the four Shastras, the Tantrasamgraham, Yukti-Bhasa, Carana Padhati, and Sadratnamala", *Transactions,* 3 (1835): 509–523, p 521

70. C N Srinivasiengar, *The History of Ancient Indian Mathematics,* Calcutta, World Press, 1967, p 146

71. C N Srinivasiengar, *The History of Ancient Indian Mathematics,* op cit, p 145

72. C T Rajagopal and M S Rangachari, "On an Untapped Source of Medieval Keralese Mathematics", *Archive for History,* 18 (1978): 89–102, p 102

73. K Venkateswara Sarma, *A History of the Kerala School,* op cit, p 12

74. C T Rajagopal and Mukunda Marar, "On the Hindu Quadrature of the Circle", *Journal of the Royal Asiatic Society (Bombay branch),* 20 (1944): 65–82

75. See, for example, Josef Wicki, *Documenta Indica,* op cit, vol X, p 239, 834, 835, 838, and 845

76. Josef Wicki, *Documenta Indica,* op cit, vol XV, p 224

77. Josef Wicki, *Documenta Indica,* op cit, vol XV, p 7*

78. Josef Wicki, *Documenta Indica,* op cit, vol XV, p 667

79. A. Balasubramaniam: 'Establishing Transmissions: Some Methodological Issues' in G.G. Joseph (Ed) Medieval Kerala Mathematics: The Possibility of its Transmission to Europe, KCHR Publication, forthcoming.

CHAPTER 15

THE PHILOSOPHY OF MATHEMATICS, VALUES, AND KERALESE MATHEMATICS

Paul Ernest
University of Exeter

WHAT IS THE BUSINESS OF THE PHILOSOPHY OF MATHEMATICS?

Traditionally, in Western philosophy, mathematical knowledge has been understood as universal and absolute knowledge, whose epistemological status sets it above all other forms of knowledge. The traditional foundationalist schools of formalism, logicism and intuitionism sought to establish the absolute validity of mathematical knowledge by erecting foundational systems. Although modern philosophy of mathematics has in part moved away from this dogma of absolutism, it is still very influential, and needs to be critiqued. So I wish to begin by summarising some of the arguments against Absolutism, as this position has been termed (Ernest 1991, 1998).

My argument is that the claim of the absolute validity for mathematical knowledge cannot be sustained. The primary basis for this claim is that mathematical knowledge rests on certain and necessary proofs. But proof

Critical Issues in Mathematics Education, pages 189–204
Copyright © 2009 by Information Age Publishing
All rights of reproduction in any form reserved.

in mathematics assumes the truth, correctness, or consistency of an under-lying axiom set, and of logical rules and axioms or postulates. The truth of this basis cannot be established on pain of creating a vicious circle (Lakatos 1962). Overall the correctness or consistency of mathematical theories and truths cannot be established in non-trivial cases (Gödel 1931).

Thus mathematical proof can be taken as absolutely correct only if certain unjustified assumptions are made. First, it must be assumed that absolute standards of rigour are attained. But there are no grounds for assuming this (Tymoczko 1986). Second, it must be assumed that any proof can be made perfectly rigorous. But virtually all accepted mathematical proofs are infor-mal proofs, and there are no grounds for assuming that such a transformation can be made (Lakatos 1978). Third, it must be assumed that the checking of rigorous proofs for correctness is possible. But checking is already deeply problematic, and the further formalizing of informal proofs will lengthen them and make checking practically impossible (MacKenzie 1993).

A final but inescapably telling argument will suffice to show that absolute rigour is an unattainable ideal. The argument is well-known. Mathematical proof as an epistemological warrant depends on the assumed safety of axi-omatic systems and proof in mathematics. But Gödel's (1931) second incom-pleteness theorem means that consistency and hence establishing the cor-rectness and safety of mathematical systems is indemonstrable. We can never be sure mathematics theories are safe, and hence we cannot claim their cor-rectness, let alone their necessity or certainty. These arguments are necessar-ily compressed here, but are treated fully elsewhere (e.g., Ernest 1991, 1998). So the claim of absolute validity for mathematical knowledge is unjustified.

The past two decades has seen a growing acceptance of the weakness of absolutist accounts of mathematical knowledge and of the impossibility in establishing knowledge claims absolutely. In particular the 'maverick' tradition, to use Kitcher and Aspray's (1988) phrase, in the philosophy of mathematics questions the absolute status of mathematical knowledge and suggest that a reconceptualisation of philosophy of mathematics is needed (Davis and Hersh 1980, Lakatos 1976, Tymoczko 1986, Kitcher 1984, Er-nest 1997). The main claim of the 'maverick' tradition is that mathemati-cal knowledge is fallible. In addition, the narrow academic focus of the philosophy of mathematics on foundationist epistemology or on Platonistic ontology to the exclusion of the history and practice of mathematics, is viewed by many as misguided, and by some as damaging.

RECONCEPTUALIZING THE PHILOSOPHY OF MATHEMATICS

Although a widespread goal of traditional philosophies of mathematics is to reconstruct mathematics in a vain foundationalist quest for certainty, a number of philosophers of mathematics agree this goal is inappropriate. "To confuse description and programme—to confuse 'is' with 'ought to be' or 'should be'—is just as harmful in the philosophy of mathematics as elsewhere." (Körner 1960: 12), and "the job of the philosopher of mathematics is to describe and explain mathematics, not to reform it." (Maddy 1990: 28). Lakatos, in a characteristically witty and forceful way which paraphrases Kant indicates the direction that a reconceptualised philosophy of mathematics should follow. "The history of mathematics, lacking the guidance of philosophy has become blind, while the philosophy of mathematics turning its back on the . . . history of mathematics, has become empty" (1976: 2).

Building on these and other suggestions it might be expected that an adequate philosophy of mathematics should account for a number of aspects of mathematics including the following:

1. **Epistemology**: Mathematical knowledge; its character, genesis and justification, with special attention to the role of proof
2. **Theories**: Mathematical theories, both constructive and structural: their character and development, and issues of appraisal and evaluation
3. **Ontology**: The objects of mathematics: their character, origins and relationship with the language of mathematics, the issue of Platonism
4. **Methodology and History:** Mathematical practice: its character, and the mathematical activities of mathematicians, in the present and past
5. **Applications and Values**: Applications of mathematics; its relationship with science, technology, other areas of knowledge and values
6. **Individual Knowledge and Learning**: The learning of mathematics: its character and role in the onward transmission of mathematical knowledge, and in the creativity of individual mathematicians (Ernest 1998)

Items 1 and 3 include the traditional epistemological and ontological focuses of the philosophy of mathematics, broadened to add a concern with the genesis of mathematical knowledge and objects of mathematics, as well as with language. Item 2 adds a concern with the form that mathematical knowledge usually takes: mathematical theories. Items 4 and 5 go beyond the traditional boundaries by admitting the applications of mathematics and human mathematical practice as legitimate philosophical concerns, as well as its relations with other areas of human knowledge and values. Item 6 adds a concern with how mathematics is transmitted onwards from one generation to the next, and in particular, how it is learnt by individuals, and the dialectical relation between individuals and existing knowledge in creativity.

The legitimacy of these extended concerns arises from the need to consider the relationship between mathematics and its corporeal agents, i.e., human beings. They are required to accommodate what on the face of it is the simple and clear task of the philosophy of mathematics, namely to give an account of mathematics.

CHALLENGING EPISTEMOLOGICAL ASSUMPTIONS AND VALUES

The challenge to the traditional philosophy of mathematics to broaden its epistemological goal, as indicated above, raises some critical issues. In particular, if providing ironclad foundations to mathematical knowledge and mathematical truth is not the main purpose of philosophy of mathematics, has this fixation distorted philosophical accounts of mathematics and what is deemed valuable or significant in mathematics? To what extent is the philosophical emphasis on mathematical proof and deductive theories justified? I want to argue that the emphasis on mathematics as made up of rigorous deductive theories is excessive, and this focus in fact existed for only two periods totaling possibly less than ten percent of the overall history of mathematics as a systematic discipline, and then only in the West.[1]

The first of these two periods was the ancient Greek phase in the history of mathematics which reached its high point in the formulation of Euclid's *Elements*, a systematic exposition of deductive geometry and other topics. The second period is the modern era encompassing the past two hundred years or so. This second period was first signaled by Descartes' modernist epistemology, with its call to systematize all knowledge after the model of geometry in Euclid's *Elements*. However, fortunately, his injunction was not applied in the practices of mathematicians for the next two hundred years, which was instead a period of great creativity and invention in the West. Only in the 19th century did the newly professionalized mathematicians turn their attention to the foundations of mathematical knowledge and systematize it into axiomatic mathematical theories. The contributions of Boole, Weierstrass, Dedekind, Cantor, Peano, Hilbert, Frege, Russell and others in this enterprise up to the time of Bourbaki are well known.

I am not claiming that all or even most mathematical work was foundational during these two exceptional periods. But the foundational work is what caught the attention of philosophers of mathematics, and in the spirit of Cartesian modernism has become the epistemological focus of modern philosophy of mathematics, as well as the touchstone for what is deemed to be of epistemologically valuable. I do not want to detract from either the magnificence of the achievement in the foundational work carried out by mathematicians and logicians, nor from the pressing nature of the problems that made attention to it so vital in the early part of the 20th century. Nevertheless, the legacy of this attention has been to overvalue the philosophical significance of axiomatic mathematics at the expense of other dimensions of mathematics. Two underemphasized dimensions of mathematics are calculation and problem solving. All three of these aspects of mathematics involve deductive reasoning, but axiomatic mathematics is valued above the others as the supreme achievement of mathematics.

There is another feature shared by the two historical periods that emphasised axiomatic mathematics, namely a purist ideology involving the philosophical dismissal or rejection of the significance of practical mathematics. The antipathy of the ancient Greek philosophers to practical matters including numeration and calculation is well known. This aspect of mathematics was termed 'logistic' and regarded as the business of slaves or lesser beings. In the modern era, calculation and practical mathematics have been viewed as mathematically trivial and philosophically uninteresting. The fact that philosophers have been concerned with ontology and the nature of the mathematical objects has engendered little or no interest in the symbolism of mathematics, or calculations and transformations that convert one mathematical object (or rather its name, a term) into another. Such a view is typified by Platonism, which concerns itself primarily with mathematical truths and objects. These are presumed to exist in an unearthly and idealized world beyond that which we inhabit as fleshy and social human beings, such as Popper's (1979) objective World 3.

Of course at the same time as these modern developments were taking place applied mathematics and theoretical or mathematical physics were making great strides, but this was not considered to be of interest to philosophers of mathematics (however much interest it was to philosophers of science), because of their purist ideology. Even in British public schools, during the late Victorian era, mathematics was taught in with ungraduated rulers because graduations implied measurement and practical applications, which was looked down upon for the future professional classes and rulers of the country. (Admittedly some of the rationale was that Euclid's geometry only requires a straight-edge and a pair of compasses as drawing instruments).

What I have described here (in order to critique it) is an ideological perspective that elevates some aspects of mathematics above others, but typically does not acknowledge that it is based on a set of values, a set of choices and preferences to which no necessity or logical compulsion is attached. Furthermore, it appears that such values have only been prominent during a small part of the history of mathematics.

In order to strengthen my critique of these values I want to point out that mathematical proof, the cornerstone of axiomatic mathematics, and calculation in mathematics, are formally very close in structure and character. In Ernest (forthcoming) I have argued that mathematical topic areas (e.g., number and calculation) can be interpreted as being made up semiotic systems, each comprising (1) a set of signs, (2) rules of sign production and transformation, and (3) an underpinning (informal) meaning structure. Such signs include atomic, i.e., basic, signs and a range of composite signs comprising molecular constellations of atomic signs. These signs may be alphanumeric (made up of numerals or letters) or figural (e.g., geometric figures) or include both (e.g., figures with labels and the types of inference employed). The use of semiotic systems is primarily that of sign production in the pursuit of some goal (e.g., solving a problem, making a calculation, producing a proof for a theorem). I want to claim that most recorded mathematical activity concerns the production of sequences of signs (within a semiotic system). Typically these are transformations of an initial composite sign (S_1), resulting after a finite number (n) of transformations, in a terminal sign (S_{n+1}), satisfying the requirements of the activity. This can be represented by the sequence: $S_1 \rightarrow S_2 \rightarrow S_3 \rightarrow \ldots \rightarrow S_{n+1}$. Each transformation (represented by $\rightarrow$) constitutes the application of one of the rules of the semiotic system to the sign, resulting in the derivation of the next sign in the sequence. More accurately these should be represented by $\rightarrow_i$, with $i = 1, \ldots, n$, since each transformation in the sequence is potentially different.

My claim is that this formal (semiotic) system describes most mathematical domains and activities. If the initial sign is the statement of a problem, the sequence represents the derivation of a solution to the problem. I will not dwell on this case as there are many complications involved in problem solving, such as the use of multiple representations, branching solution attempts,[2] etc. and some of the transformations (such as interpreting an initial problem formulation and constructing a problem representation) are neither easily made explicit nor fully formalizable. Furthermore, there is no simple characterization of the relationship between the transformational rules and the underlying informal meaning structure, for the transformations are partly structure preserving morphisms, and partly calculational.

More significantly in the present context, such transformational sequences can represent a deductive proof for a theorem. In this case it consists of a sequence of sentences, each of which is derived from its predecessors by the deductive rules of the system (including the introduction of axioms or other assumptions). The final sign in the sequence is the theorem proved. The meaning structure underpinning the rules of proof is based on the principle of the preservation of the truth value of sentences in each deductive step, and hence along the length of the proof sequence (which is why axioms can be inserted, and why proofs 'work', i.e., do what they are designed to do.)[3]

In the case of a calculation, the initial sign is usually a compound term. Subsequent terms are derived by calculational rules and typically each is a simplification in some sense of its predecessor. The final term in a calculation is the simplified numerical 'answer' to the problem. With the introduction of algebra and other functions and operations such as trigonometrical functions, the answer may instead be a simplified but non-numerical term (i.e., a function). Thus calculations are sequences of terms, each derived from predecessors by the rules of the system. The meaning structure underpinning the rules of calculation is based on the principle of the preservation of numerical value.[4]

Thus there is a strong analogy between the semiotic systems based on calculation and those based on deductive proof. The transformations of terms and sentences are based on the underlying principles of value preservation, namely numerical value or truth value, respectively, as I have demonstrated. In addition, calculation concerns terms and proof concerns sentences (or formulas), and both of these are defined similarly. Terms (sentences) are defined recursively as follows. An atomic term (sentence) consists of a constant or a variable (an n-place predicate applied to constants or variables, respectively). A compound term (sentence) is defined as the result of applying a function or operation (a logical connective or quantifier, respectively) to one or more terms (sentences, respectively), to make a new term (sentence, respectively). Thus structurally terms and sentences are very similar, defined analogously by induction.

The sequential and rule-based nature of calculation is something that precedes the development of the deductive proof of theorems by at least two thousand years. My contention is that without the long and ancient tradition of rule following in sequences of calculations, and the entrenchment of the grammatical and value preservational features noted above, the development of proof would not be possible. As I have indicated here, there is a striking analogy between calculations and deductive proofs of theorems, rarely if ever remarked upon, that puts into question the claimed superiority of proof.

Furthermore, proof and calculation are formally equivalent, in modern foundational terms. Calculations utilize the term as a basic unit of meaning (and as that which is transformed), whereas deductive proofs use the sentence (including formulas or open sentences) as a basic unit. However, there are equivalence transformations between calculations and proofs. A calculation sequence of the form $t_1, t_2, t_3, \ldots, t_n$, where each t_i $(1 \le i \le n)$ is a term, can be represented as a deductive proof of the form $t_1 = t_2, t_2 = t_3, t_3 = t_4, \ldots, t_{n-1} = t_n$ in which each identity asserts that numerical values of adjacent terms are preserved identically in the calculation. By an extended or repeated application of the transitivity of identity ($x = y$ and $y = z \to x = z$, for all x, y, and z), $t_1 = t_n$ is derived, thus equating the initial term of the calculation and the terminal term, the 'answer'.

Likewise, a deductive proof of the form $S_1, S_2, S_3, \ldots, S_n$, can be represented as a series of terms. These are the values of the truth value function f defined on numerical representations of true and false sentences to give the values 1 and 0, respectively. For a valid proof these values must be $f(S_1 \to S_2) = f(S_2 \to S_3) = \ldots = f(S_{n-1} \to S_n) = 1$. The formal details are messy and omitted here (see Gödel 1931 for the introduction of arithmetization of logic, and Kleene 1952) but the principle is both simple and sound. It is well known that f is a morphism mapping $<S, \to>$ onto a Boolean algebra $<f(S), \leq>$.[5]

The very strong analogy and structural similarities between proof and calculation, including their inter-convertibility, challenges the preconception often manifested in philosophical and historical accounts of mathematics that proof is somehow intellectually superior to calculation in mathematics.[6] Looking in detail at the technical and structural aspects of proof and calculation reveals that they cannot be so easily attributed to different epistemological domains as is often claimed. It is not defensible to say that proof alone in mathematics pertains to the true, good, beautiful, to wisdom, 'high-mindedness' and the transcendent dimensions of being, and that calculation is only technical and mechanical, pertaining to the utilitarian, practical, applied, and mundane; the lowly dimensions of existence. Such assertions are part of an ideological position incorporating a set of values that overvalues pure proof-based mathematics as having epistemological significance, and undervalues calculation and applied mathematics as having only practical significance; going back to the social divisions of ancient Greek society, as noted above, and the prejudices and ideology to which it gave rise.

This preconception or prejudice is used as the basis for asserting that the contributions of some cultures and civilizations are intellectually superior to others in history of mathematics. It also undervalues the solving of problems, calculations and other local applications of deduction in mathematics (including proof, see Joseph 1994). Thus the mathematics of ancient Egypt, Mesopotamia and India, as well as other countries outside of the Greco-European tradition, is viewed as inferior and immature. Part of the argument is that only cultures that produce axiomatic proof in mathematics achieve the highest levels of abstract intellectual achievement.

I have argued that philosophical dispositions and values have underpinned a prejudice against ascribing value to certain forms of mathematical activity. In particular, that axiomatic systems are greatly valued over less systematic forms of deduction including problem solving, calculation and unsystematized proofs. Furthermore, this prejudice also maintains the contrast between and overvalues any form of proof, including unsystematized and unaxiomatized proofs over any form of calculation or problem solving.

These two levels of prejudice, these two value-based distinctions and preferences are frequently elided in the history and philosophy of mathematics. Thus the contributions of the ancient Greeks of the Euclidean type, and the modern focus on axiomatics of the past two centuries are seen to characterize the superior forms of thought of what is purported to be a Greco-European tradition. Furthermore, the unsystematized and unaxiomatized proofs and methods characterizing the official European history of mathematics from the late-Renaissance to the beginning of the Nineteenth century are seen as also reflecting the superior methods and concepts and higher forms of thought of the modern European tradition in their nascent phase, whose superiority and value is demonstrated in the subsequent flowering of the axiomatic tradition in Europe.

Through this elision, there arises the discounting of the proof-based contribution of cultures and civilizations outside of the 'Greco-European' tradition. Thus although there is a tradition of convincing demonstration or proofs, known as Upapatis, originating around two millennia ago in India, these proofs are discounted as intellectually inferior (Joseph 1994). Admittedly there are significant differences between the ancient Greek and the early Indian concepts of proof. Joseph (2000) has convincingly argued that ancient Indian mathematics was at least partially shaped by linguistic and grammatical conceptions of knowledge, based on the contributions of Panini; whereas ancient Greek mathematics was shaped by developments in philosophical thinking. So there are differences in the epistemological basis for different forms of proof that have emerged in different cultures and civilizations. However, the current challenges to the philosophy of mathematics discussed in the beginning of this paper, legitimate challenges to the traditional univocal and absolute conceptions of mathematics, knowledge and proof. From the perspective of the new fallibilist or social constructivist philosophies of mathematics, there is no ultimate or uniquely correct form of proof. Rather the forms of proof accepted within any culture or civilization during any epoch are a function of the historically contingent conceptual history and epistemological preconceptions that emerge and are accepted by the relevant geographico-historical communities of scholars. So there is no basis for elevating certain cultural forms of proof and demoting others on epistemological grounds alone. Each must be judged within the cultural contexts of its geographico-historical location.

EUROCENTRISM IN THE HISTORY AND PHILOSOPHY OF MATHEMATICS

The above discussion raises the question of why informal and unsystematized proofs and demonstrations that occur in the mathematical histories of certain cultures are valued more than those of others. Why, for example, are the unsystematized proofs, methods and results of post-Renaissance

European mathematics regarded as superior to antecedent developments in Kerala of comparable character? To answer this it is necessary to turn to another dimension of ideological prejudice at work in the history and philosophy of mathematics. This is eurocentrism, the racist bias that claims that the European 'mind' and its cultural products are superior to those of other peoples and races. Thus Bernal (1987) has argued that during the past two hundred years or so, Ancient Greece has been 'talked up' as the starting point of modern European thought, and the 'Afroasiatic roots of Classical Civilisation' have been neglected, discarded and denied.

Against this backdrop it is not surprising that that mathematics has been seen as the product of European mathematicians. However, there is now a widespread literature supporting the thesis that mathematics has been misrepresented in a eurocentric way, including Almeida and Joseph (2004), Joseph (2000), Powell and Frankenstein (1997) and Pearce (undated). A common feature of eurocentric histories of mathematics is to claim that it was primarily the invention of the ancient Greeks. Their period ended almost 2000 years ago, which was followed by the 'dark ages' of around 1000 years until the European renaissance triggered by the rediscovery of Greek learning led to modern scientific and mathematical work in Europe (and its cultural dependencies). This trajectory is illustrated in Figure 15.1.

Some accounts have acknowledged the impact of lower level Egyptian and Babylonian mathematics on ancient Greek developments, as well as the later minor contributions of Indian and Arabic mathematicians (often seen primarily as custodians of Greek knowledge) on the history of mathematics in Europe (i.e., *The* history of mathematics). This is shown in the Modified Eurocentric model (Figure 15.2).

Figure 15.1 Eurocentric chronology of mathematics history (from Pearce, undated).

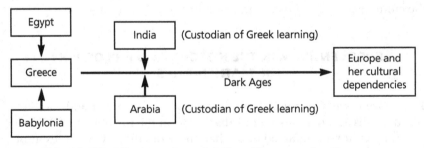

Figure 15.2 Modified Eurocentric model (from Pearce, undated).

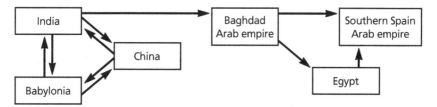

Figure 15.3 Non-European mathematics during the dark ages (from Pearce, undated).

Pearce, Joseph and others go on to argue that in the so-called 'dark ages' and beyond, from 5th–15th centuries, a great deal of mathematical work continued. Further the relationships between different regions and countries was complex and multidirectional and "A variety of mathematical activity and exchange between a number of cultural areas went on while Europe was in a deep slumber." (Joseph, 2000: 9). In Figure 15.3 I reproduce Pearce's diagram of interrelationships in the development Non-European mathematics during the dark ages.

Thus out of ignorance or prejudice arising from ideologically based values and preconceptions, eurocentric histories of mathematics neglect the 'Non-European roots of mathematics' (to quote the subtitle of Joseph, 2000). There is a small but growing impact of such critical ideas in the history and philosophy of mathematics as indicated here. However, in my view, there is still an under-emphasis on the vital role of pre-Hellenic civilizations in providing the conceptual basis for modern mathematics through calculation, problem solving, etc.

MATHEMATICS OF THE INDIAN SUBCONTINENT AND KERALA

One of the major casualties in the Eurocentric view of mathematics has been the ignoring or undervaluing of the contributions to mathematics of the Indian subcontinent. The long presence of deductive proofs in mathematics from this region has already been noted (Joseph 1994, 2000). Although the invention of zero by mathematicians of the Indian subcontinent has long been acknowledged, the significance of this as the lynchpin of the decimal place value system is often underestimated. Rotman (1987) presents a view of this innovation that puts its significance as reaching far beyond mathematics, right at the heart of European cultural and intellectual development in the Renaissance and early modern times. Pearce (undated) argues the Indian development of decimal numeration together with the place value system is the most remarkable development in the history of mathematics, as well as being one of the foremost intellectual productions in the overall history of humankind. I have indicated above how both philosophically and in the published histories of mathematics, calculation and numeration have traditionally been downplayed as epiphenomena of what is perceived to be the much more important Platonic conception of number. This is a misrepresentation of the intellectual significance of these developments without which the modern conceptions of number (including its computerization, with all of the applications this brings) would not be possible.

In the history of mathematics in the Indian subcontinent, much attention has been given to very large numbers, including powers of ten up to near 50. Whether these were contributors to or results of the development of the decimal place value system is for historians to say. Likewise it is tempting to speculate as to whether the extension of the decimal place value system into decimal fractions helped in the conceptualisation and formulation of the remarkable series expansions developed in Kerala. Although there is no unequivocal historical basis for this, it is convincingly claimed that floating point numbers were used by Kerala mathematicians to investigate the convergence of series (Almeida et al. 2001).

This brings me to one of the most remarkable and most neglected episodes in the history of mathematics, and the focus of the conference at which this paper was first presented.[7] This is the fact that Keralese mathematicians discovered and elaborated a large number of infinite series expansions and contributed much of the basis for the calculus, which is traditionally attributed to 17th and 18th century European mathematicians. Furthermore, this is not a case of simultaneous discovery in Kerala, for the work in Kerala took place two centuries before that in Europe.

Pearce (undated), Joseph (2000) and others attribute to Madhava of Sangamagramma (c. 1340–1425), the Keralese mathematician-astronomer, the important step of moving on from the finite procedures of ancient mathematics to treat their limit, the passage to infinity, the essence of modern classical analysis. He is also thought to have discovered numerous infinite series expansions of trigonometric and root terms, as well as for π, for which he calculated the value up to 13 (some say 17) decimal places (Pearce, undated). These inestimably important results anticipate some of the discoveries attributed to or named after the great mathematicians Gregory, Maclaurin, Taylor, Wallis, Newton, Leibniz and Euler.

Joseph (2000: 293) claims that "We may consider Madhava to have been the founder of mathematical analysis. Some of his discoveries in this field show him to have possessed extraordinary intuition." Almeida *et al.* (2001) have argued that Keralese contributions as a whole anticipate developments in Western Europe by several centuries in work on infinite series for numerical integration results.

In addition, these results are very possibly not just the anticipations of unacknowledged genius in the Indian subcontinent, and as such a very remarkable case of independent discovery. There is the very real possibility that these Keralese discoveries were transmitted to Europe by Jesuit missionaries and 'appropriated' by European mathematicians as their own (Almeida and Joseph 2004). The arguments for this transmission and appropriation are strong, although they remain circumstantial. Certainly the mathematicians of Renaissance Europe are known to have been secretive about their methods and knowledge, and if they had 'purloined' the foundational results of calculus from Kerala would conceal and deny their origins. However, to date, no evidence of direct transmission has been found in the extensive correspondence of Jesuits analysed so far. Furthermore, Stedall (2005) has argued that there are no miraculous or unexpected 'jumps' in the development of calculus and infinite series in seventeenth-century Europe that one might have expected to appear in the record following the appropriation of results from Kerala. However, the issue is still an open one.

As someone who is not an historian of mathematics, I find this new recognition of the major Keralese and Indian subcontinent contributions to the history of mathematics remarkable. The fact that traditional histories of mathematics fail to acknowledge these and other non-European contributions is partly due to ignorance, for until recently it was difficult to find proper sources on this in standard texts. But there is much more to this, as there have been reports of the anticipations in the literature for almost two centuries which have been disparaged or ignored (Almeida and Joseph 2007). Instead there are two sets of entwined ideological presuppositions that have led to this denial and blindness. The first is the epistemological prejudice towards a certain style of mathematics, namely the axiomatic theories and purist ideas discussed above as well as favoring proofs over calculation al and applied mathematics. Through the lenses of these modern prejudices the historical contributions of non-Eurocentric traditions has been minimized and trivialized. The second set of ideological presuppositions is more sinister. This is the racial prejudice of eurocentrism. Namely, that only the 'Western mind' (i.e., the Caucasian or European) is capable of the pure thought and insights required in the highest forms of mathematics. Thus the contributions of African, Asian, Indian subcontinent, and Oriental peoples is discounted and minimized, because by the presupposed 'very inferior nature' of these peoples they are incapable of the high levels of thought involved. Hence any results that contradict these prejudices is *ab initio* incorrect. Thus such discoveries are minimized as intellectually inferior, or doubted and attributed to the transmission and copying of ideas from West to East, or in the last resort, challenged with regard to their chronology.

CONCLUSION

So what is the philosophical significance of the Keralese and Indian sub-continent contribution to history of mathematics? Identifying the most accurate genesis and trajectory of mathematical ideas in history that current knowledge allows should be the goal of every history of mathematics, and is consistent with any philosophy of mathematics. However, I have argued that a broader conceptualization of philosophy of mathematics is needed than the traditional emphasis on scholastic enquiries into epistemology and ontology. For such an emphasis has been associated, though I add need not necessarily be so, with an ideological position that devalues non-European contributions to history of mathematics. The philosophy of mathematics needs to be broad enough to recognise the salient features of the discipline it reflects upon, namely mathematics. As Lakatos (1976) indicated in the quote given above, the philosophy of mathematics has become empty by ignoring the history of mathematics.

It is no little charge to claim that the history and philosophy of mathematics have in effect become infused with error and a racist ideology, through the implicit and unacknowledged values and prejudices. Elsewhere, as well as above, I have argued that it is the business of mathematics and the philosophy of mathematics to take the issue of values seriously (Ernest 1998), and it is no longer enough to claim that these are outside of its proper subject matter. After all, ethics is just another branch of philosophy and I can see no grounds for its *a priori* exclusion. All human activities, however rarefied and abstruse' are part of the vast cultural project of humankind, and as such none has the right to claim exemption from awareness of values and social responsibility (provided that this is not used as an excuse to limit freedom of thought and critique).

REFERENCES

Almeida, D. F. and Joseph, G. G. (2004) Eurocentrism in the History of Mathematics: the Case of the Kerala School, *Race & Class*, Vol. 45, No. 4, 2004: 5–59.

Almeida, D. F. and Joseph, G. G. (2007) 'Kerala Mathematics and its Possible Transmission to Europe', *The Philosophy of Mathematics Education Journal* No. 20 (June 2007), accessible via <http://www.people.ex.ac.uk/PErnest/>.

Almeida, D. F., John, J. K. and Zadorozhnyy, A. (2001) Keralese Mathematics: Its Possible Transmission to Europe and the Consequential Educational Implications, *Journal of Natural Geometry* Vol. **20**: 77–104.

Bernal, M. (1987) *Black Athena, The Afroasiatic roots of Classical Civilisation*, Vol. 1, London: Free Association Books.

Davis, P. J. and Hersh, R. (1980) *The Mathematical Experience*, Boston: Birkhauser.

Ernest, P. (1991) *The Philosophy of Mathematics Education*, London, Falmer Press.

Ernest, P. (1998) *Social Constructivism as a Philosophy of Mathematics*, Albany, New York: SUNY Press.

Ernest, P. (2006). A semiotic perspective on mathematical activity: The case of number. *Educational Studies in Mathematics, 61*, 67–101.

Gödel, K. (1931) translation in Heijenoort, J. van Ed. (1967) *From Frege to Gödel: A Source Book in Mathematical Logic*, Cambridge, Massachusetts: Harvard University Press, 592–617.

Høyrup, J. (1980) Influences of Institutionalized Mathematics Teaching on the Development and Organisation of Mathematical Thought in the Pre-Modern Period, Bielefeld, extract reprinted in Fauvel, J. and Gray, J., Eds, *The History of Mathematics: A Reader*, London, Macmillan, 1987, 43–45.

Høyrup, J. (1994) *In Measure, Number, and Weight*, New York: SUNY Press.

Joseph, G. G. (1994) Different Ways of Knowing: Contrasting Styles of Argument in Indian and Greek Mathematical Traditions, in P. Ernest, Ed. *Mathematics, Education and Philosophy*, Falmer Press, London, 1994: 194–207.

Joseph, G. G. (2000). *The Crest of the Peacock, non-European roots of Mathematics*. Princeton and Oxford: Princeton University Press.

Kitcher, P. (1984) *The Nature of Mathematical Knowledge*, Oxford: Oxford University Press.

Kitcher, P. and Aspray, W. (1988) An opinionated introduction, in Aspray, W. and Kitcher, P. Eds. (1988) *History and Philosophy of Modern Mathematics*, Minneapolis: University of Minnesota Press, 3–57.

Kleene, S. C. (1952). *Introduction to metamathematics*. Amsterdam: North Holland.

Körner, S. (1960) *The Philosophy of Mathematics*, London: Hutchinson.

Lakatos, I. (1962) Infinite Regress and the Foundations of Mathematics, *Aristotelian Society Proceedings*, Supplementary Volume No. 36, 155–184 (revised version in Lakatos, 1978).

Lakatos, I. (1976) *Proofs and Refutations*, Cambridge: Cambridge University Press.

Lakatos, I. (1978) *The Methodology of Scientific Research Programmes* (Philosophical Papers Volume 1), Cambridge: Cambridge University Press.

MacKenzie, D. (1993) Negotiating Arithmetic, Constructing Proof: The Sociology of Mathematics and Information Technology, *Social Studies of Science*, Vol. 23, 37–65.

Maddy, P. (1990) *Realism in Mathematics*, Oxford: Clarendon Press.

Pearce, I. G. (undated) Indian Mathematics: Redressing the balance, web material consulted on 18 November 2005 at <http://www-history.mcs.st-andrews.ac.uk/history/Projects/Pearce/index.html>

Popper, K. R. (1979) *Objective Knowledge* (Revised Edition), Oxford: Oxford University Press.

Powell, A. B. and Frankenstein, M. (1997) *Ethnomathematics: Challenging Eurocentrism in Mathematics*, Albany, New York: SUNY Press.

Rotman, B. (1987) *Signifying Nothing: The Semiotics of Zero*, London: Routledge.

Stedall, J (2005) 'Calculus and infinite series in seventeenth-century Europe', presented at AHRB Workshop *Kerala Mathematics: History and the Possibilities of Transmission*, Trivandrum, India, 15–16 December 2005 (Organizers George G. Joseph & Dennis F. Almeida).

Tymoczko, T. Ed. (1986) *New Directions in the Philosophy of Mathematics*, Boston: Birkhauser.

ACKNOWLEDGMENT

An earlier version of this paper is to appear in *Kerala Mathematics: History and the Possibilities of Transmission,* the proceedings of AHRB Workshop held in Trivandrum, India, 15–16 December 2005, edited by George G. Joseph, as well as in *The Montana Mathematics Enthusiast,* Vol. 4 no. 2 (2007) accessible via <http://www.montanamath.org/TMME/>.

NOTES

1. I take the beginning of disciplinary mathematics to be around 2500–3000 BCE, following Høyrup (1980) and (1994).
2. Clearly branching derivations can occur in virtually all mathematical processes or activities including those involved in problem solving, deriving mathematical proofs, and mathematical calculations. However, they are mostly eliminated from the transcriptions of successfully completed activities.
3. Note that I have not distinguished between the two analogous forms of proof which employ equivalence or deductive consequence as the transformational relationship at each step. In the latter the truth value derived is greater than or equal to is precedent value, in the former it is equal to it. But in each case (in bivalent logic), since the initial truth value in the sequence must be 1 the whole sequence of truth values including the final term, the theorem proved, is 1.
4. The preservation of one of the four inequality relations along the sequence is possible variation, where an upper or lower bound on the value of the term is determined
5. Technically the truth value function f can simply be defined on the domain of sentences under a given interpretation provided that there is an effective procedure for determining whether each sentence is true or false (thus giving values 1 or 0, respectively) under the given interpretation of the underlying theory or formal language.
6. Joseph (1991) is among the few to note the importance of algorithms and calculation in the history of mathematics and to note their almost universal devaluation by other commentators.
7. AHRB Workshop *Kerala Mathematics: History and the Possibilities of Transmission,* Trivandrum, India, 15–16 December 2005 (Organizers George G. Joseph & Dennis F. Almeida).

SECTION 3

MATHEMATICS, EDUCATION, AND SOCIETY

CHAPTER 16

VALUES AND THE SOCIAL RESPONSIBILITY OF MATHEMATICS

Paul Ernest
University of Exeter

MATHEMATICS AND VALUES

Mathematicians have a strong set of values that are constitutively central to mathematics. For example, Hardy (1941) argues that the most important mathematical theorems have both beauty and seriousness, which means they have generality, depth and contain significant ideas. Within mathematics, problems, concepts, methods, results and theories are routinely claimed to be deep, significant, powerful, elegant and beautiful. These attributions cannot be simply dismissed as subjective, because mathematicians and epistemological absolutists claim that at least some of these properties reflect objective features of the discipline.

Human interests and values play a significant part in the choices of mathematical problems, methods of solution, the concepts and notations constructed in the process, and the criteria for evaluating and judging the resulting mathematical creations and knowledge. Mathematicians choose which of infinitely many possible definitions and theorems are worth pur-

Critical Issues in Mathematics Education, pages 207–216

suing, and any act of choice is an act of valuation. The current upsurge of computer-related mathematics represents a large scale shift of interests and values linked with social, material and technological developments, and it is ultimately manifested in the production of certain types of knowledge (Steen 1988). The values and interests involved also essentially form, shape and validate that knowledge too, and are not merely an accidental feature of its production. Even what counts as an acceptable proof has been permanently changed by contingent technological developments (De Millo *et al.* 1979, Tymoczko 1979).

Overt or covert values can thus be identified in mathematics underlying the choice of problems posed and pursued, the features of proofs, concepts and theories that are valorized. Deeper still, values underpin the conventions, methodologies and constraints that limit the nature of mathematical activity and bound what is acceptable in mathematics (Kitcher 1984). These values are perhaps most evident in the norms that regulate mathematical activity and the acceptance of mathematical knowledge. At times of innovation or revolution in mathematics, the values and norms of mathematics become most evident when there are explicit conflicts and disagreements in the underlying values. Consider the historical resistance to innovations such as negative numbers, complex numbers, abstract algebra, non-Euclidean geometries and Cantor's set theory. At root the conflicts were over what was to be admitted and valued as legitimate, and what was to be rejected as spurious. The standards of theory evaluation involved were based on meta-level criteria, norms and values (Dunmore 1992). Since these criteria and norms change over the course of the history of ideas, they show that the values of mathematics are human in origin, and not imposed or acquired from some timeless source.

The Values of Absolutist Mathematics

A further broad selection of values can be identified in mathematics, including the valuing of the abstract over the concrete, formal over informal, objective over subjective, justification over discovery, rationality over intuition, reason over emotion, generality over particularity, and theory over practice. These may constitute many of the overt values of mathematicians, but are introduced through the traditional definition of the field. Only that mathematics which satisfies these values is admitted as *bona fide*, and anything that does not is rejected as inadmissible.[1] Thus warranted mathematical propositions and their proofs are legitimate mathematics, but the criteria and processes used in warranting them are not. The rules demarcating the boundary of the discipline are positioned outside it, so that no discussion of these values is possible within mathematics. Once meta-rules are established in this way, mathematics can be regarded as value free. In fact, the values lie behind the choice of the norms and rules. By concealing the underpinning values absolutism makes them virtually unchallengeable. It legitimates only the formal level of discourse as mathematics (i.e., axiomatic theories, not metamathematical discussion), and hence it relegates the issue of values to a realm which is definitionally outside of the discipline.

An absolutist can reply that the list in the preceding paragraph is not a matter of preferred values, but rather of the essential defining characteristics of mathematics and science, which may subsequently become the values of mathematicians. Thus, the content and methods of mathematics, by their very nature, make it abstract, general, formal, objective, rational, theoretical and concerned with justification. There is nothing wrong with the concrete, informal, subjective, particular or the context of discovery, they are is just not part of the character of justified mathematical or scientific knowledge (Popper 1972).

Absolutist philosophies of mathematics thus have internalist concerns and regard mathematics as objective and free of ethical, human and other values. Mathematics is viewed as value-neutral, concerned only with structures, processes and the relationships of ideal objects, which can be described in purely logical language. Any intrusion of values, tastes, 'flavors' or 'coloring' is either an inessential flourish, or a misrepresentation of mathematics, due to human fallibility.

Similar ideas dominated the thought of the British Empiricists, who sharply distinguished primary and secondary qualities (Morris 1963). Primary qualities were the quantifiable properties describable in the mathematics of the day, comprising the properties attributed to things-in-themselves. These include mass, shape and size; the mathematical attributes of a mechanistic world-view. Secondary qualities include colour, smell, feel, taste and sound, and are understood to be the human subject's responses to things, and not the inherent properties of objects or the external world. By analogy, human values, tastes, interests can be factored out from 'objective' knowledge as human distortions and impositions. This resembles Frege's (1892) view, as I showed in chapter 6. Thus there is a long intellectual tradition which distinguishes values and interests from knowledge, and which dismisses the former as subjective or human responses to the latter.

The claim that mathematicians and their practices display implicit or explicit values is not denied by absolutist philosophies of mathematics. It is acknowledged that mathematicians have preferences, values, interests, and that these are reflected in their activities, choices, and even in the genesis of mathematics. But mathematical knowledge once validated is objective and neutral, based solely on logic and reasoning. Logic, reason and proof discriminate between truth and falsehood, correct and incorrect proofs, valid and invalid arguments. Neither fear, nor favour, nor values affect the court of objective reason. Hence, the absolutist argument goes, mathematical knowledge is value-neutral.

However, the fallibilist view is that the cultural values, preferences and interests of the social groups involved in the formation, elaboration and validation of mathematical knowledge cannot be so easily factored-out and discounted. The values that shape mathematics are neither subjective nor necessary consequences of the subject. Thus at the heart of the absolutist neutral view of mathematics, fallibilism claims to locate a set of values and a cultural perspective, as well as a set of rules which renders them invisible and undiscussable.

The Values of Social Constructivist Mathematics

One of the central thrusts of social constructivism as a philosophy of mathematics (Ernest 1998) is to challenge a number of traditional dichotomies as they pertain to mathematics. Thus it is argued that a number of domains cannot be kept disjoint, including the philosophy of mathematics and the history of mathematics, the context of justification and the context of discovery, mathematical content and rhetoric, epistemology and objective knowledge as opposed to the social domain. An outcome of the breakdown of these traditional dichotomies is that mathematics ceases to be independent of values and interests and therefore cannot escape having a social responsibility.

Although the traditional absolutist stance is that values are excluded in principle from matters logical and epistemological, and hence can have no relevance to mathematics or the philosophy of mathematics, social constructivism argues that this claim is based on assumptions that can be critiqued and rejected. Mathematics may strive for objectivity, and consequently the visibility of values and interests within it is minimized. However, if the pairs of categories in the above oppositions cannot be regarded as entirely disjoint, it follows that mathematics and mathematical knowledge reflect the interests and values of the persons and social groups involved, to some degree. Once conceded, this means mathematics cannot be coherently viewed as independent of social concerns, and the consequences of its social location and social embedding must be faced.

In particular, a strong case has been made for the ineliminably contingent character of mathematical knowledge. If the direction of mathematical research is a function of historical interests and values as much as inner forces and logical drives, mathematicians cannot claim that the outcomes are inevitable, and hence free from human interests and values. Mathematical development is a function of intellectual labour, and its creators and appliers are moral beings engaged in voluntary actions. Mathematicians may fail to anticipate the consequences of the mathematical developments in which they participate, but this does not absolve them of responsibility. Mathematics needs to be recognised as a socially responsible discipline just as much as science and technology.

This does not merely apply to the genesis of mathematics and the context of discovery or the applications of mathematics. As has been argued, it applies equally to logic and the context of justification. For logic, reason and proof have also been shown to be contingent. They include historically accidental features, reflecting the preferences or inclinations of their makers and their source cultures. Thus even the most objective and dispassionate applications of logic and reasoning incorporate a tacit set of values and preferences.

THE SOCIAL RESPONSIBILITY OF MATHEMATICS

Social constructivism regards mathematics as value-laden and sees mathematics as embedded in society with social responsibilities, just the same as every other social institution, human activity or discursive practice. Precisely what this responsibility is depends on the underlying system of values which is adopted, and social constructivism as a philosophy of mathematics does not come with a particular set of values attached.[2] However, viewed from this perspective mathematicians and others professionally involved with the discipline are not entitled to deny on principle social responsibility for mathematical developments or applications.

The same problem of denial of responsibility does not arise in the contexts of schooling. It has always acknowledged that education is a thoroughly value-laden and moral activity, since it concerns the welfare and treatment of young persons. If, as in many education contexts, social justice values are adopted, then additional responsibility accrues to mathematics and its related social institutions to ensure that its role in educating the young is a responsible and socially just one. In particular, mathematics must not be allowed to be distorted or partial to the values or interests of particular social groups, even if they have had a dominant role in controlling the discipline historically. There is an extensive literature treating criticisms of this sort, concerning gender and race in mathematics education. It has been argued that white European-origin males have dominated mathematics and science, in modern times, and the androcentric values of this group have been attributed to knowledge itself (Ernest 1991, Walkerdine 1988, 1989, Bailey and Shan 1991). To the extent that this complaint can be substantiated, and there is significant confirmatory evidence, there is a value imbalance in mathematics and science which needs to be rectified. Such distorted values may not only permeate teaching, but also the constitution of the subjects themselves in their modern formulations. I will not pursue this further here except to remark that social constructivism, by admitting the value-laden nature of mathematics and its social responsibility, necessitates the facing and addressing of these issues.

Social constructivism links together the contexts of schooling and research mathematics in a tight knowledge reproduction cycle. This cycle is concerned with the formation and reproduction of mathematicians and mathematical knowledge, and thus is deliberately mathematics-centred. But if the reproduction of the social institution of mathematics were to be adopted as the leading aim for schooling in the area of mathematics, the outcome could be an educational disaster. For although in most developed and developing countries virtually the whole population studies mathematics in school, less than one in one thousand go on to become a research mathematician. Letting the needs of this tiny minority dominate the mathematical education of everyone would lead to an ethical problem, not to mention a utilitarian one. This problem is further exacerbated by the fact that the needs of these two groups are inconsistent. I will not go into the educational details here except to say that what is needed is differentiated school mathematics curricula to accommodate different aptitudes, attainments, interests and ambitions. Such differentiation must depend on balanced educational and social judgements rather than exclusively on mathematicians' views of what mathematics should be included in the school curriculum. Mathematicians are often concerned only with improving the supply of future mathematicians, and many of their interventions in the mathematics curriculum have been to increase its content coverage, conceptual abstraction, difficulty and rigor. Whilst this may be good for the few, the outcomes are often less beneficial for the many (Ernest 1991).

A further issue is that of the social import of differing perceptions of mathematics. There is an interesting division between the mathematicians' and the public's understanding and perceptions of mathematics. Mathematicians typically regard mathematics as a rational discipline which is highly democratic, on the grounds that knowledge is accepted or rejected on the basis of logic, not authority, and potentially anyone can propose or criticize mathematical knowledge using reason alone. Social standing, wealth or reputation are immaterial to the acceptability of mathematical proposals, so the argument goes. This claim is an idealization, and I indicate in chapter 6 of Ernest (1998), that social standing does matter to some extent with regard to the acceptance of new knowledge. Furthermore it is not pure, context-free reason that serves as the basis for proposing or criticizing mathematical knowledge, but the context-embedded reason and rhetoric of the mathematics community. This caution must be borne in mind in evaluating the mathematicians' claims. However, given this caveat, the claim of the mathematicians' is largely correct. The mathematical community greatly values reason and attempts to minimize the role of authority in mathematics.

In contrast, a widespread public perception of mathematics is that it is a difficult but completely exact science in which an élite cadré of mathematicians determine the unique and indubitably correct answers to mathematical problems and questions using arcane technical methods known only to them. This perception puts mathematics and mathematicians out of reach of common-sense and reason, and into a domain of experts and subject to their authority. Thus mathematics becomes an élitist subject of asserted authority, beyond the challenge of the common citizen. Such absolutist views attribute a spurious certainty to the mathematizations employed, for example, in advertising, commerce, economics, politics and in policy statements. The use of carefully selected statistical data and analysis is part of the rhetoric of modern political life, used by political parties of all persuasions to further their sectional interests and agendas. An absolutist perception of mathematics helps to prevent the critical questioning and scrutiny of such uses of mathematics in the public domain, and is thus open to anti-democratic exploitation. If democratic values are preferred over authoritarian ones, as presumably they are in most countries of the world, then part of the social responsibility of mathematics is to support a climate of critical questioning and scrutiny of mathematical arguments by the public. Although superficially this is consonant with mathematicians' views of the rationality of mathematical knowledge, it does not always fit with the public image of mathematics they communicate. Being aware of this, I wish to claim, is part of the social responsibility of mathematics. There are also important implications for schooling (Ernest 1991, Frankenstein 1983, Skovsmose 1994).

Social constructivism has adopted conversation as an underlying metaphor for epistemological reasons, to enable the social aspects of mathematical knowledge to be adequately treated within the philosophy of mathematics. But ethical consequences also follow, and it is worth dwelling a moment on the ethics of conversation. Part of the difference between monological and dialogical argument in conversation is the space and respect given to voices other than the proponent's. I have shown how modern proof theory in trying to remain close to mathematical practice—not for ethical reasons—has admitted the legitimacy of the multiple voices of both proponent and opponent. Habermas (1981) draws upon the dialogical logic of Lorenzen and his Erlangen school in order to ground the philosophical theory of communicative action in human actions and conversation. Much of his motive is ethical in that he posits an ideal speech community in which the alternating voices of persons are listened to with respect. Habermas's arguments for the ethical import of conversation offer support to the position defended here. The analogy with democracy is also clear, namely the legitimacy of multiple voices seeking to persuade, as opposed to the imposed monologue of the dictator.[3]

A feminist critique of logic and the Western philosophical tradition claims that, first of all, the traditional monologic has been the male voice of authority and power, which denies the legitimacy of challenge except from others speaking with the same voice (Nye 1989). This may seem to be an extreme reading, but it shares a number of points in common with the argument presented above. The social constructivist position is to acknowledge that to be heard, a voice must be from someone who is already a participant within a shared language game. But discursive practices grow and change and one generation's peripheral or excluded voice can become central in the next.[4] The very 'natures' of rationality and logic are shifting. There is a reflexive point to be made here too, in that social epistemologies like social constructivism have traditionally been excluded from philosophy and the philosophy of mathematics. However their voices are now being listened to and the traditional boundaries of the philosophy of mathematics are being widened to admit the 'maverick' tradition.

A second feminist criticism is that even where contestation is admitted, the adversary method of traditional philosophy seeks not to employ the dialectics which counterpoises the voices of proponent and critic in the quest for a shared higher synthesis. Instead the adversary method is a

> model of philosophic methodology which accepts a positive view of aggressive behaviour and then uses it as the paradigm of philosophic reasoning.... The adversary method requires that all beliefs and claims be evaluated only by subjecting them to the strongest, most extreme opposition. (Harding and Hintikka 1983: xv)

These authors, with Moulton (1983), offer in contrast to the adversary method, the *elenchus*, the method of refutation employed in the Socratic dialogues of Plato, to lead people to see their views are perceived as wrong by others. Thus the ethics of conversation requires turn-taking in respectful listening as well as in talking.

Social constructivism takes the primary reality to be persons in conversation; persons engaged in language games embedded in forms of life. These basic social situations have a history, a tradition, which must precede any mathematizing or philosophizing. We are not free-floating, ideal cognizing subjects but fleshy persons whose minds and knowing have developed through our bodily and social experiences. Only through our antecedent social gifts can we converse and philosophize. I have argued for epistemological fallibilism and relativism, but instead of rendering social constructivism groundless and rootless, its grounds and roots are to be found the practices and traditions of persons in conversation. In addition to providing a reply to the criticism that social constructivism is relativistic, this basis is also one that is irrevocably moral. For ethics arises from the ways in which persons live together and treat each other. Thus an outcome of social constructivism as a philosophy of mathematics is that questions of inter-human relations and ethics cannot be avoided and must be addressed. What is now needed, I wish to claim, is an ethics of mathematics, one which acknowledges the social responsibility of mathematics and how it is implicated in the great issues of freedom, justice, trust and fellowship. It is not that this need follows *logically* from social constructivism. It follows *morally*.

In his *Ethics of Geometry*, Lachterman (1989, back-cover), quotes Salomon Maimon's emblematic dictum: "In mathematical construction we are, as it were, gods." I have argued that in the social construction of mathematics we act as gods in bringing the world of mathematics into existence. Thus mathematics can be understood to be about power, compulsion and regulation. The mathematician is omnipotent in the virtual reality of mathematics, although subject to the laws of the discipline; and mathematics regulates the social world we live in, too. This perspective perhaps resolves the enigma, mystery and paradox of mathematics that I wrote of in the introduction to Ernest (1998). But in accepting this awesome power it also behooves us to strive for wisdom and to accept the responsibility that accompanies it.

REFERENCES

Bailey, P. and Shan, S. J. (1991) *Multiple Factors: Classroom Mathematics for Equality and Justice*, Stoke-on-Trent: Trentham Books.

De Millo, A., Lipton, R. J. and Perlis, A. J. (1979) Social Processes and Proofs of Theorems and Programs, *Communications of the ACM*, Vol. 22, No. 5, 271–280. Reprinted in Tymoczko (1986a).

Dunmore, C. (1992) Meta-level revolutions in mathematics, in Gillies (1992), 209–225.

Ernest, P. (1991) *The Philosophy of Mathematics Education*, London: Falmer.

Ernest, P. (1998) *Social Constructivism as a Philosophy of Mathematics*, Albany, New York: SUNY Press.

Frankenstein, M. (1983) Critical Mathematics Education: An Application of Paulo Freire's Epistemology, *Journal of Education*, Vol. 165, No. 4, 315–339.

Frege, G. (1892) On Sense and Reference, *Translations from the Philosophical Writings of Gottlob Frege* (Eds. P. Geach and M. Black), Oxford: Basil Blackwell, 1966, 56–78.

Habermas, J. (1981) *The Theory of Communicative Action*, 2 volumes, Frankfurt am Main: Suhrkamp Verlag. Translated by T. McCarthy, Cambridge: Polity Press, 1987 & 1991.

Harding, S. (1991) *Whose Science? Whose Knowledge?*, Milton Keynes: Open University Press.

Harding, S. and Hintikka, M. B. Eds (1983) *Discovering Reality; Feminist Perspectives on Epistemology, Metaphysics, Methodology, and Philosophy of Science,* Dordrecht: Reidel.

Hardy, G. H. (1941) *A Mathematician's Apology,* Cambridge: Cambridge University Press.

Kitcher, P. (1984) *The Nature of Mathematical Knowledge,* Oxford: Oxford University Press.

Lachterman, D. (1989) *The Ethics of Geometry,* New York and London: Routledge.

Morris, C. R. (1963) *Locke Berkeley Hume,* Oxford: Oxford University Press.

Moulton, J. (1983) A Paradigm of Philosophy: The Adversary Method, in Harding and Hintikka (1983), 149–164.

Nye, A. (1989) *Words of Power,* London: Routledge.

Popper, K. (1972/1963) *Conjectures and Refutations.* Revised fourth edition 1972, London: Routledge and Kegan Paul.

Popper, K. R. (1979) *Objective Knowledge.* Revised Edition, Oxford: Oxford University Press.

Steen, L. A. (1988) The Science of Patterns, *Science,* Vol. 240, No. 4852, 611–616.

Tymoczko, T. (1979) The Four-Color Problem and its Philosophical Significance, *The Journal of Philosophy,* Vol. 76, No. 2, 57–83.

Walkerdine, V. (1988) *The Mastery of Reason,* London: Routledge.

Walkerdine, V. (1998) *Counting Girls Out: Girls and Mathematics* (Revised 2nd edition), London: RoutedgeFalmer.

ACKNOWLEDGMENT

This paper is an adapted selection from from chapter 8 of Ernest, P. (1998) *Social Constructivism as a Philosophy of Mathematics,* Albany, New York: SUNY Press.

NOTES

1. These values and the associated powers of exclusion figure in a number of powerful feminist critiques of mathematics and science, see for example, Ernest (1991), Harding (1991), Walkerdine (1988).
2. Elsewhere in the context of education I explored the consequences of a set of values largely consonant with social constructivism (Ernest 1991). Nevertheless, any such set of educational values is independent of the philosophical position.
3. I argued in chapter 6 of Ernest (1998) that the emergence of proof in Ancient Greek mathematics, at least in part, reflected the prevailing democratic forms of social organization of the time.
4. Sometimes an excluded voice becomes strong in a second language game and then is listened to in the first on the basis of that strength.

CHAPTER 17

CLASSROOM RESEARCH

Impact and Long Term Effect *versus* Justice, Liberation and Empowerment?

Simon Goodchild
Agder University College

ABSTRACT

This paper considers the coherence between the research goals of making an impact and having a long-term effect with the researcher's concern to contribute towards justice, liberation and empowerment. The fundamental issue of how the researcher can be sure that her or his research will not result in injustice and oppression, and maybe worse, is considered and some criteria are suggested that might be used to guard against harmful effects. These criteria are then used to examine some key issues in classroom research, with illustrations from an example of the author's own classroom research into students goals.

Critical Issues in Mathematics Education, pages 217–235
Copyright © 2009 by Information Age Publishing
All rights of reproduction in any form reserved.

INTRODUCTION

I believe it goes without saying that a basic principle of educational research is that it should make an impact. As Stenhouse (1979/1985) remarked:

Research may be broadly defined as systematic enquiry *made public.* (p. 120, my emphasis).

And more recently Bassey (1995) has asserted:

I believe a definition like this needs to be nailed to the door and printed on the letterhead of everyone who claims to be an educational researcher! **Educational research aims critically to inform educational judgements and decisions in order to improve educational action.** (p. 39, bold type in original)

But my title offers the suggestion of the possibility that making an impact might be contrary to the goal of contributing to justice, liberation and empowerment. The question that I want to address in this paper is: how can we ensure that our research into teaching and learning in classrooms makes a positive contribution to the development of mankind? My concern with the issues of justice, liberation and empowerment were stimulated recently whilst reading an account of the genocide that took place in Rwanda in the 1990s (Rusesabagina, 2006). Rusesabagina traces the interracial conflict in Rwanda back to some rather bad anthropology reported one hundred and fifty years earlier by the British explorer, John Hanning Speke, this is the man who is credited with identifying the source of the river Nile. I was struck by the train of events and injustices that were linked to Speke's journal and I reflected on my own work. How can I be sure that my research activity will not eventually lead to dreadful consequences? How can I ensure that any impact from my research will be for good?

The simple answer to this question is that I must ensure that my research attains the *highest standards of scientific rigour.* In this paper I want to share my thoughts about what 'highest standards of scientific rigour' might mean in the context of classroom research.

ETHICAL CHARACTERISTICS OF EDUCATIONAL RESEARCH

Before considering the scientific standards of classroom research I want to suggest some ethical characteristics of educational research which I believe are essential if any impact or long term effect will also contribute to justice, liberation and empowerment? [Note, I am using the word characteristics rather than 'standards' here because my intention is not to argue a set of ethical standards, this has been attempted elsewhere, (e.g., British Educational Research Association, 2004). My intention is merely to identify some of the things that I can apply to the key issues of classroom research]. These characteristics are, in no particular order: honesty, openness, critical reflection, rationality, impact on informants/participants, and voice.

- *Honesty:* in reporting data and evidence; in working with our informants and participants; in recognising the limitations of any claims to knowledge.
- *Openness:* in reporting all issues and being prepared to admit error and failure; to criticism and suggestions.
- *Critical reflection:* in examining everything done at each stage of the research, to try to expose better ways of doing things, alternative explanations and interpretations, and trying to expose weaknesses and limitations; to ensure awareness of how our own knowledge, experience values, attitudes and emotions shape the research process.
- *Rationality:* in identifying the reasons for what we do, in terms of our existing knowledge, existing theory, research questions, data collection, interpretation and style of reporting.
- *Impact on informants/participants:* this is normally the main focus of an ethical risk assessment, everything already mentioned is part of an embracing 'ethical framework' but there are other issues regarding the impact upon the well-being our informants/participants.
- *Voice:* in making sure that the authentic voice of informants and participants is audible through the reporting of the research.

CLASSROOM RESEARCH

I worked for 16 years in secondary schools before moving into higher education. I then worked for a further sixteen years training secondary teachers. My whole professional life has been focused on the activity that takes place in mathematics classrooms. It follows that if I have anything to offer the wider community it will surely be from my experience and knowledge of teaching and learning mathematics in school. Consequently all that I present here is related primarily to mathematics classroom research. I hope that all I discuss is relevant to classroom research in general, irrespective of the subject, but I do accept that the subject matter is a significant focus of classroom research and this will inevitably mean differences in the research carried out.

I want to emphasise that I do not believe there is a 'specialism' that might be called 'classroom research', educational research includes classroom research, good classroom research is good research and the principles of classroom research are the same as the principles of any other type of educational research. On the other hand I do think classrooms are very special places in which to do research and they present researchers with great challenges, not least because of the complexity of the classroom. As Ernest (2001) observes:

> the mathematics classroom is a fiendishly difficult object to study. For the mathematics classroom involves the actualised relationships between a group of students... and a teacher... with a variety of material and semiotic resources in play within a set of temporally and geographically delimited spaces. Furthermore, each student, teacher, classroom, school and country has a life history with antecedent and concurrent events and experiences which impinge on the thin strand chosen for study within all this complexity: periodic mathematics lessons. (p. 7)

This complexity requires us to be very careful in researching classrooms, it is too easy to make mistakes and draw wrong conclusions, and make an impact that is contrary to the aims of justice, liberation and empowerment. Shulman (1987) summarises one concern:

> When policymakers have sought "research-based" definitions of good teaching to serve as the basis for teacher tests or systems of classroom observation, the lists of teacher behaviours that had been identified as effective in the empirical research were translated into the desirable competencies for classroom teachers. They became items on tests or on classroom-observation scales. They were accorded legitimacy because they had been "confirmed by research." While the researchers understood the findings to be simplified and incomplete, the policy community accepted them as sufficient for the definitions of standards. (p. 6)

I am not sure that we can ever protect our work from misuse; my first concern is that anything that I report is trustworthy. In the consideration of the foundations of trustworthiness, and the ethical characteristics that I have listed I want to address the following issues in relation to classroom research: types of classroom research, paradigm, theory, unit of analysis, methodology, operationalisation, method, disturbance, and concern for informants.

In addressing these issues I will make reference to my own ethnographic style case study of a year ten mathematics classroom. Briefly, I joined the class for every mathematics lesson for very nearly one complete year; I had conversations with the students while they were engaged in the tasks given to them by their teacher. My purpose was to expose the "Students' Goals" (Goodchild, 2001) in their classroom activity.

I start from the assumption that classroom research can be any inquiry into what happens in classrooms, into teaching and learning, and into teachers' and students' experiences, values, beliefs, attitudes etc. The data collection may take place wholly within classrooms—such as with observation, or wholly outside, as in the completion of questionnaires, or, indeed part in and out as with design research. In my own experience I have engaged in action-research within my own classroom, I have explored teaching and learning through observation, and engaged in curriculum evaluation—using pre and post tests. Currently I am working with teachers in a development-research project.

PARADIGM

A researcher needs to be conscious of the paradigm within which he or she is working. It is necessary to be clear within oneself whether, for example one believes in unproblematic cause-effect relationships that can be modelled in generalisable laws as in a scientific/positivist paradigm, or conversely that human behaviour is so complex and dynamically related simultaneously to a range of social, historical, emotional and physical phenomena that such generalisable rules are unknowable, even if they exist. As Lincoln and Guba (1985) summarise:

> *The possibility of causal linkages*
>
> *Positivist version*: Every action can be explained as the result (effect) of a real cause that precedes the effect temporally (or is at least simultaneous with it).
>
> *Naturalist version*: All entities are in a state of mutual simultaneous shaping so that it is impossible to distinguish causes from effects. (p. 38, italics in original)

It will become apparent that in my research into students goals I was heavily influenced by Lincoln and Guba's work. One of the weaknesses that I now recognise in my own work was that I was not sufficiently critical of their work and allowed it to exert an undue influence in all that I did. Researchers have a responsibility to understand the philosophical foundation upon which they base their claims for knowledge and to reflect on these and consider the consequences in practice. Researchers need to be clear about the paradigm within which they are working and that of any interlocutor because, the chances are that when one engages in dialogue with a researcher working in a different paradigm the likelihood is that a mutual understanding will be unattainable and any sense of agreement will be illusory.

Nevertheless, it is necessary also to act within a given paradigm in a critical fashion. Pring (2000) recognises that different social groups interpret the world differently, (I assert that this difference is a fundamental issue to be considered by classroom researchers) nevertheless, Pring argues that we can only be aware of and understand these differences because of their 'enduring features' which 'enable generalizations to be made' (p. 56). He asserts:

> The qualitative investigation can clear the ground for the quantitative—and the quantitative be suggestive of differences to be explored in a more interpretive mode.

> Understanding human beings, and thus researching into what they do and how they behave, calls upon many different methods, each making complex assumptions about what it means to explain behaviours and personal and social activities. (pp. 56–57)

I find it reassuring that an eminent educational philosopher argues so clearly for a stance that I feel, from the perspective of research practice, is sensible.

THEORY

> Theory is critical to the production of research knowledge, and to work more generally. (Boaler, 2002, p. 4)

I place my own research in a naturalistic paradigm, and I pursued and critically reflected on my work in exploring students' goals within a framework set out by Lincoln and Guba. Ten years on from when I completed that work I am rather more critical of the framework, especially when related to classroom research. For example, Lincoln and Guba (1985) argue for a grounded theory approach asserting:

> N (the naturalist) prefers to have the guiding substantive theory emerge from (be grounded in) the data because no a priori theory could possibly encompass the multiple realities that are likely to be encountered ... (p. 41, emphasis in original)

I certainly agree that theory must be grounded in data and tested empirically. Empirical evidence is essential in the verification of theory. However, I am sceptical of research based on 'the *discovery* of theory from the data' (Glaser & Strauss, 1967/2006, p. 1, my emphasis). I must emphasise that my critique here is focused on the notion of '*discovery*' that seems to ignore the existence of informal theory that resides subconsciously in the mind of the researcher. I believe that in classroom research it is fundamentally necessary to start from a priori, guiding, substantive theory. It is impossible to come to classroom research as, in Lave's (1988) description of ethnographers, 'ignorant strangers':

Ethnographers are nonmembers of the cultures they study, being observant strangers whose ignorance they themselves take to be a condition for eliciting from informants explicit accounts of the obvious and basic aspects of culture and everyday practice. (p. 185)

Apart from the small number of people taught at home or by a private tutor everyone has a range of experiences in the classroom. If we do not come to the classroom with a consciously held and clearly articulated theory then we come with a subconscious theory—theory helps the researcher to 'see' through the filter of her or his own preconceptions.

Additionally, as we approach our inquiry into classrooms from a theoretical perspective we can be more sure of making a contribution to the development of scientific knowledge, that is we avoid the creation of producing discrete pieces of information that are not connected to the wider body of knowledge. Theory also helps us to focus on the issues that concern our research and professional community.

A good theory should help researchers understand what is going on in the classroom, it is essential for framing the research, it is essential in interpreting the evidence, it is essential in the development of knowledge. Consequently I treat with suspicion research reports that 'excuse' an apparent denial of existing relevant theory by claiming a 'grounded theory' approach. I want reports to be clear—if it is meant that no explicit theory has guided the research then perhaps my time will be better spent elsewhere. If on the other hand it means that the testing and development of theory is grounded within the data then I want to applaud—but I do want that theory to be clearly articulated in the report.

This raises questions about what theory is appropriate to classroom research. In my research into students' goals I took three separate theoretical perspectives. I argued that my familiarity with the classroom, and my own personal theory of teaching and learning needed to be enlightened and challenged by trying to view and explain things from different perspectives. I chose to take a social constructivist perspective, an activity theory perspective, and a situated cognition perspective. I acknowledged the inconsistencies between these perspectives and avoided any suggestion of trying to combine them into one 'super theory'. In a way the approach was successful, the different perspectives allowed me to focus on different levels within the classroom, the individual student, the student in socially mediated activity, the student as a member of a community of practice. These different foci led to complementary accounts of what was happening in the classroom. However, ten years on, I believe the approach was naïve and possibly a lazy way of engaging with the complexity of the classroom. I no longer believe that it is necessary to use several theories, no matter how complex the classroom, each theory is sufficient in itself to address the range of individual, social and cultural issues. If I had used just one theory, perhaps my work would have been stronger and made a greater 'scientific impact', nevertheless, I think what I attempted is original and has some merit in that. However, Boaler (2002), in the same article from which the quotation at the head of this section is drawn, argues for the knowledge generating potential of drawing connections between theories—using Andrew Wiles approach to solving Fermat's last theorem as an example. Perhaps my work allows such connections to be made and there is greater value in the use of complementary theories than I have argued.

UNIT OF ANALYSIS

The discussion of theory naturally leads on to the unit of analysis. A unit of analysis is 'the minimal unit of "evidence" that preserves the properties of the whole' (Davydov & Radzikhovskii, 1985, p. 50) object that is being studied. So if we are considering the properties of 'classrooms', that is, of teaching and learning mathematics in classrooms, it is necessary to have a unit of analysis that makes possible a study of the complexity of relationships between teacher, pupils, resources, history, culture, etc. The complexity of the classroom is a challenge to classroom researchers, and it challenges us in our choice of a suitable unit of analysis.

In my work on students' goals I used three units of analysis, one relating to each of the theories I was using. Within the social constructivist theory I used Neisser's perceptual cycle (Neisser, 1976) in this the individual's mental schema directs his or her exploration which samples the object or available information (a mathematical task or problem, say) as a result of these processes the individual modifies his or her schema. Most components of this unit of analysis are hidden from view, in particular the individual's schema and the processes of direction and modification. Those elements concealed from the direct exploration of the researcher have to be inferred by careful and critical examination of the range of evidence relating to the individual's action. Within activity theory I used Engeström's (1987) model of an extended activity system. In this it is the dialectical mediation of tools and signs, community, rules and division of labour that come between the individual and the object of his or her activity that are the focus. From the perspective of situated cognition I used Lave's (1988) model of dialectical relations between students acting, the classroom arena, students in activity and the task setting. Each model characterises the student's activity differently and draws attention to different features of the context of his or her work on a task. Nevertheless it is the same data that is used, transcripts of conversations, student's writing, copies of resources used within the classroom, etc. from these pieces of information a theoretical account of the classroom can be developed, and the theory can be challenged—when events and observations defy explanation. Each unit of analysis is a product of the theoretical perspective, and as with the theories I did not try to combine the units of analysis.

It bothers me when I read in a research report, that is taking a *single* theoretical stance, of a number of different 'units of analysis'. Either the unit of analysis is 'the minimal unit of evidence' or it is not.

METHODOLOGY

Methodology is much more than method. Ernest (1994) defines methodology as:

> A theory of which methods and techniques are appropriate and valid to use to generate and justify knowledge, given the epistemology (of the research). (p. 21)

When one engages in research it is necessary to make decisions about what data to collect and how to collect it and analyse it. These decisions will rest on a number of issues including: the paradigm and theoretical framework; the nature of the research question; what research has been done in this area before; our understanding of the nature of the subject—or our research participants or informants; what possible obstacles might exist for collecting trustworthy data in an ethically justifiable manner and so on.

I want to credit Leone Burton (2002) for opening my eyes to these issues, she writes:

> I am asking for researchers to be clear to themselves about the values, beliefs, and attitudes that are driving the study that they propose to do and to make that clarity visible to the reader.... Second, I am making the assumption that research in mathematics education is emancipatory in that its intentions are to empower—pupils, teachers, curriculum designers, policymakers—those who could be users or affected by use of the research. (p. 4)

> ...the choice of which method(s) is best, in order to gather the data necessary to the exploration of a particular question, is always a function of the theoretical stance adopted by the researcher(s) together with, of course, the research context and the related research questions, the informants, and so forth. This, in itself is a product of the attitudes, beliefs, and values underlying that stance. (pp. 7,8)

OPERATIONALISATION

By operationalisation I mean the translation of theoretical constructs into phenomena that can be observed and thus used as evidence within the inquiry. Here again, in my research into students' goals I followed Lincoln and Guba (1985) who assert that operationalisation is:

> not meaningful or satisfying...too shallow—depending on sensations (not) meanings or implications...results in a meaningless splintering of the world (pp. 26, 27)

In 'students' goals I provided working definitions of the 'goals' I was seeking evidence of but I did not operationalise these by articulating beforehand how they would emerge in the data I would collect, or later before starting analysis. On the one hand this meant that the account of students' goals was created by my informants (and my interpretation!) rather than constrained by preconception. Additionally the approach avoided Brousseau's topaz effect of which Jaworski (1994) writes:

> Beware the topaz effect—'the more explicit you are about what you want, the more likely you are to get that because it's perceived that you want it, not because it is actually the case.' [(Jaworski's) paraphrasing of Brouseau, 1984]. (p. 139)

[It should be noted that for Jaworski the point is a didactic issue, that 'getting it turns 'it' into a mechanistic routine, rather than the deeply thoughtful learning process that is desired' (personal communication, October, 26, 2006)]. From my perspective, as researcher the issue is about trustworthiness, if I lead my informants/participants so carefully that they 'give' me what I want then I am only getting my own 'theory' confirmed. However, when I eventually set about analysing the data I do not believe the working definitions were sufficient to overcome my own, mostly subconscious and maybe inconsistent, notions of what might constitute evidence. The working definitions were not sufficient to transform the utterances in the conversations into clearly identifiable evidence of students' goals, the transformation took place largely within my own reflective activity and it was only after the analysis that I felt able to provide sharp definitions. Thus there is circularity in my interpretation of the data and I fall prey to the same criticism that I level at those who take a strictly grounded theory approach to their research.

In research we seek, as far as is possible, trustworthy accounts that balance the objectivity provided by theory with the acknowledged and known subjectivity of the researcher. The key issue is for the researcher is to make his or her subjectivity explicit and avoid subconscious or unconscious application of preconceptions and prejudice. I now believe, in my research into students' goals that my failure to operationalise, explicitly, the constructs I was seeking undermines my claims for the trustworthiness of my interpretation.

METHOD

The next question that I want to address concerns whether some research methods are more appropriate to classroom research than others. The simple answer is 'no'! My remarks on methodology form the basis of my assertion that the method adopted will be determined by a range of factors including the theoretical framework, the question to be researched and the nature or characteristics of the informant. Nevertheless my use of Lave's description of an ethnographer suggests to me that classroom 'ethnography' is not a reasonable option—even though I admit Woods (1990) ethnographic account of pupils in secondary school is powerful and informative. I describe my own study of students' goals as 'ethnographic style', and there are other notable research reports in mathematics education that can be similarly characterised, for example Jaworski's Investigating Mathematics Teaching (1994), and Boaler's Experiencing School Mathematics (1997). It is evident that the techniques used in ethnography have the potential to answer the questions we ask about activity in classrooms.

We have to admit that we as researchers act as filters and make decisions that will impact upon the nature and quality of the evidence we use—even before we begin to analyse the data and make interpretations. In researching students' goals most of my data arose through conversations I had with students or from discussions between students when I was present. I can only wonder what the discussions between the students would have been like if I were not present. I made a decision to collect data from the students while they were engaged in their tasks because, I argue, if I had collected it outside the classroom it would be 'coloured' by their rationalisations—it seems that I am suggesting that my presence had a smaller impact on the goals I was able to expose than their own reasoning or rationality. In my present work, more of which later, I am using 'naturally occurring data', that is recordings of events that are made in the course of meetings and workshops without the intervention of activity done purely for the research element of the project. But there is a vast amount of data and I am making choices about which pieces to use and how it will be analysed thus my influence is still present.

The point I want to make is that, in classroom research it is necessary to consider the position and voice of the teacher and students, and, naturally, the subject content of lessons. Wagner (1997) identifies three types of relationship between researchers and teachers. Data extraction agreements, where the researcher goes into the field, collects data, and returns to her or his office to analyse and interpret. Clinical partnerships, in which teachers share in the research activity in both deciding what will be done and reflecting on the outcomes, but the emphasis is on learning about what happens in classrooms, about teaching and learning—the purpose is for the researcher to inform her or his scientific community and policy makers, and the teacher about practices in the classroom. The third type is a co-learning agreement in which the researcher enters the field alongside the teacher and acknowledges that the joint activity with the teacher has the potential to inform the research about her or his role as much as the teacher about classroom practices.

In my research into students' goals my relationship with the teacher was that of a data extraction agreement. I remain deeply grateful to the teacher and students for allowing me such freedom in their classroom but as far as I can recall they showed little interest in what I was doing with the data I was collecting or what I was learning—and I did not try to interest them. The teacher did read the final report and made some comments but I am not aware that it had any impact upon her practice—I don't know! And that is the point.

In classroom research the place that, perhaps, the research has the potential to make an impact are the classrooms, and the practices of the teachers, who collaborate in the research. In this I mean a symmetrical relationship in which teacher and researcher come together, each with their own experience and specialised knowledge, each prepared to learn about their own practice in addition to learning about teaching and learning. That is, in a co-learning agreement, anything short of this is making 'use' of another person and their practice, and the researcher ends up taking away or changing the voice of the teacher. A clinical partnership has the potential to empower a teacher in her or his practice, albeit with the teacher and researcher in asymmetrical relationship and the teacher being given the researchers' voice. A co-learning agreement has the potential to empower and the teacher's voice to be heard.

As an example, I am currently working with a very experienced and talented primary school teacher. I am learning a great deal about his practice and the way that he manages teaching and learning in his classroom. We have written one paper together and I hope this will be the first of many, in the paper I reflect on what I am learning from this classroom, I describe events that I see and how I interpret these from my perspective as an experienced secondary teacher. However, the teacher I am working with has the opportunity to ensure that his own voice is heard and the balance of reporting fits with his view of his classroom. In fact he is in a powerful position because even though I have done the greatest part of the drafting of the paper he has translated it into Norwegian. I have a basic understanding of

Norwegian but it is insufficient to judge the nuances of language that can make so much impact on meaning.

DISTURBANCE

Is inevitable! Almost any form of data collection is likely to disturb the subject. I have remarked that I am now using naturally occurring data, this is available because we record, audio or video, every event that takes place within the project. Within our project we have become so accustomed to the presence of a recorder or camera that we forget it is there—I guess up to the moment that we begin to think very carefully about what we say because we know there will be a recording!

Gallagher (1995) writes about an instance in her research when a student asked her for help, she struggled with her role as 'detached observer' and her feelings as person and teacher. She decided to help the student. Following the event she felt that she had violated the tenets of good research and resolved to remain detached in future. Gallagher goes on to regret that decision believing that she might have been able to have a positive impact on the teacher's practice. In researching students' goals it was inevitable that I would disturb the students, given that I was sitting next to them and talking to them during the course of their activity. I also found it very difficult to 'switch off' the teacher within me, so at times my 'research conversation' took the form of a 'teaching conversation' in which I seem as interested to expose to the student the mathematics in which they are engaged as I am to expose for myself the goals they are working towards.

Wiske (1995) reports from research in which teachers and researchers worked together in clinical interviews of students and describes how the teacher's view of the interview differed from the researcher.

> The teacher regarded the clinical interview as an educational experience for the student. She wanted the child to be treated as the teacher would have treated her in class, not allowed to feel stupid or discouraged by a prolonged period of ignorance unlike anything the teacher would willingly sustain in class. (p. 203)

My sympathy here is with the teacher. I think it is an abuse of our informants or participants and the trust that they put in us when they agree to take part in our research if our actions lead them to feel 'stupid' or 'discouraged'.

While researching students' goals I felt reasonably pleased with myself that the students did not view me as a teacher. I recall, for example, an occasion near the beginning of my field work when the class had a substitute teacher, we were relocated in an unfamiliar and uncomfortable room which unsettled the class and resulted in an unusual degree of poor be-haviour. It seemed to make no difference to the students that I was sitting next to the paper aeroplane production and launch site. Later in the year, towards the end of my field work a student asked me if I was intending to become a mathematics teacher. However my feelings of satisfaction were diminished very near to the end of my work. I had analysed 90% of the data and produced my account of students' goals. I then turned to the 10% of data that I had selected randomly for archiving and later testing the theory I built. One of these archived conversations took place in the first lesson in which I started talking to the students. In this conversation I observed a new phenomenon. In all the conversations I had used to build the theoretical account of the classroom I had been able to identify goals towards which students were working. However, in this archived conversa-tion I could find no evidence of any goal, in fact it seemed that the student had no goal, because prior to the conversation the student had been inac-tive. One conclusion I drew from this was that my presence in the classroom had sharpened the students' awareness to the possibility of me coming to them and asking them about what they were doing and why. I believe, part of the honesty that I assert underpins research that contributes to justice, empowerment and liberation is to admit the possibility that we may have it wrong! Unfortunately this does not go down too well with the policymakers who want definite answers, hard facts and relationships. To admit to the possibility of error weakens our case and undermines the impact that our research might have.

REGARD FOR INFORMANTS AND PARTICIPANTS

The whole of this paper is, I believe, an ethical statement. But there are a few recognisably ethical considerations that I want to add. First, we take for granted that we must seek permission from our informants before we collect data of any kind and we are careful to explain how the data might be used. I wonder how aware teachers and students are of the risk they are taking when they give permission for research to take place. Teachers in particular: the research might expose parts of their practice that make them feel uncomfortable and perhaps even a loss of confidence; I think this is almost inevitable when we lift the lid on a person's practice and ex-amine carefully what is going on. Some research entails the teacher trying something new; this also puts the teacher at risk because it moves her or him from their established patterns of behaviour. When we seek permission for research from teachers I wonder how careful we are to offer them a risk analysis—and if we did whether they would still participate.

One of the realities of our lives as researchers is that our work is often determined by the time frames of short-term funded projects. Within a project we establish a working relationship with teachers, complete our research agenda and then have to walk away because the funds no longer exist to sustain the relationship, we have to turn our attention to the next project. I do not think this is being fair to the teachers and their students, unless we build into the project some mechanism by which any impact of the research may be sustained after the project ends.

Cooper and McIntyre (1996) offer three characteristics of the behaviour of researchers towards informants: being willing and able to empathise with informants; unconditional regard for participants, that is liking and being interested in informants as individuals; and congruence, that is the researcher is honest and authentic in his or her relationships with informants. I would like to believe that anyone involved in classroom research will have had experience as a teacher which will ensure empathy. I know that during my 16 years teaching in school I will have demonstrated both good and mediocre practice, things I am proud of and things I am ashamed of. And it is not the case that all the bad came at the beginning and the end was only good. The opportunity to reflect on one's own practice in the past should prevent one being judgemental of another teacher in the present. I think if one does not have a liking and interest in teachers and students then one should not even begin to research classrooms. Being honest and authentic also requires some humility, to be clear about one's own mistakes and weaknesses.

VOICE

All that I have argued so far needs to be reported with the same care as the research is carried out, thus the report will be characterised in the same way by honesty, openness, critical reflection and 'voice'. I want to say something in particular about giving the teacher and students 'a voice'. Keitel (2004) writes about using my report of Students' Goals with a group of students on an initial teacher training course:

> I confronted [a] group with some of the excerpts of students' interviews from Goodchild's study and asked them to recall their own experiences in school time and search for similar situations, or make notes about other students' experiences. As they usually are asked to write a biographical essay, they complemented their excerpts with examples of stories about getting stuck, complete lack of understanding, hard debates with other students about getting meaning and significance of a given task and so on. In short, the excerpts encouraged them to look for their own stories and remember their school mates' struggle as well as their own. They were mostly fascinated that a researcher

and teacher are giving students a voice and really were interested to listen to them—they argued that this might be an almost necessary condition for becoming a teacher, but almost nobody among them had experienced such a 'listening' mathematics teacher. (p. 276)

Naturally I am pleased to see the report being used but I do wonder if I am really 'giving students a voice' as suggested. I think it is more likely that I am using their voice to convey my story. I made selections from my data, using those bits that I thought were interesting and 'useful' for my account. I presented these within the context of my account, with my interpretation alongside. I did the same with the teacher, presenting my account of the classroom using extracts of the teacher's explanations from an extended interview I had with her when I had completed the classroom observations. This has the appearance of giving the teacher and students a voice but in reality I am using their voice to speak my meaning. Perhaps this is more dishonest than giving no voice at all.

Sometimes we will be allowed into classrooms and observe a consistency of bad practice that causes us alarm. For example, we might observe students getting a severely impoverished experience of mathematics and a teacher struggling with subject content knowledge, or pedagogical content knowledge. What do we do in these cases? If our relationship with the teacher is of the data extraction type, as it was with students' goals I think there is little that we can do. The teacher did not allow me into her classroom to criticise her practice and it would have been a violation of our agreement if I had. If I went past the teacher and described her poor practice publicly then the chances are I would never be allowed into another classroom. This provides me with one more reason for entering into at least a clinical partnership with a teacher and ideally a co-learning agreement in which both I and the teacher engage in the activity prepared to learn about our own practices.

CONCLUSION

Returning to the title of this talk: in classroom research I do not believe it is possible to engage in respectable research that will contribute to justice, liberation and empowerment and produce easy recipes for teaching and learning that are likely to appeal to policy makers and thus make a substantial or long term impact. In this respect I can see that impact and long term effect are possibly contrary to the goals of justice, liberation and empowerment. Nevertheless, I do think it is possible to have an impact, immediately and in the long term, not as I did with my research into students' goals but rather by entering into co-learning agreements with teachers and working with them in developmental research as in the Learning Communities in Mathematics Project (LCM) in which I am now engaged. [LCM is supported by the Research Council of Norway (Norges Forskningsråd): Project number 157949/S20]

LCM is a development-research project in which the aim is for didacticians to enter co-learning agreements with teachers. The project is established on a notion of 'inquiry as a way of being' (Jaworski, 2004, p. 26). Inquiry is seen to be empowering. At the heart of the development part of the project the aim is for teachers, working in school teams to design tasks. The teachers' role in the design process respects their knowledge. A condition of joining the project was that at least three mathematics teachers in any school should participate; this is part of our attempt to ensure sustainability beyond the end of the project.

We aim to facilitate teachers expressing their own voice. One such instance is the joint article I have mentioned earlier, and at the moment we are busy preparing a project book which will include chapters written by school teams as well as didacticians—and some eminent mathematics educators who have made a special contribution to the project. At the beginning of September this year, as we came to the end of the second phase of the project we held a major conference with participants coming from all parts of Norway, teachers and didacticians made contributions. The conference helped to convince the didacticians that the project was having and impact on the professional practice of the teachers. It also gives us hope that the impact might be felt beyond our small project group.

We are now entering the third phase and final year of the project. One aim this year is to consolidate the developments that are taking place in schools, but we have been pleased to be given more funds from the Research Council of Norway for a new project which aims to widen the scope in two ways. First there is the possibility of more schools being drawn in, but also the new project is based on a consortium of mathematics educators in university colleges in different regions of Norway who will pursue their own projects within the common theoretical framework of communities of inquiry that has been established within LCM.

In these projects I want to assert that it is possible to achieve impact and long term effect alongside the goals of justice, liberation and empowerment.

REFERENCES

Bassey, M. (1995). Creating education through research: A global perspective of educational research for the 21st century. Edinburgh, Scotland: British Educational Research Association.

British Educational Research Association (2004). Revised ethical guidelines for educational research. Retrieved March, 01, 2006, from http://www.bera.ac.uk/publications/pdfs/ETHICA1.PDF

Boaler, J. (1997). Experiencing school mathematics: Teaching styles, sex and setting. Buckingham, England: Open University Press.

Boaler, J. (2002). Exploring the nature of mathematical activity: Using theory, research and 'working hypotheses' to broaden conceptions of mathematics knowing. Educational Studies in Mathematics, 51, 3–21.

Brousseau, G. (1984). The crucial role of the didactical contract in the analysis and construction of situations in teaching and learning mathematics. In Theory of Mathematics Education Occasional Paper 54, November, (pp. 110–119). Institut für Didaktich der Mathematik IDM, Universität Bielefeld,.

Burton, L. (2002). Methodology and methods in mathematics education research: Where is the why? In S. Goodchild & L. English (Eds.), Researching mathematics classrooms: a critical examination of methodology (pp. 1–10). Westport, CT: Praeger.

Cooper, P. and McIntyre, D. (1996). Effective teaching and learning: Teachers' and students' perspectives. Buckingham, England: Open University Press.

Davydov, V. V. & Radzikhovskii, L. A. (1985). Vygotsky's theory and activity-oriented approach in psychology. In J. V. Wertsch (Ed.), Culture, Communication and Cognition: Vygotskian Perspectives (pp. 35–65). Cambridge, England: Cambridge University Press.

Engeström, Y. (1987). Learning by expanding: An activity-theoretical approach to developmental research. Helsinki, Finland: Orienta-Konsultit.

Ernest, P. (1994). An introduction to research methodology and paradigms. Exeter, England: University of Exeter School of Education Research Support Unit

Ernest, P. (2001). Foreword. In S. Goodchild, Students goals: a case study of activity in a mathematics classroom (pp. 7–8). Bergen, Norway: Caspar Forlag.

Gallagher, D. J. (1995). In search of the rightful role of method: reflections on conducting a qualitative dissertation. In T. Tiller, A. Sparkes, S. Kårhus & F. Dowling Næss (Eds.), The Qualitative Challenge: Reflections on Educational Research. (pp. 17–35). Bergen, Norway: Caspar Forlag.

Glaser, B. G. & Strauss, A. L. (2006). The discovery of grounded theory: Strategies for qualitative research. New Brunswick, NJ: Aldine Transaction. (Original work published 1967)

Goodchild, S. (2001). Students goals: A case study of activity in a mathematics classroom. Bergen, Norway: Caspar Forlag.

Jaworski, B. (1994). Investigating mathematics teaching: A constructivist enquiry. London: Falmer Press.

Jaworski, B. (2004). Grappling with complexity: Co-learning in inquiry communities in mathematics teaching development. In M. Johnsen Høines & A. B. Fuglestad (Eds.), Proceedings of the 28th Conference of the International Group for the Psychology of Mathematics Education: Vol. 1 (pp. 17–36). Bergen, Norway: Bergen University College.

Keitel, C. (2004). Book review. Journal of Mathematics Teacher Education, 7, 269–277.

Lave, J. (1988). Cognition in practice: Mind, mathematics and culture in everyday life. Cambridge, England: Cambridge University Press.

Lincoln, Y. S. and Guba, E. G. (1985). Naturalistic inquiry. London: Sage.

Neisser, U. (1976). Cognition and reality: Principles and implications of cognitive psychology. San Francisco: W. H. Freeman and Co.

Pring, R. (2000). Philosophy of educational research. London: Continuum.

Rusesabagina, P. (with Zoellner, T.) (2006). An ordinary man: The true story behind 'Hotel Rwanda'. London: Bloomsbury.

Shulman, L. S. (1987). Knowledge and teaching: Foundations of the new reform. Harvard Educational Review, 57, 1–22.

Stenhouse, L. (1979). Research as a basis for teaching: inaugural lecture, University of East Anglia, February 1979. In L. Stenhouse (1983), Authority education and emancipation London: Heinemann Educational pp. 177–195. This print from J. Rudduck & D. Hopkins, (Eds.) (1985), Research as a basis for teaching: readings from the work of Lawrence Stenhouse (pp. 113–128). London: Heinemann Educational.

Wagner, J. (1997). The unavoidable intervention of educational research: A framework for reconsidering researcher-practitioner cooperation. Educational Researcher, 26, 13–22.

Wiske, M. S. (1995). A cultural perspective on school-university collaboration. In D. N. Perkins, J. L. Schwartz, M. M. West & M. S. Wiske (Eds.), Software goes to school: Teaching for understanding with new technologies (pp. 187–212). Oxford, England: Oxford University Press.

Woods, P. (1990). The happiest days? How pupils cope with school. London: Falmer Press.

CHAPTER 18

WHAT HAS POWER GOT TO DO WITH MATHEMATICS EDUCATION?

Paola Valero
Aalborg University

Whenever I meet a new person and we come to talk about my work people tend to be surprised. Mathematics and power? Mathematics and democracy? Those things do not go together! For many people it is astonishing that mathematics can be thought in relation to something "social" such as power relationships, political affairs and actions, and values and forms of living such as democracy. After a long conversation and many examples some people may come to see my point. However, it is difficult to break, all of a sudden, the view of mathematics as numbers, rules and procedures, which have no relation with people and their every day lives in society. Such a view is deeply entrenched as a result of people's own school experiences and of their understanding of the public ways of talking about mathematics in the media and in society in general. This view is not only shared by those who dislike mathematics (and have probably had a "traumatic" school experience with it) but also by many of those who have been successful and like it. Part of the view is based on the assumption that mathematics has a

Critical Issues in Mathematics Education, pages 237–254
Copyright © 2009 by Information Age Publishing
All rights of reproduction in any form reserved.

life on its own, independently from the human beings that both have invented and used it.

Critique to such an image has emerged from recent work in the philosophy of mathematics—as for example in the work of Paul Ernest (1998b) and his analysis of mathematics as a socially constructed knowledge—and from research in mathematics education. When mathematics is considered as one of the many social activities and practices that human beings carry out, the entrenchment of mathematics into cultural, political and economic phenomena becomes more evident. Now, if we enter mathematics education, we definitely have to pay attention to society, not only because the content of this education is considered to be important in society, but specially because mathematics acquires an important part of its social life through educational relationships in schools. This is precisely the viewpoint I adopt and from which I would like to reflect about mathematics, education and power. My reflections are inscribed in a perspective concerned with the political dimension of mathematics education. This perspective has not been so widespread in the field, although it has been a growing trend (as shown in, for example, Valero & Zevenbergen, 2004; and Walshaw, 2004).

In what follows I will discuss my view of mathematics education as a field of study and as a field of practice and will point to the necessity of broadening dominant definitions of these fields in order to grasp their socio-political complexity. I will also discuss three notions of power and how they have permeated mathematics education research. Finally I will present some remarks about the challenges that adopting a socio-political viewpoint for an understanding of the practices of mathematics teaching and learning poses to research.

MATHEMATICS EDUCATION: BROADENING THE SCOPE

The term "mathematics education" can refer to two different domains. On the one hand, mathematics education names a *field of practice*, which is the space where people actually engage in the activities connected to the teaching and learning of mathematics. On the other hand, we mathematics education refers to the *field of study*, which is the space of scientific inquiry on and theorization about the field of practice (Ernest, 1998a). It is interesting to consider the relationship between these two fields, in particular, the way in which they are dialectically defined.

A dominant definition of mathematics education as a scientific field of study is that of the discipline "covering the practice of mathematics teaching and learning at all levels in (and outside) the educational system in which it is embedded" (Sierpinska & Kilpatrick, 1998, p. 29). In this field, "[...] mathematics and its specificities are inherent in the research questions from the outset. One is looking at mathematics learning and one cannot ask these questions outside of mathematics."(p. 26). This way of defining the field of study highlights the *didactic triad*—that is, the relationships between teacher, learners and mathematical content—as the privileged space of enquiry. This definition determines both what is taken as legitimate research questions and approaches in the field of study. Studies in mathematics education are mainly interested in characterizing, theorizing and explaining the practices that are clearly inscribed in the didactic triad. This definition of the field of study frames the possible meanings and definitions of the field of practice. Mathematics education practices are defined, then, as all the activities revolving around teachers' instruction of a given content to some students who are engaged in the learning of that content. As a result, mathematics education practices also get inscribed into the closeness of the didactic triad that defines the field of study.

Concerning this narrowness of scope, Gómez (1999) argues that "mathematics education research production is centered mainly on cognitive problems and phenomena; that it has other minor areas of interest; and that it shows very little production on those themes related to the practices that influence somehow the teaching and learning of mathematics from the institutional or national point of view" (p. 2–3). Chassapis (2002) also argues that little and almost insignificant attention has been paid in 30 years of research production to the issue of who are the mathematics learners and how the learners' background influences mathematical learning. This lack of attention contributes to a lack of comprehension about the social, political and cultural complexity of mathematics education and the factors involved in it. Vithal and Valero (2003) also raise a critique to the lack of attention that dominant mathematics education research has paid to the impact of social context —and its conflictive nature— on the study of the micro-processes of mathematical teaching and learning. This restrictive frame for what is considered to be legitimate objects of study has an impact in conceptions of the practices of mathematics education in society and, correspondingly, in conceptions about ways of undertaking the study of those practices. These arguments also resonate with Apple's (1995) comment that most of the discussions about mathematics education have left apart "critical social, political, and economic considerations"; they have limited their scope to the individual realm and, therefore, "lost any serious sense of the social structures, race, gender, and class relations" that constitute individuals; and finally, they have not been situated "in a wider social context that includes larger programs for democratic education and a more democratic society" (p. 331). The field of study defines the field of practice and itself in a closed, internalistic fashion. This has as a consequence that connections to the external environment of the teacher-learners-mathematics triad are not researched, neither taken into account as a part of practice.

Dominant definitions of mathematics education as a field of study and their corresponding definitions of the field of practice are problematic. It is not possible to think about neither the field of study, nor the field of practice, as spheres existing independently from the social relations and conditions where they occur. It is clear that there are many other factors affecting mathematics education than those considered in the didactic triad. Research in mathematics education within the "social turn" (Lerman, 2000) has pointed to some of the shortcomings of internalistic definitions of the fields of study and practice. Are there other definitions that open for other possible understandings of these two fields?

Let us think about mathematics education as a field of practice covering the net-work of social practices carried out by different social actors and institutions located in different spheres and levels, which constitute and shape the way mathematics is taught and learned in society, schools and classrooms (Valero, 2002). This means that, besides the three basic elements of the didactic triad, there is a series of actors with their practices that contribute shaping the mathematics education practices occurring in the relationship between teacher, students and a content displayed in their interaction. One could consider, for example, the role of the group of mathematics teachers and administrators in the educational institution, in connection to the teachers' practices inside the classroom. Textbook writers with their materials definitely shape teachers' practices. One could also take into account how teacher-educators impact practices through the construction and provision of certain teacher qualification. Community's and of labor market's expectations and demands about the mathematical competencies of students as potential working force can not be discarded as relevant contributors to shape mathematics education practices. This is also connected to the influence of policy-makers in mathematics education, and to the politicians' inputs and expectations about the outputs of schooling practices. In summary, mathematics education as practices are not restricted to the sphere of the classroom, but transcend it by including the practices of different social actors and institutions, and the interconnection between those across levels.

This broader definition of the field of practice evidences the social, political, cultural and economic dimensions that are a constitutive element of mathematics education practices. This means that in practice, students' cognitive processes of a particular subject matter are only one of the many activities that take place in classrooms, and only one of the components of the larger net-work of practices. The implications of this view for a definition of mathematics education research, that is, the field of study, are enormous. The theories, the methodologies, the positioning of the researcher, the criteria for judging the quality of research and even the forms of communicating research are open to discussion (Vithal & Valero, 2003). Although the intention of this paper is not to go into the details of those implications, let me just say that opening the scope of mathematics education research challenges the established ways of working in the discipline and invites researchers to engage in a more uncertain process of knowledge construction and sense making about the practices of mathematics education.

Having this approach, I would like to concentrate on the political dimension of mathematics education practices and on the ways research has tackled it through defining the meaning of power in relation to mathematics and mathematics education.

POWER IN MATHEMATICS EDUCATION RESEARCH

Authors such as Lerman (2000) have argued that the growing attention during the last 20 years to the social aspects of mathematics education could be rooted in the political concerns of some researchers who saw that "inequalities in society were reinforced and reproduced by differential success in school mathematics" (p. 24). I have argued elsewhere (Valero, 2004b) that this initial political awareness has not necessarily led to the constitution of a socio-political approach in research. One of the reasons for this is that the notion of power has not been extensively and systematically analyzed neither operationalized in research. In what follows I would like to examine some of the existing literature in search of the meaning given to the notion in different mathematics education research discourses.

The Intrinsic Power of Mathematics and Mathematical Learning

In the recent *Handbook of International Research in Mathematics Education,* English (2002) invited contributing authors to think about the issue of "access to *powerful* mathematical ideas." In her text, English provides meaning to this phrase and to the term *powerful,* in the following way:

> [...] the lack of access to a quality education—in particular, a quality mathematics education—is likely to limit human potential and individual economic opportunity. Given the importance of mathematics in the ever-changing global market, there will be increased demands for workers to possess more advanced and future-oriented mathematical and technological skills. Together with the rapid changes in the workplace and in daily living, the global market has alerted us to rethink the mathematical experiences we provide for our

students in terms of content, approaches to learning, ways of assessing learning, and ways of increasing access to quality learning. (p. 4)

She follows her explanation about "powerful mathematical ideas" in the following terms:

Students are facing a world shaped by increasing complex, dynamic, and powerful systems of information and ideas. As future members of the workforce, students will need to be able to interpret and explain structurally complex systems, to reason in mathematically diverse ways, and to use sophisticated equipment and resources. [...] Today's mathematics curricula must broaden their goals to include key concepts and processes that will maximize students' opportunities for success in the 21st century. These include, among others statistical reasoning, probability, algebraic thinking, mathematical modeling, visualizing, problem solving and posing, number sense, and dealing with technological change. (p. 8)

Let me examine English' words. In the first quotation there is established a connection between the quality of the mathematical education of a person and the person's potential and economic opportunity. This seems to imply that good mathematics education gives "power" to a person because it gives people mathematical skills that are of paramount importance in current social processes. English also establishes a connection between mathematics (and mathematics education) with current economic and productive processes. The power of mathematics and mathematics education is also brought in relation to a person's participation in a global economy. The demands of the global economy should make educators rethink the kind of mathematical experience provided to all students. In the second quotation English makes more explicit the demands of the global economy to people's performance. Powerful mathematical ideas are those that will allow people to think in ways that secure their success as working force in the 21st century, that is, in the global economy.

I take English words as representative of a type of discourse about power and mathematics education. Her definition of powerful mathematical ideas does in fact resonate with the way in which the term power features in most literature in mathematics education, where the term power appears in association with statements such as: "Since mathematics is a *powerful* knowledge in our society, then it is important to improve the access of as many students as possible to a quality mathematics education so that they get *empowered*." Such a statement brings together two basic ideas: on the one hand, that mathematics has power, and that, therefore, mathematics can empower those who acquire it, on the other hand. These assumptions are sometimes explicit, but most of the times they remain tacit. When remaining tacit, the assumptions do not differ substantially from the also tacit concern of hard-core psychological research in mathematics education, where it is assumed that there is an intrinsic resonance between the goodness of mathematics and all the positive contributions of mathematics education both to the individual and to society (see Skovsmose & Valero, 2001 for details on this discussion).

These assumptions also rely on a notion of power rooted in a liberal functionalist tradition. In many analysis of power within this trend (for example, Weber, 1964), the concept is defined as the capacity of an actor *A* to influence the behavior of another actor *B*. *A* has power over *B* if *A* can modify *B*'s actions and therefore the results of *B*'s actions. If power is such capacity, then *A* is in possession of a form of control over other people or situations. *B* accepts *A*'s influence on the grounds of *B*' acknowledgement of the legitimacy and desirability of *A*'s influence. The public recognition of *A*'s capacity allows *A* to exercise influence despite possible disagreement or even opposition from *B*'s side. Furthermore, on the grounds of *A*'s authority and legitimacy *A* can empower *B*, if desired. Power can be passed on the will of the powerful and the acceptance of the empowered.

When translated into an educational arena, this view of power has led to view education as a powerful process where the teacher has power not only because s/he can modify the student's behavior, but mostly because s/he possesses a capacity that allows him/her to control students. Such a capacity is normally associated with teachers' knowledge. When one says that teachers can empower students, it is further assumed that the capacity that makes teachers powerful (in this case knowledge) can be transferred. Teachers transfer knowledge to students and as a result students acquire power. It is in this way that education is an empowering process. Knowledge allows students to think and therefore act in appropriate and desirable ways in the society in which they live. Students have gained power, which they can later exercise in relation to other people and other situations inside and outside the school.

In mathematics education this assumption is even stronger: mathematics teachers transfer a very special and in itself powerful knowledge. The traditional idea that mathematics education is important because it develops the brain and thinking functions of people due to its dealing with ideas and structures (see Niss, 1996) is in line with this view. Once mental structures are in place then individuals can engage in legitimate, credible actions such as describe, count, measure, control, predict, argue, communicate, etc., in order to influence their environment. All of these activities are possible thanks to the possession of mathematical knowledge, abilities, competencies, etc. Teachers, the possessors of knowledge, transfer mathematics to students who then become empowered by the acquisition of a knowledge that allows students to exercise powerful actions.

This conception of power and of power in relation to mathematics and mathematics education is problematic. First of all, saying that mathematics is powerful is equivalent to asserting that mathematics exerts power. Saying that mathematics exerts power implies that mathematics *can do* something in itself. That is, mathematics is given the status of a social actor who can perform actions. In this way mathematics is given a life of its own. Supposing that mathematics has a life of its own (independently of people) implies a reliance on Platonist philosophies of mathematics that conceive mathematics as ever existing objects. Such a view is incompatible with the social constructivist ontologies of mathematics which are at the base of socio-political approaches to mathematics education. Here we fall in a contradiction, which may easily lead to an in internalistic conception of both mathematics and mathematics education.

Second, this conception supposes that there is transference in education from the structures of mathematics to mental structures, and from the potentialities of mathematics to people's capacities. The issue of transference of power has been questioned from poststructuralist viewpoints (such as Foucault, 1972). I'll come back top this view later on. Suffice to say by now that the constitution of power in social practice is much more complex that what this view of power supposes, and therefore it is not possible to assume that empowerment (or transference of A's capacity to B) can take place in such an unproblematic way. The issue of transference in learning, particularly the transference of schemes of thinking from one situation from another, has also been criticized by situated cognition theories that emphasize the dialectical relationship between social practice and its setting, and thinking and learning (as in, for example, Lave, 1989). That is, if thinking and learning happen in the constitutive relation between a person's action, a social setting and activity, then it is not possible to assume that people can always manage to transfer thinking from one situation to another. Thus, it is not possible to assume either that the ways of thinking involved in the development of the discipline of mathematics can be transferred to children in school, since the way in which children in school develop their thinking is related to the social practices happening in school settings, and those setting and practices are different in time, space and activity from those in which the thinking of mathematicians develops. Furthermore, it is not plausible to suppose that, once school children have developed one or another way of mathematical thinking, they will in transfer that way of thinking into any other field of practice, in particular everyday life settings (Boaler, 1997).

This type of conceptualization of power in relation to mathematics and mathematics education, I have argued, does not bring us further in an understanding of the functioning of mathematical knowledge and of school mathematics education in the current Modern, Western world. Rather, it leads us to some contradictions and shortcomings.

Power As Structural Imbalance of Knowledge Control

In the work of Marilyn Frankenstein (see, for example, 1995) there is a different way of talking about power in relation to mathematics education. She says:

> So, I argue that mathematics education in general, and mathematics in particular, will become more equitable as the class structure in society becomes more equitable. Since I also contend that working-class consciousness is an important component in changing class inequities, developing that consciousness during teaching could contribute to the goal of ensuring equity in mathematics education. [...] I think that mathematical disempowerment impedes an understanding of how our society is structured with respect to class interests. (p. 165)

A first concern of Frankenstein is the existence of deep class inequalities in society that are also present in school and that permeate the way in which mathematics is taught. Students' awareness of these class inequalities is essential in a move towards a more equitable society. Mathematics education (of certain kind) can help students gaining class-consciousness since it can make visible the way in which mathematical calculations are implicated themselves in the production of those inequalities. Mathematics education empowers students to gain this awareness. A lack of mathematical capacities—mathematical disempowerment—blocks the gaining of class consciousness.

I take Frankenstein words as representative of a different way to conceive power in mathematics education research. In this perspective there are new elements associated to the meaning of power. First of all, there is a clear assumption about society—an unequal, class-divided society—which differs from the kind of global, market society to which English (2002) refers to. Frankenstein's perspective is in line with Marxist interpretations of the capitalist society. The general inequalities in society are reproduced in the ideological apparatus of the State, which include schools, and within them, mathematics classrooms. Second, there is also a definition of power rooted in the Marxist tradition. Power is the capacity of the owners of productive resources to alienate others from such resources including their own working force, and, as a result, to create a situation of oppression and dispossession for the latter. These inequalities produced through the production system and made visible in the divisions of class are structurally reproduced through practices in many other fields of social action, particularly in those fields where ideology is constituted. Schools are a particular space for that reproduction, and there power is exercised by some people at the expense of others. Although this definition, so formulated, may misrepresent the depth of its theoretical lineage, it is important to highlight that the essence of such a definition is a struggle between those who are structurally 'included' and those who are 'excluded'. This struggle represents a relation in which the powerful tend to win—although there are may be chances of resistance on the side of the excluded, or initiatives of critical people to help the excluded break their alienation and gain power. Third, mathematical empowerment is seen as the capacity that an individual gains, via the learning of mathematics, to see the way in which mathematics operates in society and contributes to perpetuate an unequal class distribution. Its opposite, mathematical disempowerment, contributes to the general alienation of people as part of the operation of the capitalist system. Empowerment, though, is not a result of an individual enlightening process but rather a social process in which the disempowered are assisted by others in order to gain consciousness.

Although the discourse on society and the structural misdistribution of access to resources is different in the first and second perspectives, the discourse around mathematical power in this Marxist perspective does not seem significantly different from the one in the Liberal perspective. The idea that mathematics gives students or people a capacity to act in the social world is similar and therefore these two perspectives may fall in the contradiction of ascribing mathematics the role of a social actor. In both perspectives mathematics empowers students. However, they differ in their view of the kind of actions that can be undertaken with the use of mathematics. While in the liberal position mathematics is seen as a positive constructive tool, in this Marxist, critical position it is seen as a tool that both can be used in constructive and in destructive ways.

Another example of this perspective is to be found in the political challenge posed by ethnomathematics to the reign of Western, white mathematics. A fundamental critique by D'Ambrosio (1993) is the uncontested imposition of mathematics as the privileged form of thinking of human beings. Because of this high, culturally given status in the Western world, mathematics 'is positioned as a promoter of a certain model of exercising power through knowledge' (p. 24, my translation). In the historical development of the Western world, which has as well impacted the transformation of the rest of other peoples, mathematics imposes the rationality of the dominant power over all other kinds of forms of thinking and expression in non-Western, indigenous, colonized cultures. Powell (2002, p. 17) also highlights that ethnomathematics departs from forms of thought that privilege 'European, male, heterosexual, racist, and capitalistic interests and values'. This essential critique to mathematics as a tool of ideological domination is incorporated in research and in the pedagogical proposals derived from it, such as in the work of Powell (2002).

One element that emerges clearly from this type of definition of power—in association with the use of Critical Theory (see, for example, Held, 1980) and Marxist approaches—is the necessity of questioning both mathematics and mathematics education practices. In the case of an ethnomathematical program it is clear that any reformulation of mathematics education as social and cultural practices should look critically at the goods and evils of the uses of mathematics within the social structures in which they emerge. In the case of Marilyn Frankenstein's criticalmathematics education, mathematics is implicated in the creation of unequal social structures by means of the way it is used in society. The 'uses of mathematics' here do not only refer to the concrete applications of mathematics in the development of technological devices—as Skovsmose (1994) emphasizes—but also the 'functionality' that people give to it in the construction of social relations and culture.

A risk in adopting this definition of power in mathematics education could be to adhere to the thesis of the *dissonance* between mathematics education, power and democracy (Skovsmose & Valero, 2001). This risk would equate with seeing no possible alternative to break the 'intellectual oppression' exercised by the imposition of mathematics and mathematics education over other possible human rationalities. The "destructive" effect of power may be emphasized to a point where it becomes impossible to think about the "constructive" effects of power.

Power as Distributed Positioning

In his plenary address to the Third Mathematics Education and Society Conference, Thomas Popkewitz (2002) presented the pillars of his analysis of mathematics education as a school subject. He says:

> The mathematics curriculum [...] is an ordering practice analogous to creating a uniform system of taxes, the development of uniform measurements, and urban planning. It is an inscription devise that makes the child legible and administrable. The mathematics curriculum embodies rules and standards of reason that order how judgments are made, conclusions drawn, rectification proposed, and the fields of existence made manageable and predictable.
>
> I consider mathematics education in this manner not only because mathematics education is one of the high priests of modernity. Mathematics education carries the salvation narrative of progress. The narratives are of the enlightened citizen who contributes to the global knowledge society. The story of progress is also told about a pluralism of the diverse people who come to school. Yet while the speech is about a universal child who is not *left behind* and

all children will learn, some children are never even brought to the table! How does that happen? What are the concrete cultural practices in the curriculum that produce the distinctions and divisions that qualify some and disqualify others? (p. 35)

In Popkewitz' word mathematics education is seen as a social practice which, together with other sets of practices, contributes to the governance of citizens. That governance is carried through the instauration of systems of reason, that is, socially constructed and accepted forms of characterizing and organizing the world, which frame what is possible, desirable and appropriate and that, therefore, constitute the basis of classification of individuals in a society. The mathematics curriculum and the teaching of mathematics are not devices and processes in charge of the transmission of a highly valued knowledge. They are social practices that, through the transformation of knowledge from one field of practice to another field of practice, helps regulating the action of students, their thinking frames and their possibilities of participation and exclusion from participation in the social world. Mathematics education operates as part of broader mechanisms which determine what is valued, what is right and what is normal in society. Mathematics education are practices through which social relations of classification and regulation are established, and through which some social actors use particular resources in particular situations to position themselves and others in those socially defined categories and norms.

I take Popkewitz' formulations as being representative of a third view of power. Popkewitz' perspective is highly inspired by Foucault's analysis of the microphysics of power in modern societies. In this view, power is a relational capacity of social actors to position themselves in different situations, through the use of various resources. This definition implies that power is not an intrinsic and permanent characteristic of social actors; power is relational and in constant transformation. This transformation does not happen directly as a consequence of open struggle and resistance, but through the participation of actors in social practices and in the construction of discourses. In this sense power is not openly overt but subtly exercised. This also means that power is both a constructive and destructive force, and that duality is always present in any social situation. When power is defined in these terms, it becomes possible to enter into a very fine grained analysis of how mathematics and mathematics education are used in particular discourses and of the effects of those discourses in people's lives.

This way of defining power has not been so popular among mathematics education researchers. However this type of definition could bring new insights in research because it finds resonance not only with the advance of postmodern ideas in education (e.g., Popkewitz & Brennan, 1998) and in mathematics education (for example, Ernest, 2004) but also with new possibilities of reinterpreting many of the theories that have been at the core of the discipline.

In the recent book *Mathematics Education within the Postmodern* (Walshaw, 2004), there is a series of articles adopting this perspective of power. Hardy (2004), for example, presents to the reader a toolkit, a series of notions coming from Foucault (1972), which have helped her seeing how in mathematics classrooms power is exercised in the relationship between students, a teacher and school mathematics activities. Though the examination of a video excerpt from a teacher training material published by the UK government as part of the National Numeracy Strategy, she presents an interpretation of the interaction between teacher and students in which the teacher's pedagogical techniques are in operation. From her perspective the teacher creates a situation of surveillance in which students' actions are exposed to the control of the teacher, who publicly approves and disapproves students' answers to calculations. Students are not only "answering" to the teacher's demands, they are being identified with an answer and are learning to identify themselves with an accepted (or rejected) behavior and thinking. The teacher's way of managing the classroom discourse plays with the double strategy of individualizing (that is, making noticeable in public an individual action) and totalizing (that it, hiding individuals within a collectivity) through her constant distinction between particular students (with proper name) and the collectivity of the class (the "we" referring to "all" in the classroom). This strategy is used in systematic ways: individualization is used to publicly correct wrong answers and to reward right answers and by this creating a clear differentiation between those who cannot and can do the mathematics; while totalization is used to give a collective legitimacy to what the teachers considers to be appropriate behavior. With this analysis Hardy illustrates that the power dynamics of a classroom go deeper than the expected mathematical empowerment assumed by the views of power presented in the two previous sections of this paper.

Meaney (2004) also uses Foucault's idea of power as embedded in social actors' relationships in order to analyze her role as a white expert consultant when working with a Maōri community, socially positioned as a disadvantaged community, in the development of a mathematics curriculum. In her analysis of the changing positions that both her and the community acquired during the inquiry process, she highlights that what comes to be considered as valid knowledge and truth is deeply dependent on the way in which the relationship among the project participants evolved. She argues that power fluctuated among participants in their differential use of strategies to argue for and give meaning to the knowledge being constructed in their relationship.

Both Hardy and Meaney, as well as other authors such as Cotton (2004) and Valero (2004a), argue that an analysis of power in these terms is not restricted to the practices of teaching and learning where school mathematics is implicated. The analysis should also extend to the way in which research is produced. Researchers, in their privileged position as active constructors of knowledge, and with it, of discourses about what is valid true, participate in the consolidation of certain systems of reason. As Popkewitz (2004, p. 259) argues, "intellectual traditions of research construct ways of thinking and ordering action, conceive of results, and intern and enclose the possibilities imagined." In this sense, researchers' discursive practices are not a neutral search for truth but an active engagement in opening/closing possibilities for phrasing and giving meaning to the social world. Therefore, this view of power opens for an examination of the way in which researchers are also implicated in the social distribution of power.

A SOCIO-POLITICAL APPROACH IN MATHEMATICS EDUCATION RESEARCH

Valero and Matos (2002) have discussed some of the dilemmas that researchers may face when exploring the broad field of mathematics education practices, as defined at the beginning of this paper. The *dilemma of the mathematical specificity* illustrates the tension between a traditional focus on the mathematical content in educational interactions and the opening of scope that makes that content one of the many other aspect at stake such as language, students' backgrounds and foregrounds for learning, cultural conflicts between the school culture and out-of-school culture, etc. The *dilemma of the scope* addresses the issue of navigating in an open field of investigation instead of researching highly specialized, well-delimited problems. The *dilemma of the scientific distance* points to the positioning of the researcher and how much criteria of objectivity and distancing between the researcher and the research participants influence the whole process of data collection. This contrasts with a view in which the researcher is viewed as an actor whose activity influences the situation to be researched, even if the researcher does not choose to play the role of an active participant. This is precisely the point being illustrated by Meaney (2004) in her discussion of power/knowledge in a research project. The *dilemma of relevance* addresses the issue of for whom and from whose perspective a given research in mathematics education is seen as being of importance. Somehow this dilemma allows to question whether the discourse of empowerment and improvement set in place by definitions of power such as the liberal one discussed above really holds for those on whom research is traditionally been carried out. Does the narrative of "salvation"—illustrated by English's (2002) formulations, and criticized by Popkewitz's (2002) analysis—hold its strength in the eyes of students, teachers, researchers and many of the other actors involved in the practices of mathematics education? Furthermore, which implications does such a discourse have in the way in which research names and defines the role of mathematics education in society and in the lives of individuals? A socio-political approach to research, that is the adoption of theories and forms of enquiry that place power in the center of mathematics education practices, invites researcher into discussions about the very same epistemological and ontological basis of the process of knowledge production.

Vithal and Valero (2003) have also pointed that an opening of scope in research to grasp the social and political complexity of mathematics education practices demands a re-examination of research questions and agendas, of theories and methodologies and of criteria for judging the quality of research. A consideration of power in mathematics education invites to question what has been taken for granted in the historical development of the field of study. Without such critical stance mathematics education research risks either preaching a modernist narrative of salvation with little impact on students' lives, or closing the possibilities to reach a more nuanced and richer understanding of mathematics education in our current societies.

ACKNOWLEDGMENT

A preliminary version of this paper was presented at 4th Dialogue on Mathematics Teaching Issues: Social and Cultural Aspects of Mathematics Education. Aristotle University of Thessaloniki–Primary Education Department. Thessaloniki, Greece, March 19–20, 2005, and was translated into Greek by Dimitris Chassapis and published in D. Chassapis (Ed.), *Proceedings of the 4th Dialogue on Mathematics Teaching Issues: Social and Cultural Aspects of Mathematics Education* (pp. 25–43). Thessaloniki (Greece): Aristotle University of Thessaloniki–Primary Education Department. The ideas here have been further developed in collaboration with Ole Ravn Christensen and Diana Stentoft and have been published in Christensen, O. R., Stentoft, D., & Valero, P. (2008). "Power distribution in the network of mathematics education practices" and "A landscape of power distribution." In K. Nolan & E. De Freitas (Eds.), *In(ter)ventions in mathematics education.* New York: Springer. Parts of this text have also been published in Valero, P. (2008). Discourses of power in mathematics education research: Concepts and possibilities for action. *PNA. Revista de investigación en didáctica de la matemática*, 2(2), 43-60. They are reproduced here with permission of PNA.

REFERENCES

Apple, M. (1995). Taking power seriously: New directions in equity in mathematics education and beyond. In W. Secada, E. Fennema & L. Adajian (Eds.), *New directions for equity in mathematics education* (pp. 329–348). Cambridge, USA: Cambridge University Press.

Boaler, J. (1997). *Experiencing school mathematics.* Buckingham: Open University Press.

Chassapis, D. (2002). Social groups in mathematics education research. An investigation into mathematics education-related research articles published from 1971 to 2000. In P. Valero & O. Skovsmose (Eds.), *Proceedings of the Third International MES Conference* (pp. 229–237). Copenhagen: Centre for Research in Learning Mathematics.

Cotton, T. (2004). What can I say and what can I do? Archaeology, narrative, assessment. In M. Walshaw (Ed.), *Mathematics education within the postmodern* (pp. 219–238). Greenwich, CT: Information Age.

D'Ambrosio, U. (1993). *Etnomatemática. Arte ou técnica de explicar e conhecer.* Sao Paulo: Ática.

D'Ambrosio, U. (1996). *Educacão matemática: Da teoria à pràtica.* Campinas, Brazil: Papirus.

English, L.D. (Ed.) (2002). Handbook of international research in mathematics education: Directions for the 21st century. Mahwah, USA: Lawrence Erlbaum.

Ernest, P. (1991). The philosophy of mathematics education. London: Falmer.

Ernest, P. (1998a). A postmodern perspective on research in mathematics education. In A. Sierpinska; J. Kilpatrick (Eds.), *Mathematics Education as a Research Domain: A Search for Identity* (p. 71–86). Dordrecht: Kluwer.

Ernest, P. (1998b). *Social constructivism as a philosophy of mathematics.* New York: SUNY Press.

Ernest, P. (2004). Postmodernity and social research in mathematics education. In P. Valero & R. Zevenbergen (Eds.), *Researching the Socio-political Dimensions of Mathematics Education: Issues of Power in Theory and Methodology* (pp. 65–84). Dordrecht: Kluwer.

Frankenstein, M. (1987). Critical mathematics education: An application of Paulo Freire's epistemology. In I. Shor (Ed.), *Freire for the classroom: A sourcebook for liberatory teaching* (pp. 180–210). New Hampshire, USA: Boyton and Cook.

Frankenstein, M. (1995). Equity in mathematics education: Class in the world outside the class. In W. Secada, E. Fennema, & L. Adajian (Eds.), *New directions for equity in mathematics education* (pp. 165–190). Cambridge, USA: Cambridge University Press.

Foucault, M. (1972). *The archaeology of knowledge.* New York: Pantheon.

Gómez, P. (2000). Investigación en educación matemática y enseñanza de las matemáticas en países en desarrollo. *Educación Matemática, 12*(1), 93–106.

Hardy, T. (2004). "There's no hiding place." Foucault's notion of normalization at work in a mathematics lesson. In M. Walshaw (Ed.), *Mathematics education within the postmodern* (pp. 103–119). Greenwich, CT: Information Age.

Held, D. (1980). *Introduction to Critical Theory. Horkheimer to Habermas.* Oxford: Polity.

Lave, J. (1988). *Cognition in practice: Mind, mathematics and culture in everyday life.* Cambridge: Cambridge University Press.

Kilpatrick, J. (1992). A history of research in mathematics education. In D. Grouws (Ed.), *Handbook of research on mathematics teaching and learning* (pp. 3–38). New York: Macmillan.

Lerman, S. (2000). The social turn in mathematics education research. In J. Boaler (Ed.), *Multiple perspectives on mathematics teaching and learning* (pp. 19–44). Westport, CT: Ablex Publishing.

Meaney, T. (2004). SO what's power got to do with it? In M. Walshaw (Ed.), *Mathematics education within the postmodern* (pp. 181–200). Greenwich, CT: Information Age.

Niss, M. (1996). Goals of mathematics teaching. In A. Bishop, K. Clements, C. Keitel, J. Kilpatrick, & C. Laborde (Eds.), *International Handbook of Mathematics Education* (pp. 11–47). Dordrecht: Kluwer.

Popkewitz, T. (2002). Whose heaven and whose redemption? The alchemy of the mathematics curriculum to save (please check one or all of the following: (a) the economy, (b) democracy, (c) the nation, (d) human rights, (d) the welfare state, (e) the individual). In P. Valero & O. Skovsmose (Eds.), *Proceedings of the Third International MES Conference, Addendum* (pp. 1–26). Copenhagen: Centre for Research in Learning Mathematics.

Popkewitz, T. (2004). School subjects, the politics of knowledge, and the projects of intellectuals in change. In P. Valero & R. Zevenbergen (Eds.), *Researching the Socio-political Dimensions of Mathematics Education: Issues of Power in Theory and Methodology* (pp. 251–268). Dordrecht: Kluwer.

Popkewitz, T. & Brennan, M. (Eds.) (1998). *Foucault's challenge. Discourse, knowledge and power in education.* New York: Teachers College.

Powell, A. (2002). Ethnomathematics and the challenges of racism in mathematics education. In P. Valero & O. Skovsmose (Eds.), *Proceedings of the Third International MES Conference* (pp. 15–28). Copenhagen: Centre for Research in Learning Mathematics.

Sierpinska, A. & Kilpatrick, J. (Eds.) (1998). *Mathematics education as a research domain: A search for identity.* Dordrecht, The Netherlands: Kluwer.

Skovsmose, O. (1994). *Towards a philosophy of critical mathematics education.* Dordrecht, Netherlands: Kluwer.

Skovsmose, O. & Valero, P. (2001). Breaking political neutrality. The critical engagement of mathematics education with democracy. In B. Atweh, H. Forgasz & B. Nebres (Eds.), *Sociocultural research on mathematics education: An international perspective* (pp. 37–55). Mahwah, NJ: Lawrence Erlbaum.

Skovsmose, O. & Valero, P. (2002). Democratic access to powerful mathematical ideas. In L.D. English (Ed.), *Handbook of international research in mathematics education: Directions for the 21st century* (pp. 383–407). Mahwah, NJ: Lawrence Erlbaum.

Valero, P. (2002). *Reform, democracy, and secondary school mathematics.* Copenhagen (Denmark), The Danish University of Education, Ph.D. dissertation.

Valero, P. (2004a). Postmodernism as an attitude of critique to dominant mathematics education research. In M. Walshaw (Ed.), *Mathematics education within the postmodern* (pp. 35–54). Greenwich, CT: Information Age.

Valero, P. (2004b). Socio-political perspectives on mathematics education. In P. Valero & R. Zevenbergen (Eds.), *Researching the Socio-political Dimensions of Mathematics Education: Issues of Power in Theory and Methodology* (pp. 5–24). Dordrecht: Kluwer.

Valero, P. & Matos, J.F. (2000). Dilemmas of Social / Political / Cultural research in Mathematics Education. In J.F. Matos & M. Santos (Eds), *Mathematics Education and Society* (pp. 394–403). Lisboa: Centro de Investigação em Educação da Faculdade de Ciências da Universidade de Lisboa.

Valero, P. & Zevenbergen, R (Eds.) (2004). Researching the socio-political dimensions of mathematics education: Issues of power in theory and methodology. Dordrecht: Kluwer.

Vithal, R., & Valero, P. (2003). Researching mathematics education in situations of social and political conflict. In A. Bishop et al. (Eds.), *Second International Handbook of Mathematics Education* (pp. 545–592). Dordrecht: Kluwer.

Walshaw, M. (Ed.) (2004). *Mathematics education within the postmodern.* Greenwich, CT: Information Age.

CHAPTER 19

FORMATTING POWER OF "MATHEMATICS IN A PACKAGE"

A Challenge for Social Theorising?

Ole Skovsmose and Keiko Yasukawa

ABSTRACT

In this chapter we, firstly, want to give an example of "mathematics in a package" to which we relate most of the rest of our considerations. The package we will open and look into is an encryption package used for the secure transfer of files and emails over the Internet. In particular, we will locate some of its mathematical content. In this way we want to clarify the question *"What is in the package"*

Secondly, we want to emphasise that this package is embedded in reality unlike many of mathematical theorem and results normally do. Although the package represents knowledge, and mathematical knowledge in particular, it does not operate simply as a cluster of theoretical knowledge. It becomes relevant to ask: *"Whose package is it?"* as it has become part of our social and economic reality.

Critical Issues in Mathematics Education, pages 255–281
Copyright © 2009 by Information Age Publishing

255

Thirdly, we want to clarify the question *"What could be done by means of the package."* In the clarification we will be more specific about the thesis of the formatting power of mathematics, which claims that mathematics operates almost everywhere, and that this operation has a social significance. The aspects of the formatting power which we will examine are: (1) By means of mathematics it is possible to provide new spaces for socio-technological action. (2) By means of mathematics it is possible to investigate details of a hypothetical situation. (3) Mathematics becomes "locked in" in reality and becomes inseparable from other aspects of society.

Fourthly, we want to discuss the question: *"What do these observations mean for social theorising?"* In particular, we will consider what our observations related to the formatting power of mathematics may have of implications for social theorising in general. An understanding of how mathematics may be operating is not only of relevance for the philosophy and the sociology of mathematics, but also for social theorising in general. In particular, notions like "trust," "risk" and "reflexivity" must include the study of mathematics based technological actions in order to interpret the social and political dynamics of the "information" of "network" society.

INTRODUCTION

People wake up to a world where their level of health can be determined for them by quantified comparators such as weight, height, blood cholesterol level, and other measures of their physiological machinery. Furthermore, they are able to "fix" some of their physiological deficiencies by taking appropriate amounts of vitamins and other drugs correlated mathematically to their particular physiological indicators. For many working people across the industrialised world, the working day is determined by some form of industrial award which specifies the number of hours, if not which hours, they are expected to work. Increasingly, the work done is translated into performance or productivity measures, which in turn can be calculated into performance based pay. The business risks that organisations decide to take are based on a range of quantitative cost-benefit analyses. Out of ones daily or fortnightly pay, a certain amount is set aside for superannuation whose rate of contribution and returns are based on complex actuarial calculations. Coming home from work, people sort through their mail, and perhaps they will find that the rates are due on their property—perhaps there has been an increase due to a higher valuation of their property. In addition to this excessive quantification, which have a direct impact on one's life and identity, there is the mathematically determined wider environment in which people operate. Thus, models of the national economy determine "acceptable" levels of unemployment, immigration quota, inflation rates, interest rates; assessment of public risks, such as pollution levels, that are deemed acceptable or otherwise through some "scientific" models. And so on.

Investigators of the social studies of mathematics have examined society's increased propensity for measuring and quantifying social phenomena, reliance on mathematically based "intellectual technologies" to replace human decision-making, and the use of "inscription devices" to abstract and categorise physical and non-physical features.[1] Such studies focus on the processes or effects of quantifying aspects of social and physical phenomena, and how this may in turn mediate social interactions. This chapter also examines manifestations of mathematics in society. In particular we shall try to condense these observations into a theses of the formatting power of mathematics. Furthermore, we shall argue that these observations are essential to any social theorising.

What is of interest to us is the sharp contrast between the impotence of mathematics claimed by the mathematician G.H. Hardy and the central role that his very area of mathematics, Number Theory, has played in the emergence of a fundamental social condition in the new electronic environment. We let Hardy represent the extreme position that mathematics can be considered as "gentle and clean" as it has no social impact, however, we contest Hardy's claim:

> If the theory of numbers could be employed for any practical and obviously honorable purpose, if it could be turned directly to the furtherance of human happiness or the relief of human suffering, as physiology or even chemistry can, then surely Gauss nor any other mathematician would have been so foolish as to decry or regret such applications. But science works for evil as well as for good (and particularly, of course, in time of war); and both Gauss and lesser mathematicians may be justified in rejoicing that there is one science at any rate and that their own, whose very remoteness from ordinary human activities should keep it gentle and clean.[2]

In contrast to Hardy's reflections, we find that mathematics operates almost everywhere, and that it operates as an integral part of the socio-technological structures of today's society. The notion of technology we are going to use is broad, including social, economic, cultural, military devises. When we want to emphasise that by technology we not only thing of its mechanical aspects but also of its manifestations in forms of operations and decision making, we shall talk about technological actions. Sometimes we will also talk about socio-technological actions in order to emphasise the broad implications of such actions. For example, we shall examine the construction of "trust" from fundamental results in Number. We shall conduct our analysis by discussing the following four questions.

First, we want to give an example of "mathematics in a package" to which we relate most of the rest of our considerations. The package we will open and look into is an encryption package used for the secure transfer of files and emails over the Internet. In particular we will locate some of its mathematical content. In this way we want to clarify the question "What is in the package?"

Secondly, we want to emphasise that this package operates in a way more "real" than clusters of mathematical theorem and results normally do. Although the package represents knowledge, and mathematical knowledge in particular, it does not operate simply as a cluster of theoretical knowledge. It becomes relevant to ask: "Whose package it it?" as it has become part of our social and economic reality.

Thirdly, we want to clarify the question "What could be done by means of the package." In the clarification we will be more specific about the thesis of the formatting power of mathematics, which condense the idea that mathematics operates almost everywhere, and that this operation has a social significance. The aspects of the formatting power which we will examine are: (1) By means of mathematics it is possible to provide new spaces for socio-technological action. (2) By means of mathematics it is possible to investigate details of a hypothetical situation. (3) Mathematics becomes "locked in" in reality and becomes inseparable from other aspects of society.

Fourthly, we want to discuss the question: "What do these observations mean to social theorising?" In particular we will consider what implications our observations related to the formatting power of mathematics may have for social theorising in general. An understanding of how mathematics may be operating is not only of relevance for the philosophy and the sociology of mathematics, but also for social theorising in general. In particular, notions like "trust," "risk" and "reflexivity," which play a role in the works of Giddens and Beck, represent aspects of the formatting power of mathematics. In developing a social theory of new social networks on *terra-silica*, one must be aware that the nature of the social interactions which can occur is founded as much on how mathematical capabilities are appropriated as it is on the patterns of interactions which people are used to on *terra-firma*. As a consequence, a part of the sociological study of the dynamics of virtual networks and of the "informational society," as coined by Castells, must include the study of mathematical theories and results as key "actors" in that socio-technical environment.

WHAT IS IN THE PACKAGE?

Encryption, the process of encoding messages to achieve confidentiality, is a technique which could most readily be associated with warfare—coding of secret strategic information about enemy movements, dissemination of false information to fool enemy intelligence, leakage of intelligence information, and so on. However, interest in cryptography, the study of encryption systems and methods, has now expanded far beyond the military sphere into the wider social spheres of commerce, social movements and personal communications. Thus, a particular encryption software system as PGP (Pretty Good Privacy), can offer new socio-technological possibilities. This in turn can disturb alter forms of social relations in potentially profound ways.

PGP is a software package that was released in 1991 by a private individual Phil Zimmerman (rather than a commercial software firm) in the USA to provide electronic mail and file storage security. PGP is one of many security packages that incorporate encryption systems. PGP supports the basic cryptographic services expected of an electronic mail security system, namely, confidentiality and authentication. Confidentiality means protection from unintended parties reading the contents of the message or file being transmitted, and authentication means the assurance of the correct identity of the message source and the integrity of the message (that is, that the message has not been changed).[3]

Some Background on Encryption

Before we start examining PGP in detail, we will make some comments about cryptography in general. Suppose we call the original message or the "plaintext" P. Then the secret message or the "ciphertext" C is produced by a transformation of P by an encryption function, say E. The relationship between P and C can be represented by:

$$C = E(P)$$

The encryption function should have the property that

$$P_1 \neq P_2 \Rightarrow E(P_1) \neq E(P_2),$$

that is, different messages are encrypted into different ciphertexts so that ambiguity does not occur. "Decryption" is the process of recovering the plaintext from the ciphertext through some function D:

$$D(C) = P$$

The functions E and D should have the property that

$$D \circ E(P) = P. \quad [4]$$

Thus, if the plaintext *P* is encrypted and the result decrypted, the result should be the plaintext P itself. Stated in another way, the function *D* is the inverse function of *E*: the decryption function reverses the action of the encryption function.

To give an example, Ludwig Wittgenstein was known to use a simple cryptographic tool when he wrote his private remarks: the first letter of the alphabet *a* was substituted by the last letter *z*, the second *b*, by the second to the last, *y*, and so on. In this way the message: *P*: "this is a secret" will be encrypted as *C* = *E*(*P*): "eqpf pf z ftvgte." In this particle example we have the *D* being identical to *E*.

Modern cryptography refers to the studies of encryption and decryption functions that take the form of mathematically based computer algorithms. In talking about modern encryption, we are assuming a computerized process where the original message in natural language has already been converted into a suitable machine-readable representation. In fact, we can think of the machine-readable representation as a number, i.e., we can think of P as a number.

Encryption and decryption algorithms, except possibly those used in military and intelligence work, are public algorithms. One can purchase (or in the case of PGP freely download) these encryption packages. Although it may appear that making algorithms public would make encrypted communication less secure, the consensus has been on the contrary.[5] Because the algorithms themselves are public, confidentiality is established by the secrecy of a "key." The key is necessary for the execution of an encryption algorithm.

Modern cryptography can be classified into two distinct approaches: (1) conventional, single-key or symmetric cryptography, and (2) public-key or asymmetric cryptography. Symbolically, these processes of encryption and decryption can be represented by the equations:

$$C = E_{K}1(P)$$

$$P = D_{K}2(C)$$

where *EK1* is the encryption algorithm using the key *K1*, and *DK2* is the decryption algorithm which uses key *K2*. For conventional encryption the two keys *K1* and *K2* are the same, while for public-key encryption they are different.

The big step forward in cryptography was to avoid the need to use the same, or closely related, keys for encryption and decryption. Using the same key for encryption and decryption requires a system of key distribution. Somehow the key which has to be kept secret between the two communicating bodies must be safely distributed from whoever generates the key to the other. Any capture of this key by a third party compromises the confidentiality of the communication. The public-key system, on the other hand, relies on an "asymmetric," or two key system. In this system, each party has a "public key" which is known and shared by both the sender and the receiver (and which could be known to the wider public without compromising the security system), and a unique "private key." Hence, a message is encrypted by a sender A using the public key $K1$ of the receiver B; this message, however, can only be decrypted by the intended receiver B using their private key $K2$, which even the sender does not know. The strength of the public-key schemes is based on the difficulty of determining the private key $K2$ from the knowledge of the public key $K1$ and the ciphertext C alone.

PGP uses a combination of conventional encryption techniques and public key encryption techniques to achieve confidentiality of e-mail transactions. It is the public-key encryption schemes used in PGP that we will be examining in some detail.

Opening the Package

At the surface where PGP is implemented and used, PGP's mathematical artefacts are invisible. People implementing PGP would typically download from the Internet or purchase from a vendor, a whole package within which the various public-key and symmetric algorithms reside. Whilst in some cases, a version of PGP may offer options in features such as key sizes, the user would not look beyond the technical specifications of the product to determine whether or not it is the appropriate, machine compatible product to buy. Nor would the user need to be conversant with the mathematical theories behind cryptographic algorithms. The package would have a user-friendly interface which guides the users through the process of getting encryption-enabled; encrypting a message; decrypting a message; and digitally signing a message.[6] For the users, the power of the product is entrusted not in what they understand of the mathematics upon which the algorithms are constructed, but in the vendors of the package with whom they deal, and or the reputation that the whole package has established among relevant professional communities.[7]

The PGP package is made up of a collection of modular entities; for example, encryption algorithms such as IDEA (International Data Encryption Algorithm) and RSA (Rivest-Shamir-Adleman[8]) have been separately developed, tested and distributed, and have been appropriated as building blocks of PGP. The design of PGP ensures that these algorithms are "seamlessly" linked to serve the overall functions of PGP. When we ask "what mathematics makes up the package?," we are effectively asking what mathematics underpins each of the different algorithms within the package. We will focus on the mathematics which underpin the RSA public-key algorithm used in PGP, i.e., the algorithm by means of which public and private keys are generated. Effective encryption systems requires two key features. It must be easy to implement, but difficult to compromise. For public-key systems, implementation involves the generation of the public and private keys. Compromising the system involves among other things determining the private key from knowledge of the public key and possibly some ciphertexts.[9]

The class of mathematical functions upon which the designers of public-key systems have employed to achieve the requirements of the systems is what is known as "trap-door one-way functions." These are functions where the function values are easy to compute, but where, given the function and a function value, it is difficult to compute the value(s) which the function acted upon, without additional information. An example of this is a function that takes two prime numbers and computes their product; multiplying two prime numbers is simple, but determining the prime factors of an arbitrary number, especially a large number, is difficult. This prime-factorization problem is in fact the mathematical basis of the RSA public-key algorithm.

The main role in this public-key algorithm is played by two prime number p and q. As any two numbers, p and q can be multiplied, and we get $n = pq$. However, if p and q are large enough primes, it will be very difficult to determine p and q from n, even knowing that n is a product of two primes. The degree of this difficulty represents the degree of security in the PGP system. So, for the constructor of a system, it is essential to get hold of two large prime numbers. For the hacker ability to factor a very large number becomes essential.

Of particular relevance in the design of a public-key encryption system is a result dating back to the days of Euclid. There exist infinitely many prime numbers. A proof of this was constructed by Euclid, and the statement of this result is known as Euclid's Theorem. The implication of this theorem on the analysis of encryption schemes is two fold: a hacker trying to determine the private key would have to go through the process of searching for a pair of prime numbers over an infinitely large set, while those generating the key pairs have the benefit of an infinite set from which to choose a suitable pair of prime numbers. From both perspectives, number theoretical issues becomes relevant. For instance, how difficult is it to get hold of large prime numbers? The number of primes between 1 and 20 is eight, while the number of primes between 101 an 120 is five. Does the density of primes decrease dramatically, when we search among large numbers? The answer given by the Prime Number Theorem is: "yes, but not too much."[10] For the hacker, this is a discouraging result. Their search for possible prime pairs does not become easier as the search set increases in size; for those generating the keys, they can be assured that they have a fair chance of striking a suitable prime number no matter how large a prime number they wish to have.

While the search for prime factors of a given number x may be laborious, the Fundamental Theorem of Arithmetic states that any natural number can be factored into prime numbers in one and only one way (if we do not consider the order of the factors). This effectively means that once a hacker finds a set of prime factors whose product is the original compound number x, they need not search further because these are *the* prime factors. The Fundamental Theorem of Arithmetic also ensures that, when properly chosen, the properties $P_1 \neq P_2 \Rightarrow E(P_1) \neq E(P_2)$ and $D \circ E(P) = P$ hold.

In order to construct a system of encryption, two large primes p and q have to be identified. The product $n = pq$ is calculated. Using classical results from Number Theory and given the number n it is possible to de-termined two other numbers, e and d. The public key $K1$ is then the pair (e, n) and the private key $K2$ is (d, n). The process of both encryption end decryption becomes simple mathematical procedures, just involving simple arithmetic calculations. Thus, encryption is achieved by :

$$C = P_e \bmod n$$

and decryption by:

$$P = C_d \bmod n$$

Thus encryption means raising the plaintext (the number) P to the pow-er of e, and dividing the result by n.[11] The ciphertext C will then be the remainder. Decryption will mean raising the the ciphertext C to the power of d and dividing the result with n, the remainder will be the plaintext P.

The product of the two primes is no secret, as n is part of the public key (e, n). The calculation of the number d, however, will break the code. The calculation of d is in fact also simple, if we know the two primes which multiplied gives the result n. So, breaking the codes ends up in the task of making a prime number factorisation. Constructing a code, on the other hand means identifying two fairly large prime numbers.

Although factoring is a simple process to describe mathematically, it is a time consuming task when we are working with large numbers. "At present, a 200-digit number that is the product of two 100-digit numbers cannot be factorised in any reasonable time ... In fact, not so long ago, the most ef-ficient factoring algorithms on a very fast computer were estimated to take 40 trillion years, or 2000 times the present age of the universe."[12]

This result appears amazing. Thus, it takes less that three lines (using the print on the present text) to write down a 200-digit number. And this three-line number will provoke a task beyond the shared effort of all computers of the world, drawing as well upon all present mathematical knowledge.

A New Significance of Mathematical Formulae?

If we study classics in Number Theory, like *Elementary Number Theory* by Hardy and Wright, then it is clear that no kind of applications to anything external to pure mathematics are considered.[13] The study stays well within the walls guarding the purity of mathematics from the contamination of the real world. However, the significance of a mathematical theorem is relative to its contexts. When a theorem is presented in relation to the derivation of another theory or in a textbook, it might appear "clean and gentle" and insignificant as far as its social impact is concerned; but when it appears in an application package, such as PGP, its significance may be completely different. In fact, this change of significance is an essential new departure point for the philosophy of mathematics. In this section we will consider how a number of what may be regarded as classical or fundamental results in Number Theory, namely, Euclid's Theorem, the Prime Number Theorem, Euler's Theorem, and Fermat's Little Theorem have come to acquire significance well beyond the walls of pure mathematics.

The effectiveness of the RSA algorithm described earlier, and in particular, its key generation algorithm, rests heavily on the ability to find suitable pairs of primes p and q. To this end, one of what Hardy calls a "real" mathematical theorem is central, Euclid's Theorem. Hardy states: "...Euclid's theorem is vital for the whole structure of arithmetic.[14] The primes are the raw material out of which we have to build arithmetic, and Euclid's theorem assures us that we have plenty of material for the task."[15] However, Hardy assumes that as important as this theorem is in building the theories of arithmetic, it has little practical relevance: "There is no doubt at all, then of the "seriousness" of either [Euclid's or Pythagoras's] theorem. It is therefore the better worth remarking that neither theorem has the slightest 'practical' importance. In practical applications we are concerned only with comparatively small numbers...I do not know what is the highest degree of accuracy which is ever useful to an engineer—we shall be very generous if we say ten significant figures.... The number of primes less than 1,000,000,000,000 is 50,847,478: that is enough for an engineer, and he can be perfectly happy without the rest."[16] Hardy not only predicts incorrectly the uselessness of Euclid's Theorem, but also the critical importance on the size of numbers for some engineering applications. It is recommended that the size of the prime numbers for RSA encryption is in the order of 75–150 digits.[17]

However, it is not enough that Euclid's Theorem tells us that prime numbers are abundant. In RSA (and other similar algorithms), it is critical, not only that these numbers can actually be found in principle, but that they can be found efficiently. Because there is no known way of analytically generating prime numbers, a search algorithm, one being the Miller-

Rabin algorithm which involves testing the primarity (whether a number is prime or not) of a randomly selected number, is used.[18] The search is a probabilistic one, which identifies whether or not a random number has a high probability of being prime. The search for prime numbers can be compared with the search for gold and diamonds.

In order to have any confidence that a computer search algorithm has any practical potential, however, we need also to have some assurance that prime numbers occur "often enough" in the random search of the infinite set of integers. As mentioned, the Prime Number Theorem provides us with the assurance about the frequency of occurrence of primes that is needed. This theorem is another of what Hardy calls a "deeper" theorem of mathematics: "When we ask these questions, we find ourselves in quite a different position [to the shallower inquiry which can be handled with Euclid's Theorem alone]. We can answer them, with rather surprising accuracy, but only by boring much deeper, leaving the integers above us for a while, and using the most powerful weapons of the modern theory of functions. Thus the theorem which answers our questions (the so-called Prime Number Theorem) is a much deeper theorem than Euclid's or even Pythagoras's."[19] But the theorem does not simply provide us with a pure knowledge of Natural Numbers. It acquires significance not observed by Hardy.

Another theorem which takes on a new significance in the light of the implementation of the algorithm for searching for prime numbers is Fermat's Little Theorem.[20] Propositions derived from Fermat's Little Theorem give an upper bound on the probability of a number being composite; at this point in time, this is the best sort of test that is practicable.[21] Another Number Theoretic result known as Euler's Theorem forms the basis for producing the desired encryption-decryption relationship to make the algorithm achieve its cryptographic purpose.[22]

As resources for implementing a software package, "old" pieces of mathematical results are brought into play, not only to provide the basis of a central algorithm as Euclid's Theorem does for RSA. They also provide a resource for deriving efficient ways of implementing the algorithm and as an insurance that certain checks are adequate.[23] Thus, the mathematical theorems such as those mentioned above become the basis of trust that people implicitly invest in a package such as PGP to provide secure communication.

In particular the thrust is based on the difficulty of making prime factorisation of a big number. This is a mathematical observation, as mathematical research not yet has been able to provide an efficient algorithm. But it is not proved impossible to identify much more efficient hacker-friendly algorithms. In this sense the degree of security depends on the present state of mathematical research, and furthermore on how this research is conducted in terms of public dissemination of relevant results.

Although we have indicated the number of classical mathematical results that have acquired new significance in cryptography, we also need to note that mathematical results in new areas of mathematics have also been incorporated into cryptography. To give one example, bounds on how hard the problem of a brute force attack by prime factorisation is in terms of computational effort, are derived from results in a relatively new field of mathematics, complexity theory. This theory provides us with the rate of growth in computational effort when key sizes are increased, i.e., the number of decimals of the prime numbers in question. Thus not only are classical mathematical results being applied in new areas of practical significance, they are also being challenged for their "truth" value, in a practical sense.

WHOSE PACKAGE IS IT?

We have now got an impression of what is in the package. Let us now look at the package as a whole entity. By discussing the question "Whose package is it?" we will try to illustrate in what sense mathematics "materialises." Questions addressing the package is quite different from questions we can address about bits and pieces of mathematical knowledge. The classical perspective of the philosophy of science and of the philosophy of mathematics becomes inadequate. The package is a new entity.

If those pure mathematicians who generate the resources for cryptography stay at an arm's length, or further, from the applications of their results, who are the social actors who package these results into packages such as PGP? How do these people interact with each other? Do their interactions represent a new pattern of knowledge production, or are they similar to those in the traditional paradigm of mathematical research?

As already mentioned, PGP was written by Phil Zimmerman, who was fascinated by cryptography, and found a career niche in the area in college when he started to seriously research the literature and publish in the area. [24] As will become clearer, the birth of PGP is attributable to a combination of his fascination with cryptography and his ideological conflict with the U.S. government. Specifically, PGP is claimed to have been "proposed as counter-terrorism legislation which would have placed limits on the use of encryption technology by U.S. citizens." [25] The intention was that PGP would become available before this law came into effect, even though the Bill eventually failed to be passed into law.

Being "needs driven," especially where the need is related to some practical political and legal developments, the creation of PGP can hardly be compared to the derivation of the sorts of "useless" results and theorems in Number Theory. PGP is an applications package, developed with the full intention of use. Zimmerman released the package on the Internet, making it freely available to the public. This immediately constituted a breach of the US Arms Export Control Act, and constituted illegal export. In particular, PGP, being an encryption software, was one of the items listed in what is called the US State Department's Munitions list, a list of items that cannot be exported with a special license, and which includes other items such as machine guns, missiles, and bombs.[26]

Zimmerman was then further challenged by RSA Data Security, the company holding the patent for the RSA algorithm, for theft and infringement of their patent, because according to them, PGP resembled a proprietary software written by Rivest of RSA Data Security, and marketed in the mid 1980s.[27]

By releasing the technology on the Internet, Zimmerman gave the "average" U.S. citizen a powerful tool for protecting their privacy from government authorities. According to Zimmerman:

> I wrote PGP from information in the open literature, putting it into a convenient package that everyone can use in a desktop or palmtop computer. Then I gave it away for free, for the good of democracy.... This technology belongs to everybody.... Today, human rights organizations are using PGP to protect their people overseas. Amnesty International uses it. The human rights group in the American Association for the Advancement of Science uses it. It is used to protect witnesses who report human rights abuses in the Balkans, in Burma, in Guatemala, in Tibet.[28]

Certainly, mathematics is in operation.

The free distribution of the package poses a direct threat to commercial and more highly guarded systems, while providing ready access to anyone wanting to acquire and implement a secure e-mail system. The free access has in fact posed direct threats to RSA Data Security, Inc who holds the patent for RSA algorithm in the USA (and which expired in the year 2000), and Zimmerman was accused by Data Security Inc of breaching the law by exporting PG. Free export of PGP also posed a direct challenge to the USA Government, and the NSA (National Security Agency), who have closely guarded the design of the DES algorithm used in PGP by legally limiting the key size to 56 bits, while the exported PGP enabled 128 bit keys to be used; this, in terms of security for the user is more advantageous, but for Governments who may want to intercept and access transmissions between 'suspect' entities is entirely inconvenient.

Mathematical knowledge, as expressed in theorems, conjectures and the like are made public through scholarly publications, and then diffused and distilled in textbooks, university lectures and academic meetings. So how can we understand the legal disputes about license and export regulations which surrounded the initial release of PGP? In PGP, as in other encryption and mathematically based software packages, we see a fusion of mathematical results, which together achieve certain functionalities as a package. The Fundamental Theorem of Arithmetic alone does not enable powerful encryption; however, its application to cryptography, together with a number of other mathematical results and reasoning, produces a powerful algorithm such as RSA. Being a complex algorithm, with specific functionalities, it acquires commercial, as well as political value.

It has been the assumption in every (classical) philosophy of science that knowledge is public. For instance, the whole notion of the "Third World," as discussed by Karl Popper, expresses the idea that knowledge constitutes an entity which cannot be privatised. The entities of Popper's Third World are not put into any package. They are claimed to be free flowing entities. But knowledge-in-a-package is different from knowledge-in-a-free-flow. It can be patented, it can be exported, it can also be made freely available. Knowledge-in-a-package operates completely differently from knowledge-in-a-free-flow.

The question "Whose mathematics is it?" does not really make sense. The answer appear rather simple. Mathematics is public. However, the question "Whose package is it?" is a tricky question. A "wrong" answer to this question could bring a person to court. Ownership can be maintained, and Robin Hood-like actions can be carried out with packages of knowledge. This shows that we have to think differently about an entity which has left the realm of Popper's Third World and entered the real world.

WHAT IS DONE BY MEANS OF THE PACKAGE?

Mathematics can "materialise" out of its abstract entity of symbols and theorems and emerge as part of a functional entity that drives a computer package. This package provides a specific meaning to the more general claim, that knowledge and information become intellectual technology.[29] The package can be installed and implemented, and its implementation "makes a difference" on a number of fronts—social, political and technological. A package underpinned by mathematical functions in sharp contrast to any "scientific theory." It materialises into both the physical world and people's economic and social reality.

What the examples of mathematisation cited in the Introduction and other numerous examples do not fully reveal are the mechanisms by which mathematics penetrates, pervades, and constitutes reasoning in our sociotechnological sphere.[30] What interests us here is not only that mathematics is being used to model social systems, but that mathematics is being used to create new social realities. We could talk about a packet-built reality. Thus, mathematics is no longer only attempting to provide a more or less adequate picture of reality from which certain conclusions can be drawn (for example, a map of a city). The utility of mathematics in providing a picture model of physical phenomena was what gave mathematics the "marvel" which Eugene Wigner wrote about in his famous essay "The Unreasonable Effectiveness of Mathematics in the Natural Sciences."[31] But mathematics can do more than describe what is already there. Mathematics can also create and become part of reality itself, such as the socio-economic reality created and built further upon by mathematical models of an economic system. In this sense, we contest the limited notion that "mathematics and science help us to analyze existing ideas and their embodiment in 'things', but [that] these analytical tools do not in themselves give us those ideas."[32] We argue here that mathematics operates in social and technological contexts and in doing so mathematics not only represents but also constitutes reality. Mathematics is not only operating in Popper's Third World, say, as a servant for other of the inhabitants of this world as Natural Science for instance.

In this section we will concentrate more generally on how mathematics generates and constrains socio-technological actions.[33] We will see how mathematics, in particular, influences the way we construct our social realities and operates within the space of possible actions. It would be absurd to say that the space for possible socio-technological actions is constituted by mathematics alone, but we argue that mathematics plays a role in the construction of such spaces and also in the way we investigate and choose between alternatives. We will call the alternatives within a space of technological actions, hypothetical situations.

We shall try to clarify the question "What is done by means of the package" by consideing the following three ideas representing the thesis of the formatting power of mathematics: (1) By means of mathematics new technological alternatives are presented—alternatives that are not possible to grasp and identify without mathematics as a tool for analysis and construction. However, mathematics also limits the set of hypothetical situations that are presented, as mathematical construction is only one way of expressing a sociological imagination. (2) By means of mathematics we can investigate particular details of situations not yet realised. A particular strength of mathematics is to enable hypothetical reasoning which refers to reasoning about technological details of a not yet realized technological construction. However, mathematics may also produce blind spots concerning the effects of such a not yet realized construction. Those effects cannot be foreseen and they emerge only when the technology has been implemented. (3) When choices are made and the technological ideas are translated into new technological and, hence, social realities, mathematics simultaneously "enters" this reality in a concrete form. Mathematics assumes an essential functional role within technological packages, and once that happens, the mathematical influences on this reality become inseparable from the other social realities in which it is acting. Mathematics will have become "socialised."

We suggest that the three aspects are generally applicable to the role of mathematics as resources for socio-technological actions. However, we will discuss the aspects, first of all, with reference to the example of cryptography and PGP.

By Means of Mathematics it Is Possible to Provide New (Maybe Limited) Space for Technological Actions

In the past a combination of logical operations and substitutions seemed adequate tools for encryption. However, large scale industrialised cryptography is not easy to handle with any classical approach. It would not be possible to imagine the development of public-key cryptography without the development in its mathematical basis. Informed by Number Theory, a completely new approach to encryption became identified. Naturally, it is not that Euclid's Theorem, the Prime Number Theorem, Euler's Theorem and Fermat's Little Theorem to get with observations about trap-door one-way functions provide a new possibility. But social and economic interest and mathematical resources creates new spaces for technological actions. In this way we can consider mathematics as an essential resource for a technological imagination, by means of which we grasp not yet realised technological alternatives

The technology of public-key encryption is by no means a product of "natural evolution" or accident. It is a product of a deliberate effort to improve on traditional cryptography that became increasingly inadequate as the sole approach to providing confidentiality and authentication for new cryptographic needs in the so-called information society. These two trends: the mathematical developments relevant to cryptography and the emergence of social needs, were catalysts in the conceptualisation of the possibilities for new alternatives in encryption.

That mathematics can open new spaces for technological actions, we see as a general observation related to mathematics. This is not limited to cryptography. Mathematics opens possibilities for new approaches in large scale economic management by providing possibilities for experimenting with different economic policies by means of simulations models. Simulations models can in fact be used in any form of technological constructions, like design of bridges, airplanes, cars, etc. All such simulation models help to locate a space for technological actions within which to operate.

However, as mathematics opens a new space of technological actions, mathematics can also create blind spots for other openings and for different approaches to resolving threats to privacy. Mathematically based encryption only allows certain questions to be asked about the problem perceived in the present situation. If the solution space is defined within the bounds of mathematics, then fundamental changes in social relationships, for example, which are equally valid ways of addressing the perceived threat to privacy, may not be considered. The choices that are considered become limited to choices in encryption schemes and particular realizations of the schemes.

In viewing a threat to privacy as a problem of current methods of data communication, rather than a fundamental social problem, a "fix" of the technological process of communication by technological means becomes a "natural" alternative. Once the root of the problem is identified as a technological one, then this precludes a wide range of questions and possibilities by means of which the perceived problem is investigated. Thus, the notion of trust and privacy may include many other aspects than those that can be tackled by proper cryptographic procedures. But such aspects may be downplayed when the space of possible solutions are created by mathematical tools

As mathematics opens new spaces for technological action, mathematics may also create limitations, as the identified technological possibilities may come to represent the only space for technological action. What mathematics does provide is, thus, both a new area for technological development as well as a trap, which encapsulates technological imagination, and separate it from other non-mathematically based forms of sociological imagination. This is a general observations related to all situations where mathematics establishes resources for opening spaces for technological actions.

By Means of Mathematics it Is Possible to Investigate (Maybe Rather Particular) Details of a Hypothetical Situation

By a combination of mathematical tools various investigations of particular details of encryption can be carried out. Complexity theory allows some approximate bounds to be made about how easily a hacker can succeed in determining the private keys of a public-key encryption scheme. Number theoretic tools such as "number sieves" provide cryptanalytic tools to assess the security of schemes such as RSA, and facilitate investigations of new sources of methods.

The point is that all such investigations can be carried out on the basis of hypothetical situations. We need not construct any actual system in order to investigate details of hypothetical situations. In fact, it is a strength of mathematically based technological imagination that it can be carried out on the basis of hypothetical situations. Mathematics helps designers of encryption algorithms to envisage hypothetical scenarios of hacking, or computer memory requirements, and so on, which are essential to meeting the functional requirements of the package. Mathematics helps us to predict the effectiveness of the algorithm design: for example, how many bits in the ciphertext would be affected with a change in one bit of the bit sequence of the key; or how many iterations are needed to realise a certain level of "confusion" and "diffusion."[34]

However, mathematics only allows us to investigate particular details. So, when an encryption system is implemented, there might occur malfunctions, which simply have not been considered. Mathematically based hypothetical reasoning has limitations. It contains blind spots, and from these blind spot risk structures may emerge, such as the spread of a bug in the system in a potentially far-reaching way because of the Internet environment in which systems such as PGP reside. Another risk is the effect of the technologically established "web of trust" being compromised in some way.

Although there are mathematical techniques of risk assessment, analysis and management, these tools are very limited in their use in the sorts of risk scenarios described above. Mathematical concepts of risks are expressed in terms of probability of a risk event, and the "cost" determined in some way of such an event. In situations where the probability is very very low (a very rare event), there is not the basis upon which to calculate the corresponding risks in any meaningful way. Hence in terms of encryption systems such as PGP, much "trust" is invested in the mathematical rigour of the system, but mathematics is unable to resolve the risks associated with the technology going wrong.

This also bring us to a general observation. A technological imagination, supported by mathematics, can address hypothetical situations, and can in fact investigate particular details of such hypothetical situations. This is the general idea of providing any simulation model. Then it is not necessary to realise, say, the bridge construction for carrying out a detailed investigation of how it might operate in stormy weather with a maximum load of cars. And certainly such hypothetical reasoning may include blind spots. This is exactly behind the blind spots of mathematical based hypothetical reasoning addressing events that are related to not yet realised technological constructions, that the risk of the risk society emerge.[35] The carefully estimated almost not existent risk related to the operations of atomic power plans illustrates both that decisions making is related to mathematical calculations, and also that mathematical based hypothetical reasoning contains a blind sport.

Mathematics Becomes "Locked-in" in "reality" and Becomes Inseparable from Other Aspects of Society

When installed, the PGP package becomes part of a larger technological construction, and it operates within this construction. This mechanical expression of the package is of interest to us. What remains to be seen of the mathematics when the cryptographic elements are built into a package? The interfaces with the user certainly reveal no traces of, for instance, Euler's Theorem, Euclid's algorithm, or any of the fundamental results of mathematics that underpin the logic of the encryption algorithm that is used. Many users would see PGP as a package within a bigger bundle, say an e-mail service, such as the commonly used Eudora software. In keeping with the nature of these higher level applications, the mode of interactions would be "user friendly," with a pulldown menu which guides the user through the services which they might require, rather than through the underground corridors of the algorithms from where the essential services are being delivered.

Without the mathematics, there is no package.[36] Without the package there is no technological construction. As new rooms can be added to a house by using bricks and mortar, extensions can be added to the network society by using packages as bricks. The technological construction, however, is much more that the "machinery." As mentioned in the Introduction, we see technology as inccluding also organisationan matters and proceedures for decision making. It also includes the construction of security and of trust.

Trust is established and secured by a mathematically based technology. For those arenas where encryption has become the norm, trust that has been a fundamentally social attribute has been transformed into a technologically defined attribute. Within these arenas, the social realities, therefore, are intrinsically determined by the technological package that characterises the nature of the relationships within them. Trust, security, privacy and several other social phenomena become reshaped in terms of spaces for technological actions made available by means of mathematics.

The inventor of the Diffie-Hellman key exchange (an optional key exchange algorithm within PGP) claims: "As human society changes from one dominated by physical contact to one dominated by digital communication, we will have many opportunities to choose between preserving the older forms of social interaction and asking ourselves what those forms were intended to achieve. . . . The area in which technology can most clearly make a positive contribution to privacy is encryption. If we assert the individual's right to private conversation and take measures in the construction of our communications systems to protect that right, we may remove the danger that surveillance will grow to unprecedented proportions and become an oppressive mechanism of social control. Fortunately, the fight for cryptographic freedom, unlike the fight against credit databases, is a fight in which privacy and commerce are on the same side."[37]

That mathematics materialise and "socialise" is not just a phenomenon related to packages of cryptography. Mathematics has materialised in many systems for decision making. As indicated in the Introduction such decision can concern medical actions. It can become part of the design for working conditions. It is an integrated tool for the identification and measurement of risks. The estimations of "acceptable" levels of unemployment have reached a new sophistication be means of simulation models. Thus, "acceptable" no longer simply refers to moral standards addressing how big a part of the adult population which could be left at the margin. Now acceptable also includes estimations related to the gross national product, to the level of salary, which is determined also by the degree of employment. Thus the notion of "acceptability" become re-interpreted in terms of detailed cost-benefit considerations by the help of mathematics.

WHAT DO THESE OBSERVATIONS MEAN TO SOCIAL THEORISING?

Mathematics represents a powerful resource for socio-technological action. The space of technological actions is involved in rapid and unpredictable developments. New technological options are generated, but the implications for realising these options are only accessible by hypothetical reasoning. Technological developments are supported by mathematics, because mathematics often helps to establish hypothetical situations and analyse particular aspects of (some of) these situations. Eventually, mathematics becomes part of the social reality in which the technological actions are finally carried out.

We have illustrated this idea with reference to the PGP package. But we find that these three aspects of the formatting power of mathematics characterise general aspects of mathematics in action. We will now try to point out a few implications of this attempt to clarify the notion of the formatting power of mathematics. In particular, we will emphasise implications for social theorising. We find that it is essential to social theorising to pay attention to how mathematics is operating in social affairs. Any social theorising which includes notions like "trust" and "risk" are in need of a clarification of how such phenomenon is constituted also by means of mathematics. More generally, we find that overall sociological notions like "reflexivity," which refers to the feedback process which, according to Beck, provokes the risk of the risk society, needs an understanding of the formatting power of mathematics in order to obtain an adequate interpretation. An sociological interpretation of the "informational society," to use a term suggested by Castells are in need of an understanding of how mathematics operates in packages, as mathematics stuffed packages are important bricks in the network of the network society. Naturally, mathematics does not in itself provide resources to establish new social forms and relations, but it represents society's overall disposition for socio-technological action. It can produce new spaces for socio-technological actions and help to specify aspects of this space. As a consequence we find that any theorising which addresses general features of social development, and which includes consideration expressed in terms of "trust," "risk," "reflexivity" and "informational society," must pay attention to the propensities for social development established by mathematics.

The unpredictability of the development of the space for technological actions is one of the forces in the development of "trust" and "risk.". In the encryption example the unpredictability of the degree of "trust" and "risk" is closely related to the unpredictability of the results of mathematical research, for instance related to algorithm theory. The unpredictability is linked as well to the medium in which the package operates and is exchanged. The medium, that is the Internet, has a global reach, and so the Internet effectively plays a role in "selling" a new ideology. The risks that are perpetrated by the same medium are also cause for concern, and constitute another example of the sorts of incalculable risks that pervade our risk society. A fundamental problem faced by any encryption package is that when a hacker breaks the code, they will typically not let anyone else know. People using a broken encryption scheme may continue to communicate with each other, believing that the scheme is secure. Another risk, is the continued use of compromised encryption keys in public-key systems. The risk perceived by governments is that "unlawful" or politically subversive actions may be taking place without their knowledge. Any sociological understanding of the nature of the distribution of trust and risk can draw upon an awareness of their mathematical constituents.

The formatting power of mathematics is elusive, unlike economic interests and political agendas. The formatting power is invisible. But being invisible does not mean being "not-real." The powers of mathematics are able to interact with other "powers." We find that the formatting power of mathematics is real, both in a physical sense (as for example, models of magnetic and gravitational forces), and in a sociological sense (as in models of economic and political forces), precisely because mathematics can interact with both kinds of power. And one of the implications of this interaction is the emergence of new risk structures. This provides an opening not only for an understanding of the nature of the risk society, but also for a further interpretation of the remark of Beck's that "reflexivity" takes place outside the control of the democratic institutions of society (as well as outside the attention of sociology).[38] Refexivity refers to fundamental but unnoticed feedback processes where the results and consequences of certain technological actions, returns to out reliaty, however in an unexpected form. Certainly not in a form which was conceptualised as part of the planning process. We can think of forms of pollution as an example of processes of reflexivity. The creation of new risk-structures is a more general expression of this observation.

If we want to obtain an understanding of social actions including the possible implications, it becomes important to consider how the space of technological actions functions in the process of social development. Social and political decision making is related to the creation and explorations of such spaces. In particular, it becomes important to consider the role mathematics is playing as a resource for both the construction and the exploration of particular details of hypothetical situations. Sociology cannot pursue its goal of providing a basic understanding of social agency without paying a special attention to the formatting power of mathematics. How can sociology interpret a society which resides and unfolds in terms of the "informational society," without a sound appreciation of how the space of technological possibilities has come to unfold?

These observation can also be expressed in terms of suggestions for the philosophy of mathematics. This philosophy has concentrated on understanding the nature of mathematical knowledge, which has led to analyses of abstraction and formalisation.[39] Thus, analyses, as for instance carried out by Imre Lakatos, has concentrated on investigating particular mathematical ideas as part of a process of theoretical development. Lakatos fully accepted the idea that mathematical knowledge is part of Popper's Third World, and he tried to describe the rational development of this free-flowing knowledge in terms of proofs and refutations. However, observations related to the formatting power of mathematics provokes a different emphasis in the philosophy of mathematics: How can it be that we are staying in a world in which so much mathematics in fact is operating? How do we cope with the living conditions of staying in this technological constructed environment? The new significance of mathematical formulae, which we have mentioned, becomes an essential departure point for any philosophy of mathematics which does not want to consider the constructions of mathematics form within only.

The significance of mathematics changes dramatically when the mathematical ideas are investigated as part of a package. This represents a unit of particular importance for understanding the social role of mathematics. The questions: "What is in the package?," "Whose package is it?" "What technical effects does the package have?" represent new questions to be discussed in a philosophy of mathematics. It is a new task for the philosophy of mathematics to engage with mathematics-in-a-package in order to identify how spaces for technological action are constructed, and how a formatting power of mathematics might be exercised.

We have emphasised that by means of mathematics it is possible to provide new spaces for technological actions. Let us finally pay attention to the notion of "action." Mathematics is a resources for actions of a grand variety. It is not the case that mathematics provides these actions with any particular quality. These actions must therefore be addressed by critique and by ethical considerations as any other kinds of action. In this way the thesis of the formatting power of mathematics shows the necessity of opening the considerations about mathematics in two ways. Thus, the philosophy of mathematics must open its perspective and consider what actions mathematics makes part of. And in particular, social theorising must open its perspective in order to include observations of mathematics based technological action in order to provide adequate interpretation of the propensities of the informational society.

ACKNOWLEDGMENTS

We wish to thank Marten Blomhøj, Gunnar Bomann, H. C. Hansen, Mike Newman, Miriam Godoy Penteado, John Reizes, Carl Winsløw, and Warren Yates for their critical comments and suggestions for the improvement of the earlier versions of this paper.

NOTES

[1] The term 'intellectual technology' was used by Daniel Bell to describe the "substitution of algorithms (problem-solving rules) for intuitive judgements." See *The Coming of Post-Industrial Society* (New York,1973), 29. The concept of 'inscription devices' is attributed to Bruno Latour. See for example *Science in Action* (Cambridge, MA, 1987), 68.

[2] GH Hardy, *A Mathematician's Apology* (Cambridge, 1967), 120–121.

[3] Other security services offered by the software packages such as PGP are: message compression—for economically efficient storage and transmission of message; e-mail compatibility—for wide application with different e-mail systems; and message segmentation—for accommodating message size limits imposed by e-mail systems.

[4] Here, $D \circ E$ means the composition of the two functions E and D, where the function D is applied after the function E is applied to a value.

[5] By publishing the algorithm, it becomes available to a large number of users. This provides a means of testing the strength of the algorithm; the fewer successful attacks made on an algorithm used by a large number of people—the greater assurance people can have on the strength of the algorithm against attacks. The public availability of a few statistically proven encryption algorithms is also advantageous from the point of view of compatibility. One would not have to exchange a new algorithm each time one establishes confidential communication with a new partner. The weakness of this argument, however, is that if anyone were to succeed in making a successful cryptographic attack, they would keep that secret in order to be able to continue their attack on subsequent messages.

[6] P. Loshin, *Personal Encryption: Clearly Explained,* (Boston: Academic Press, 1998), 296–302.

[7] Indeed in recent training courses attended by one of the authors (Information and Network Security, and Practical Cryptography, offered by ACL Sydney, October 26 to 29, 1998) there were but a few network security pro-

fessionals who were mathematically literate in the mathematical foundations of encryption; for most, the starting point was the 'package,' and their main aims appeared to be gaining expert advice on the suitability of available products.

[8] The RSA algorithm was developed by Ron Rivest, Adi Shamir and Len Adleman at MIT in 1977, and released in 1978.

[9] Here we are considering cryptanalytic attacks, where the private key is sought by means of some systematic or analytic approach. This is in contrast to brute force attacks where a random trial and error approach is taken to determine the key, or the plaintext of a particular ciphertext.

[10] In fact, if $\pi(x)$ is the number of primes π less than or equal to a number x then there is an estimate of this number for very large x:

$$\pi(x) \sim x/\ln(x)$$

The ratio $x/\ln(x)$ approximates the number of primes less than x, as x becomes very large. The relative "prime density" will therefore be $\pi(x) = x/\ln(x)$. The Prime Number Theorem was proved in 1896 by Hadamard and de la Valleé Poissin. The distribution of primes raises many interesting and unsolved mathematical questions: Does there exist infinitely many prime twins (those pairs of primes whose difference is 2)? Can every integer: 4, 6, 8, 10,... be written as the sum of two prime numbers?

[11] The algorithm uses what is called "modular arithmetic" which is based on mapping all integers into a smaller and finite set of numbers ranging from 0 to n–1, by mapping each number with the remainder of dividing that number by n.

[12] M.R. Schroeder, *Number Theory in Science and Communication: With Applications in Cryptography, Physics, Digital Information. And Computing and Self-Similarity, Third edition,* (Berlin: Springer-Verlag, 1997), 131. In a previous version of the book Schroeder wrote: "At present (1983), a 100-digit number that is the product of two 50-digit numbers cannot be factorized in any reasonable time...."

[13] G.H. Hardy and E.M. Wright, *Introduction to the Theory of Numbers, Fifth edition,* (Oxford, 1979). See also E. Landau, *Elementary Number Theory,* (New York, 1958), and A. Baker *A Concise Introduction to the Theory of Numbers,* (Cambridge, 1984).

[14] G.H. Hardy, *A Mathematician's Apology,* (Cambridge, 1967), 91.

[15] Hardy, 99.

[16] Hardy, 101–102.

[17] J.C.A. van der Lubbe, *Basic Methods of Cryptography,* (Cambridge, 1998), 143; W. Stallings, *Cryptography and Network Security,* (New Jersey, 1999), 181; B. Schneier, *Applied Cryptography,* (New York, 1996), 467.

[18] The Miller-Rabin algorithm for testing primarity is integrated into an iterative search algorithm summarized as follows: (1) Pick an odd integer m at random (e.g., using a pseudo-random number generator). (2) Pick an integer $a < m$ at random. (3) Perform the probabilistic primarity test, such as Miller-Rabin. If m fails the test, reject the value m and go to step 1. (4) If m has passed a

Formatting Power of "Mathematics in a Package" ■ **279**

sufficient number of tests, accept *m*; otherwise go to step 2. W. Stallings, *Cryptography and Network Security,* (New Jersey, 1999), 178.

[19] G.H. Hardy, *A Mathematician's Apology,* (Cambridge, 1967), 111–112.

[20] Fermat's Little Theorem says: if *n* is a prime and $gcd(a, n) = 1$, then $an-1 \equiv 1$ mod *n*. Many primality tests are founded on this theorem. A number *n* for which the condition

$$an - 1 \equiv 1 \text{ mod } n$$

holds for a number *a*, is called 'pseudo-prime to the base *a*'. Primality testing algorithm consists of testing the probability of the primality of a number by checking whether a candidate number is a pseudo-prime with respect to a range of bases.

[21] This test, however, would be a deterministic test of primality if one were to accept what is presently a conjecture known as the 'Generalized Riemann Hypothesis', which, if proven true, makes the Miller-Rabin test a conclusive (deterministic) test of primality, hence giving even greater confidence to the effectiveness of the RSA algorithm. This is a conjecture which Hardy himself tackled and for which he in his theorem of 1915, gained a partial result. See A. Baker, *A Concise Introduction to the Theory of Numbers,* (Cambridge, 1984), 14–15.

[22] Euler's Theorem states that for all numbers *a* and *n*, which are relatively prime to each other and for which $n > 0$ and $0 < a < n$, it is true that

$$a\phi_{(n)} \equiv 1 \text{ mod } n.$$

The function ##(*n*) is know as Eulers's totient funtion. Euler's Theorem forms the basis for proving that the relationship between the decryption exponent *d* and the encryption exponent *e* and the modulus *n* in fact produces the desired encrytpion-decryption relationship.

[23] The collection of articles in J.H. Loxton, *Number Theory and Cryptography,* (Cambridge, 1990) provides us with ways in which other classical results from mathematics, diophantine approximation and number sieves, for example, have also gained renewed interest as resources for cryptanalysis of the RSA encryption scheme. The most recent 'breakthrough' in public key encryption has been the recognition that applications of the idea of elliptic curves over finite fields provide can provide an even more powerful 'one-way function' for public-key encryption than that which algorithms such as RSA are based

[24] R. D. Hoffman, 'Interview with author of PGP (Pretty Good Privacy)', 1996, http://www.animatedsoftware.com/hightexh/philspgp.htm (last accessed 1/10/98).

[25] For more details see posting by Dave Boychuk, 1999 on the site "Privacy for Online Communication: Public Key/Private Key Strong Encryption and Public Policy" at http://www. cous.uvic.ca/poli/bennett/courses/456/fm/messages/3.htm/ (last accessed 13/9/00).

[26] R. D. Hoffman, 'Interview with author of PGP (Pretty Good Privacy)', 1996, http:// www.animatedsoftware.com/hightexh/philspgp.htm.

[27] W. Diffie and S. Landau, *Privacy on the Line: The Politics of Wiretapping and Encryption,* (Cambridge, MA, 1998), 205.

[28] Testimony of Philip R. Zimmermann to the Subcommittee on Science, Technology, and Space of the US Senate Committee on Commerce, Science, and Transportation 26 June 1996, at the PGP Security web site http://www.pgp.com/phil/phil-testimony.asp (last accessed 13/9/00). The free distribution of PGP is only one example of a new phenomenon in software development known as open source software development. The most well-known open source software is the Linux operating system. See, for example, C.C. Mann, 'Programs for the People', *Technology Review: MIT's Magazine of Innovation*, Vol. 102(1), (1999), 36–43.

[29] D. Bell, 'The Social Framework of the Information Society', in T. Forrester (ed.), *The Microelectronic Revolution*, (Oxford, 1980), 500–549.

[30] Consider for example, a recent publication in the popular press by the physicist Paul Davies, where he suggests the application of mathematical chaos theory to model economic systems, thereby attempting to resolve the inadequacies of their existing mathematical models. P. Davies, 'Ants in the Machines', *The Sydney Morning Herald*, October 17, 1998, 6.

[31] E. Wigner, 'The Unreasonable Effectiveness of Mathematics in the Natural Sciences', *Communications in Pure and Applied Mathematics*. Vol. 13 (1960), 1–14.

[32] H. Petroski, Henry, *Invention by Design: How Engineers Get from Thought to Thing*. (Cambridge, MA, 1996), 2.

[33] See also O. Skovsmose, Travelling Through Education: Uncertainty. Mathematics, Responsibility (Rotterdam: Sense Publishers, 2005).

[34] Confusion is a feature sought in conventional encryption. A high level of confusion means that there is a very complex relationship between the encryption key and the ciphertext, so that a hacker is less able to capture the key from an analysis of the ciphertext. Diffusion on the other hand is a feature where each bit of the ciphertext is a function of every bit of the plaintext, so that the plaintext cannot be easily recovered from the statistical analysis of the ciphertext.

[35] This blind spot of mathematical based reasoning contributes to what Beck calls "industrialized, decision-produced incalculabilities. See his book Risk Society: Towards a New Modernity, 1992, page 22.

[36] A cursory tour of some of the FAQ (frequently asked questions) web sites for PGP confirms the invisibility of the mathematics to the users. The users and potential users are concerned with questions about implementation and availability of the package (will it work with Windows NT, how does version 5.1 differ from version 5, and where can one get the commercial version, and so on). See the PGP Security site at http://www.pgp.com/phil/default.asp (last accessed 13/9/00) as a starting point.

[37] W. Diffie and S. Landau, *Privacy on the Line*, (Cambridge, MA, 1998), 238–239.

[38] The term "risk society" is attributed to the German sociologist Ulrich Beck who published a book by that name *Risk Society* in 1992. Reflexivity refers to a process where society confronts the consequences of its own creation. A further discussion of reflexivity and risk society can be found in U. Beck "Risk

Society and the Provident State," in S. Lash, B. Szerszynski, and B. Wynne, *Risk, Environment and Modernity,* (London, 1996), 27–43.

[39] See, for example, B. Gold, "What is the Philosophy of Mathematics, and What should it be?," *The Mathematical Intelligencer.* 16(3) (1994), 20–24.

CHAPTER 20

WHAT IS MATHEMATICS ABOUT?

Paul Budnik
Mountain Math Software

ABSTRACT

As the Platonic philosophy of mathematics is increasingly being questioned, computer technology is able to approach Platonic perfection in limited domains. This paper argues for a mathematical philosophy that is both objective and creative. It is objective in that it limits the domain of mathematics to questions that are logically determined by a recursively enumerable sequence of events. This includes the arithmetical and hyperarithmetical hierarchies but excludes questions like the Continuum Hypothesis. This philosophy is creative in recognizing that Gödel's Incompleteness Theorem implies one can only fully explore this mathematics by considering an ever increasing number of incompatible possibilities without deciding which is correct. This is how biological evolution created the mathematically capable human mind.

INTRODUCTION

Mathematics began with counting and measuring as useful procedures for dealing with physical reality. Counting and measuring are abstract in that

Critical Issues in Mathematics Education, pages 283–291

the same approach applies to different situations. As these techniques were developed and refined, problems arose in connecting highly refined abstractions to physical reality. The circles that exist physically were never the same as the ideal geometric circle. The length of the diagonal of an ideal square could not be expressed in the standard way that fractional numerical values were defined as the ratio of two integers. Mathematical thought seemed to be creating an abstract reality that could never be realized physically.

Plato had a solution to this problem. He thought all of physical reality was a dim reflection of some ideal perfect reality. Mathematics was about this ideal reality that could be approached through the mind. The difficulties with connecting mathematical abstractions to physical reality often involved the infinite. It takes a continuous plane with an infinite number of points to construct the ideal circle or diagonal of an ideal square. Plato's ideal reality seemed to require that the infinite exists. The idea that infinite mathematical abstractions are an objective Platonic reality became the dominant philosophy of mathematics after Cantor seemed to discover a complex hierarchy of infinite sets.

This hierarchy has its origins in Cantor's proof that there are 'more' reals than integers. Set A is larger than set B if one cannot define a map or function that gives a *unique* member of B for *every* member of A. This is fine for finite mathematics where one can physically construct the map by pairing off members of A and B. It becomes problematic for infinite sets. If A is larger than B it is said to have larger cardinality than B. The smallest infinite set has the cardinality of the integers. Such sets are said to be countable.

PROBLEMS WITH INFINITE SETS

The definition of cardinality creates problems because it depends on what infinite maps are defined in a mathematical system. Formal mathematical systems are, in effect, computer programs for enumerating theorems[1] and thus can only define a countable number of objects. Because all possible maps from integers to reals are not countable, no formal system will contain all of these maps. Thus we have the paradoxical Löwenheim Skolem Theorem. This says that every formal system that has a model must have a countable model. Thus no matter how large the cardinals one can define in a formal system, there is some model of the formal system in which all these cardinals can be mapped onto the integers. However this cannot be done within the system itself. But when one looks at the system from outside one can easily prove this is true because a formal system is a computer program for enumerating theorems. Every proof that some set exists comes at a unique finite time and thus the collection of everything that provably exists is countable.

A major question about the hierarchy of cardinal numbers is whether the reals are the smallest cardinal larger than the integers. The conjecture that this is true is called the Continuum Hypothesis. It has been proved that both the Continuum Hypothesis and its negation are consistent with the standard axioms of set theory. Thus the question can only be settled by adding new axioms and there is nothing remotely close to agreement about how to construct such axioms. On the contrary there is increasing doubt as to whether the Continuum Hypothesis is true or false in any objective sense.

Solomon Feferman, the editor of Gödel's collected works, observed:

I am convinced that the Continuum Hypothesis is an inherently vague problem that no new axiom will settle in a convincingly definite way. Moreover, I think the Platonic philosophy of mathematics that is currently claimed to justify set theory and mathematics more generally is thoroughly unsatisfactory and that some other philosophy grounded in inter-subjective human conceptions will have to be sought to explain the apparent objectivity of mathematics.[4]

ALTERNATIVE PHILOSOPHIES

The Platonic philosophy of mathematical truth is dominant but not universal. Constructivism demands that all proofs be constructive. It disallows proof by contradiction.[1] The constructivist treats only those mathematical objects that he knows how to construct as having an objective mathematical existence. Social constructivism has recently been applied to mathematics.[3] This approach sees mathematics as a fallible social construction that changes over time. That is an accurate appraisal of the history of mathematics.

The dominant Platonic philosophy and the extreme form of social constructivism are at opposite ends of a spectrum. In Platonic philosophy there is only absolute truth that must be discovered. In extreme social constructivism all truth is relative to some cultural group that creates and recognizes 'truth' through a cultural process.

Constructivism sits between these extremes. It accepts constructive proofs as being absolute but only allows truth values to be assigned to propositions for which there is a constructive proof. It rejects the idea that all valid mathematical questions must be objectively true or false.

Rules in some form are a common element in all these approaches. Mathematics based on a Platonic philosophy depends on formal systems which are precise rules or algorithms (in effect computer programs) for enumerating provable theorems from a set of assumed axioms. Constructivists use similar formal systems with the elimination of proof by contradiction. Social constructivism emphasizes that real proofs are never carried out in complete formal detail, that there are many errors in published work and that there is no agreement about the fundamental axioms of mathematics. This suggests that a social process is the primary element in determining accepted mathematical truth at a given period of time. Nonetheless social constructivism depends on the "rules of the game" as providing a foundation for their philosophy. If there were not rules that could be enforced with some, albeit imperfect, consistency in a social milieu there could be no theory of social constructivism.

CREATIVITY VERSUS OBJECTIVITY

Platonic philosophy ignores the marvelous creativity of our universe. Reproducing molecules have evolved to the depth and richness of human consciousness and created the mathematically capable human mind. One can only gasp in dumbfounded wonder at the miracle of it all. Social constructivism minimizes the connection between objective physical reality and mathematics. It sees mathematical creativity is somewhat or mostly arbitrary like many cultural practices seem to be. Is it possible to square this circle with a philosophy of mathematics that integrates aspects of these two philosophies to produce a creative philosophy of mathematics rooted in the objectivity of physical reality and yet open to the astounding creativity that characterizes the human condition? I believe this is the direction the philosophy of mathematics should pursue.

Finite mathematics is objective because we can physically build at least some of what it talks about. Among the finite objects we can construct are precise sets of rules in the form of computer programs. One element common to all approaches to the philosophy of mathematics described here can be made, through technology, to approach the absolute perfection of Platonic philosophy. The execution of computer programs, in contrast to semi-formal mathematical proofs, obey a rigorous set of rules (defined by the characteristics of the machine they are running on) with something approaching absolute certainty. We cannot construct a perfect circle but we can compute the ratio of the circumference to the diameter of the perfect circle, π, to a million or more decimal places with a very high certainty that we have done it correctly. The same is true for the diagonal of the perfect square. We can write a program that could, if it were possible for it to run forever with no errors, eventually output each digit of π or square root of 2.

There is a basis in physical reality for the perfection (or something very close to it) that Plato first described. However, when we move beyond finite questions and procedures, things become more ambiguous. This first happens in mathematics when we ask if some recursive property is true of all integers. To be recursive the condition must be verifiable in a finite number of finite operations for each integer. The ability to verify the condition for *each* integer does not allow one to determine the question for *all* integers. Such questions are cultural creations. There is no physical reality that embodies the solution. Yet such questions are logically determined by a recursively enumerable set of events i.e., by a set of events that can be output by a single computer program that runs error free forever and has access to unlimited storage.

One can of course argue that the universe is not infinite or potentially infinite and thus such questions do not have a connection with our physical reality. Brian Rotman, a social constructivist, has written a book objecting to the idea of potential infinity. [5] We can never know if the universe is potentially infinite, but it would be hard to prove this is not the case. Throughout history, theories of the universe have given a limit to its size. Those limits have repeatedly been vastly expanded. Cosmology is, of necessity, a highly speculative science. It projects what we think we understand about physical reality over vast distances and epochs of time using very limited information. The existence of *ultimate* limits to the size of the universe will be an open question for the foreseeable future even as the dominant cosmology confidently quotes its estimate of the size of the universe.

Because of this uncertainty and more importantly because of the practical value of proofs about properties of all integers, I assume such questions are meaningful and objectively true or false even though there exists no general method for deciding them. This is where I part company with both constructivism and social constructivism. On the other hand I do not come remotely close to embracing the hierarchy of infinity in the Platonic philosophy of mathematics. For me infinity is deeply connected to the creative evolution over time that characterizes biological evolution and is the richest and most interesting aspect of existence that I know of.

OBJECTIVE MATHEMATICS

Questions about all integers have a unique existential status with their tenuous connection to physical reality. On the one hand they are logically determined by a recursively enumerable sequence of events. They can be falsified by counter examples. However there is no general way to determine if they are true. Gödel's Incompleteness Theorem established this. Gödel proved that any system that is powerful enough to embed any computer program (or equivalently embed the primitive recursive functions) must be incomplete. In particular such formal systems could not prove their own consistency unless they were inconsistent.

The consistency of any formal system is equivalent to the halting problem for some particular computer program. (The halting problems asks will a computer program, with access to unlimited storage and able to run error free forever, eventually halt.) Similarly all integers satisfy some recursive property if and only if some particular computer program halts. In the former case we can program the computer to enumerate every theorem of the formal system and test each theorem against all previous theorems. If a contradiction is discovered the program halts. Otherwise it runs forever. In the latter case we program the computer to test the condition for each integer in succession and halt if and only if the conditions fails for some integer.

The question does a computer have an infinite number of outputs has an even more tenuous connection to objective reality then does the halting problem. The latter question can be falsified by a finite event, the former cannot. To prove a program has an infinite number of outputs requires some form of induction. The proof can be trivial. A program might produce an output and then return to exactly the same state it had before the output was generated implying it will loop forever continually producing new outputs. But no finite event can establish or contradict this proof.

How far do we take this process? What is objective mathematics? It turns out that much of set theory is about questions that are determined by a recursively enumerable sequence of events. The arithmetical hierarchy includes all questions defined by a recursive relationship and a finite number of existential and universal quantifiers over the integers. This hierarchy is equivalent to the series of questions does a computer have an infinite number of outputs, does it have an infinite number of outputs such that an infinite subset of these are programs that themselves have an infinite number of outputs etc. The nth question in this hierarchy asks does the computer have an infinite number of outputs and infinite subset of which satisfy the $n-1$ condition in the hierarchy. A single computer program can nondeterministically[2] enumerate all events at any level in this hierarchy by simulating what all computer programs do for all time. To enumerate all computer programs take the programming language for any Universal Turing Machine[3] and generate all possible finite sequences in that language. Then nondeterministically simulate each of these programs to enumerate what every computer program will do at every point in time.

Even some questions that require quantification over the reals are determined by a recursively enumerable sequence of events. For example consider a computer program that accepts an integer as input and outputs either 0 or another computer program with the same definition as itself. We can apply a sequence of integers to such a program. The first integer is applied to the base program. If we get 0 we do nothing and say the path terminates. Otherwise we apply the next integer in the sequence to the computer program that was previously output. We keep doing this for every integer in the sequence until and unless we get a zero output. If applying every sequence of integers to the base computer results in a terminated path after some finite number of steps the original base computer program is said to be well founded.

Asking if a computer program is well founded requires quantification over the reals. Every question in the hyperarithmetical hierarchy is equivalent to the question is a particular computer program well founded. Yet each such question is determined by a recursively enumerable sequence of events. A single program can nondeterministically do what every computer program will do for every integer input. This includes all the events that determine if a computer program is well founded.

The property of a computer program being well founded is impredicative. Properties defined in terms of all reals or all infinite sequences of integers can be circular because they can be used to define new reals. This is an issue because such circular definitions have, in the past, led to contradictions in mathematical systems. Claiming that some impredicative definitions must be objectively true or false crosses a major boundary. None the less I think a computer being well founded is an objective property with important practical significance.

We can attribute a limited form of objectivity to any question logically determined by a recursively enumerable sequence of events. Everything that determines the outcome could happen in a finite but potentially infinite universe. This definition separates the Continuum Hypothesis (not objectively true or false) from the question of whether a computer program is well founded. However it does not constrain objective mathematics to a particular set of propositions.

A CREATIVE OBJECTIVE PHILOSOPHY

A philosophy of mathematics must deal with two opposing forces. Computer technology allows us to create, in limited ways, structures that can approach the ideal perfection of Plato's philosophy. One can never eliminate all possibility of error but, in limited domains and with enough resources, the error rate can be made arbitrarily small. Today's computers perform billions of operations a second with rare hardware failures. Application and even operating system program bugs are far more common but the basic hardware is extremely stable.

Simple programs carefully reviewed can be error free. Complex programs are another matter. However *what they produce* can be made relatively error free. The largest computer chips today have hundreds of millions of switches and can only be designed with the aid of computer programs. Those programs are not error free but the entire design process allows one to produce a chip that ultimately is error free. Furthermore one must be able to detect all manufacturing faults in every chip produced. Thus the computer chip must be designed to make such verification possible. A limited form of Plato's heaven exists today in the engineering labs of Intel and AMD.

The opposing force is Gödel's Incompleteness Theorem and its implications. The hope that there can be a precise set of rules that determine all mathematical truth has been dashed forever. There can be no general solution even to a question as basic as the halting problems for computers. For me this is a reflection of the creative reality of our existence. One cannot determine all mathematical truth, even in a potentially infinite universe, but one can *explore* all of it in such a universe. If we insist on a single approach to mathematics we will inevitably run up against a Gödelian limit. This will not be a fixed limit or specific event. Rather it will be never ending progress that continually generates new results. However the collection of all these results will be subsumed in a single mathematical truth that we will never discover or explore.

If, on the other hand, mathematics becomes a divergent process that continually explores ever more possibilities, then there is no limit to the mathematics we may explore. This may seem as fanciful as Plato's heaven or a measurable cardinal but a divergent process, biological evolution, created the mathematically capable human mind. Evolution on this planet is enormously diverse. Over a vast expanse of time this diversity has increased enormously from the first reproducing molecules to today's biosphere. It is a safe bet that without this diversity, the enormous complexity and enormous depth of the human mind could never have evolved.

This suggests to me that the stakes are much higher than what happens with our mathematical knowledge. The hierarchy of mathematical truth involves ever more complex levels of abstraction and self reflection. The evolution of the mathematically capable human mind and the evolution of the depth and richness of human consciousness both seem to depend in part on the rich and subtle powers of abstraction and self reflection that uniquely characterize human thought and awareness. We are entering a unique period in biological evolution. Evolution has become, through us, conscious of itself and is acquiring the tools to control its future destiny. Understanding the role of diversity in not limiting the potential of future evolution may be crucial to what we can become.[2]

A CULTURAL PRESCRIPTION

Ironically the key to expanding mathematical diversity lies in embracing the technology through which humanity has obtained something approaching Platonic perfection. One must turn the foundations of mathematics into an experimental science embracing computer technology as an essential research tool just as every other major branch of science has done.

There is a cultural bias in mathematics to come up with the simplest most elegant approach possible. Most mathematical research is done using pencil, paper and the mathematician's mind, limiting the complexity that can be dealt with. Computers may be used to replace pencil and paper but they are rarely used as a research tool or to verify proofs. Of course elegance and simplicity are worthy goals, but one must not insist on them to the point of failing to deal with the enormous complexity that the foundations of mathematics suggests we can explore. The strength of a formal system is determined, in large measure, by the ordinals definable within it. Notations for recursive ordinals and recursive operations on these notation can be explored experimentally using computers. Recent history of science suggests that leveraging human intuition with the combinatorial power of computers will lead to results far beyond what the unaided human mind is capable of. I do not think the foundations of mathematics will be an exception.

REFERENCES

[1] L. E. J. Brouwer. Intuitionism and Formalism. *Bull. Amer. Math. Soc.*, 20:81–96, 1913.

[2] Paul Budnik. *What is and what will be: Integrating spirituality and science.* Mountain Math Software, Los Gatos, CA, 2006.

[3] Paul Ernest. *Social Constructivism as a Philosophy of Mathematics.* State University of New York Press, 1998.

[4] Solomon Feferman. Does mathematics need new axioms? *American Mathematical Monthly*, 106:99–111, 1999.

[5] Brian Rotman. *Ad Infinitum.* Stanford Univeristy Press, 1993.

NOTES

1. It is tedious but straightforward to go from the axioms of a formal system such as set theory and the laws of logic to a computer program that would enumerate every theorem provable from those axioms and laws of logic. This however is not a practical way to generate new mathematics because most theorems would be trivial.

2. A nondeterministic computer program can simulate an infinite sequence of computer programs by simulating the first program for one time step, followed by simulating the first two programs for two time steps, followed by simulating the first three programs for three time steps, etc.

3. A Universal Turing Machine is a computer that can simulate any computer that it is possible to build. It can be a simple device, but it requires access to unlimited storage.

CHAPTER 21

APPLIED MATHEMATICS AS SOCIAL CONTRACT

Philip J. Davis
Brown University

ABSTRACT

The author takes the position that mathematical education must redefine its goals so as to create a citizenry with sufficient knowledge to provide social backpressure on future mathematizations. This can be accomplished by increasing the part of mathematical education that is devoted to the description and interpretation of the processes of mathematization and by allowing the technicalities of the formal operations within mathematics itself to be deemphasized or automated out by computer.

THIS MATHEMATIZED WORLD

As compared to the medieval world or the world of antiquity, today's world is characterized as being scientific, technological, rational and mathematized. By "rational" I mean that by an application of reason or of the formalized versions of reason found in mathematics, one attempts to understand the world and control the world. By "mathematized," I shall mean the employment of mathematical ideas or constructs, either in their theoretical

Critical Issues in Mathematics Education, pages 293–304

form or in computer manifestations, to organize, to describe, to regulate and to foster our human activities. By adding the suffix "ized," I want to emphasize that it is humans who, consciously or unconsciously; are putting the mathematizations into place and who are affected by them. It is of vital importance to give some account of mathematics as a human institution, to arrive at an understanding of its operation and at a philosophy consonant with our experience with it, and on this basis to make recommendations for future mathematical education.

The pace of mathematization of the world has been accelerating. It makes an interesting exercise for young students to count how many numbers are found on the front page of the daily paper. The mere number of numbers is surprising, as well as the diversity and depth of the mathematics that underlies the numbers; and if one turns to the financial pages or the sports pages, one sees there natural language overwhelmed by digits and statistics. Computerization represents the effective means for the realization of current mathematizations as well as an independent driving force toward the installation of an increasing number of mathematizations.

PHILOSOPHIES OF MATHEMATICS

Take any statement of mathematics such as 'two plus two equals four', or any more advanced statement. The common view is that such a statement is perfect in its precision and in its truth, is absolute in its objectivity, is universally interpretable, is eternally valid and expresses something that must be true in this world and in all possible worlds. What is mathematical is certain. This view, as it relates, for example, to the history of art and the utilization of mathematical perspective has been expressed by Sir Kenneth Clark ("Landscape into Art"): "The Florentines demanded more than an empirical or intuitive rendering of space. They demanded that art should be concerned with certezza, not with opinioni. Certezza can be established by mathematics."

The view that mathematics represents a timeless ideal of absolute truth and objectivity and is even of nearly divine origin is often called Platonist. It conflicts with the obvious fact that we humans have invented or discovered mathematics, that we have installed mathematics in a variety of places both in the arrangements of our daily lives and in our attempts to understand the physical world. In most cases, we can point to the individuals who did the inventing or made the discovery or the installation, citing names and dates. Platonism conflicts with the fact that mathematical applications are often conventional in the sense that mathematizations other than the ones installed are quite feasible (e.g., the decimal system). The applications are of ten gratuitous, in the sense that humans can and have lived out their lives without them (e.g., insurance or gambling schemes). They are provisional in the sense that alternative schemes are often installed which are claimed to do a better job. (Examples range all the way from tax legislation to Newtonian mechanics.) Opposed to the Platonic view is the view that a mathematical experience combines the external world with our interpretation of it, via the particular structure of our brains and senses, and through our interaction with one another as communicating, reasoning beings organized into social groups.

The perception of mathematics as quasi-divine prevents us from seeing that we are surrounded by mathematics because we have extracted it out of unintellectualized space, quantity, pattern, arrangement, sequential order, change, and that as a consequence, mathematics has become a major modality by which we express our ideas about these matters. The conflicting views, as to whether mathematics exists independently of humans or whether it is a human phenomenon, and the emphasis that tradition has placed on the former view, leads us to shy away from studying the processes of mathematization, to shy away from asking embarrassing questions about this process: how do we install the mathematizations, why do we install them, what are they doing for us or to us, do we need them, do we want them, on what basis do we justify them. But the discussion of such questions is becoming increasingly important as the mathematical vision transforms our world, often in unforeseen ways, as it both sustains and binds us in its steady and unconscious operation. Mathematics creates a reality that characterize our age.

The traditional philosophies of mathematics: platonism, logicism, formalism, intuitionism, in any of their varieties, assert that mathematics expresses precise, eternal relationships between atemporal mental objects. These philosophies are what Thomas Tymoczko has called "private" theories. In a private theory, there is one ideal mathematician at work, isolated from the rest of humanity and from the world, who creates or discovers mathematics by his own logico-intuitive processes. As Tymoczko points out, private theories of the philosophy of mathematics provide no account either for mathematical research as it is actually carried out, for the applications of mathematics as they actually come about, or for the teaching process as it actually unfolds. When teaching goes on under the banner of conventional philosophies of mathematics, if often becomes to a formalist approach to mathematical education: "do this, do that, write this here and not there, punch this button, call in that program, apply this definition and that theorem." It stresses operations. It does not balance operations with an understanding of the nature or the consequences of the operations. It stresses syntactics at the expense of semantics, form at the expense of meaning. A fine place to read about this is in "L'age du

Capitaine" by Stella Baruk, a mathematics supervisor in a French school system. Baruk writes:

> From Pythagoras in antiquity to Bourbaki in our own days, there has been maintained a tradition of instruction—religion which sacrifices full understanding to the recitation of formal and ritual catechisms, which create docility and which simulate sense. All this has gone on while the High Priests of the subject laugh in their corners.

How many university lecturers, discoursing on numbers, say, allow themselves to discuss where they think numbers come from, what is one's intuition about them, how number concepts have changed, what applications they have elicited, what have been the pressures exerted by applications, how we are to interpret the consequences of these applications, what is the poetry of numbers or their drama or their mysticism, why there can be no complete or final understanding of them. How many lecturers would take time to discuss the question put by Bertrand Russell in a relaxed moment: "What is the Pythagorean power by which number holds sway above the flux?"

Opposed to "private" theories, there are "public" theories of the philosophy of mathematics in which the teaching process is of central importance. Several writers in the past half century have been constructing public theories, and I should like to add a few bricks to this growing edifice and to point out its relevance for the future of mathematical education.

APPLIED MATHEMATICS AS SOCIAL CONTRACT

I shall emphasize the applications of mathematics to the social or humanistic areas though one can make a case for applications to scientific areas and indeed to pure mathematics itself (see, e.g., Spalt).

Today's world is full of mathematizations that were not here last year or ten years ago. There are other mathematizations which have been discarded (e.g., Ptolemaic astronomy, numerological interpretations of the cosmos, last years' tax laws). How do these mathematizations come about? How are they implemented, why are they accepted? Some are so new, for example, credit cards, that we can actually document their installation. Some are so ancient, e.g., numbers themselves, that the historical scenarios that have been written are largely speculative. Are mathematizations put in place by divine fiat or revelation? By a convention of Elders? By the insights of a gifted few? By an evolutionary process? By the forces of the market place or of biology? And once they are in place what keeps them there? Law? Compulsion? Inertia? Darwinian advantage? The development of a bureaucracy where sole function it is to maintain the mathematization? The development of businesses whose function it is to create and sell the mathematization? Well, all of the above, at times, and more. But, for all the lavish attention that our historians of mathematics have paid to the evolution of ideas within mathematics itself, only token attention has been paid by scholars and teachers to the interrelationship between mathematics and society. A description of mathematics as a human institution would be complex indeed, and not be easily epitomized by a catch phrase or two.

The employment of mathematics in a social context is the imposition of a certain order, a certain type of organization. Government, as well, is a certain type of organization and order. Philosophers of the 17th and 18th-century (Hobbes, Locke, Rousseau, Thomas Paine, etc.) put forward an idea, known as social contract, to explain the origin of government. Social contract is an act by which an agreed upon form of social organization is established. (Here I follow an article by Michael Levin.) Prior to the contract there was supposedly a "state of nature." This was far from ideal. The object of the contract, as Rousseau put it, was "to find a form of association which will defend the person and goods of each member with the collective force of all, and under which each individual, while counting himself with the others, obeys no one but himself, and remains as free as before." In this way one may improve on a life which, as Hobbes put it in a famous sentence, was "solitary; poor, nasty, brutish, and short." The contract itself, whether oral or written, was almost thought of as having been entered into at a definite time and place. Old Testament history; with its covenants between God and Noah, Abraham, Moses, the Children of Israel, was clearly in the minds of contract theorists. In the United States, political thinking has often been in terms of contracts, as in the Mayflower Compact, the Constitutions of the United States and of the individual states, the Establishment of the United Nations in San Francisco in 1945, and periodic proposals for constitutional amendments and reform.

It was generally assumed by the contract theorists that "Human society and government are the work of man constructed according to human will even if sometimes operating under divine guidance." That "man is a free agent, rather than a being totally determined by external forces," and that society and government are based on mutual agreement rather than on force (see Levin).

The acceptability of social contract as a historical explanation hardly lasted till the 19th century; even if political contracts continued to be entered into as instances of democratic polity. It is an instructive exercise, I believe, in order to get a grasp on the relationship between society and mathematics, to take the outline of social contract just given and replace the words "government and society" by the word "mathematization." Though it is naive to think that most mathematizations came about by formal contracts, the "contract" metaphor is a useful phrase to designate the interplay between people and their mathematics and to make the point that mathematizations are the work of man, constructed according to human will, even if operating under a guidance which may be termed divine or logical or experimental according to one's philosophic predilection.

A number of authors, some writing about theology, and others about political or economic processes, have pointed out that contracts are continuously entered into, broken, and reestablished. I believe the same is the case for mathematizations. Consider, for example, insurance. This is one of the great mathematizations currently in place, and I personally, without adequate coverage, would consider myself naked to the world. Yet I am free to throw away my insurance policies. Consider the riders that insurance companies send me, unilaterally abridging their previous agreements. Consider also that in a litigious age, with a populace abetted by eager lawyers and unthinking juries, what appears as the 'natural' stability of the averages upon which possibility of insurance is based, emerges, on deeper analysis, to incorporate the willingness of the community to adjust its affairs in such a way that the averages are maintained. The possibilities of insurance can be destroyed by our own actions.

Another example that displays the relationship between mathematics, experience, and law is the highway speed limit in the United States. Before the gas shortage in 1974, the limit was 65 miles per hour. In 1974, the speed limits were reduced to 55 miles per hour in order to conserve gasoline. As a side effect, it was found that the number of highway accidents was reduced significantly. Now (1987) the gas shortage is over, and there is pressure to raise the legal speed limit. Society must decide what price it is willing to pay for what some see as the convenience or the thrill of higher speeds. Here is mathematical contract at work.

The process of contract maintenance, renewal or reaffirmation, in all its complexities, is open to study and description. This is a proper part of applied mathematics and I shall argue that it should be a proper part of mathematical education.

WHERE IS KNOWLEDGE LODGED?

There is an epistemological approach to the interplay between mathematics and society and that is to look at the way society answers the question that heads this section. According to how we answer this question, we will mathematize differently and we will teach differently

Where, then, is knowledge lodged? (Here I follow an article by Kenneth A. Bruffee.) In the pre-Cartesian age, knowledge was often thought to be lodged in the mind of God. Those who imparted knowledge authoritatively derived their authority from their closeness to the mind of God, evidence of this closeness was often taken to be the personal godliness of the authority

In the post-Cartesian age knowledge was thought to be lodged in some loci that are above and beyond ourselves, such as sound reasoning or creative genius or in the 'object objectively known'.

A more recent view, connected perhaps with the names of Kuhn and Lakatos, is that knowledge is socially justified belief. In this view, knowledge is not located in the written word or in symbols of whatever kind. It is located in the community of practitioners. We do not create this knowledge as individuals but we do it as part of a belief community. Ordinary individuals gain knowledge by making contact with the community of experts. The teacher is a representative of the belief community.

In my view, knowledge as socially justified belief provides a fair description of how mathematical knowledge is legitimized but we must keep clearly in mind that perceptions of what 'is', theory formation, validation, and utilization are all part of a dynamic and iterative process. Knowledge once thought to be absolute, indubitable, is now seen as provisional or even probabilistic. Science is seen as a search for error as much as it is a search for truth. Eternally valid knowledge, may remain an ideal which we hold in our minds as a spur to inquiry This view fits with the idea of applied mathematics as social contract, with the contractual arrangements being concluded, broken, and renegotiated in endless succession.

Another view of the locus of knowledge, not yet elevated to a philosophy, is that knowledge is located in the computer. One speaks of such things as 'artificial intelligence', 'expert systems', and more than one theoretical physicist has opined that all the essentials are now known (despite the fact that the same was asserted 100 years ago and 200 years ago) and that the computer can fill in the details and derive the consequences for the future.

Advocates of this view have asserted that while education is now teacher oriented, in the full bloom of the computer age, education will be knowledge oriented. These two contemporary views are not necessarily antithetical, provided we accommodate the computer into the community of experts, clarify whether 'belief' can reside in a computer, and decide whether mankind exists for the sake of the computers or vice versa.

MATHEMATICAL EDUCATION AT A HIGHER METALEVEL

A mathematized and computerized world brings with it many benefits and many dangers. It opens many avenues and closes many others. I do not want to elaborate this point as I and my co-author Reuben Hersh have done so in our book "Descartes' Dream," as have numerous other authors.

The benefits and dangers both derive from the fact that the mathematical/computational way of thinking is different from other ways. Philosopher and historian Sir Isaiah Berlin called attention to this divergence when he wrote "A person who lacks common intelligence can be a physicist of genius, but not even a mediocre historian." For the mathematical way to gain ascendancy over other modes is to create an imbalance in human life.

The benefits and dangers derive also from the fact that mathematics is a kind of language, and this language creates a milieu for thought that is hard to escape. It both sustains us and confines us. As George Steiner has written of natural language (1986): "The oppressive birthright is the language, the conventions of identification and perception. It is the established but customarily subconscious unargued constraints of awareness that enslave." One can assert as much for mathematics as a language. The subconscious modalities of mathematics and of its applications must be made clear, must be taught, watched, argued. Since we are all consumers of mathematics, and since we are both beneficiaries as well as victims, all mathematizations ought to be opened up in the public forums where ideas are debated. These debates ought to begin in the secondary school.

Discussions of changes in mathematics curricula generally center around (a) the specific mathematical topics to be taught, e.g., whether to teach the square root algorithm, or continued fractions or projective geometry or Boolean algebra, and if so, in what grade, and (b) the instructional approaches to the specific topics, e.g., should they be taught with proofs or without; from the concrete to the abstract or vice versa, what emphasis should be placed on formal manipulations and what on intuitive understanding; with computers or without; with open ended problems or with "plug and chug" drilling.

Because of widespread, almost universal computerization, with hand-held computers that carry out formal manipulations and computations of lower and higher mathematics rapidly and routinely, because also of the growing number of mathematizations, I should like to argue that mathematics instruction should, over the next generation, be radically changed. It should be moved up from subject oriented instruction to instruction in what the mathematical structures and processes mean in their own terms and what they mean when they form a basis on which civilization conducts its affairs. The emphasis in mathematics instruction ought to be moved from the syntactic-logico component to the semantic component. To use programming jargon, it ought to be "popped up" a metalevel. If, as some computer scientists believe, instruction is to move from being teacher oriented to knowledge-oriented—and I believe this would be disastrous—the way in which the role of the teacher can be preserved is for the teacher to become an interpreter and a critic of the mathematical processes and of the way these processes interact with knowledge as a database. Instruction in mathematics must enter an altogether new and revolutionary phase.

THE INTERPRETIVE COMPONENT
OF MATHEMATICAL EDUCATION

Let me begin by asking the question: to what end do we teach mathematics? Over the millennia, answers have been given and they have differed. Some of them have been: we teach it for its own sake, because it is beautiful; we teach it because it reveals the divine; because it helps us think logically; because it is the language of science and helps us to understand and reveal the world; we teach mathematics because it helps our students to get a job either directly, in those areas of social or physical sciences that require mathematics, or indirectly, insofar as mathematics, through testing, acts as a social filter, admitting to certain professional possibilities those who can master the material. We teach it also to reproduce ourselves by producing future research mathematicians and mathematics teachers.

Ask the inverse question: what is it that we want students to learn? We may answer this by citing specific course contents. For example, we may say that we now want to emphasize discrete mathematics as opposed to continuous mathematics. Or that we want to develop a course in non-standard analysis on tape so that joggers may learn about hyperreal numbers even as they run. Or, we may decide for ourselves what the characteristic, constitutive ingredients of mathematical thought are: space, quantity, deductive structures, algorithms, abstraction, generalization, etc., and simply assure that the student is fed these basic ingredients, like vitamins. All of these questions and answers have some validity, and tradeoffs must occur in laying out a curriculum.

Within an overcrowded mathematical arena with many new ideas competing for inclusion in a curriculum, I am asking for a substantial elevation in the awareness of the applications of mathematics that affect society and of the consequences of these applications. If formal computations and manipulations can be learned rapidly and performed routinely by computer, what purpose would be served by tedious drilling either by hand or by computer? On-the-job training is certainly called for, whether at the supermarket checkout counter or on the blackboards of a hi-tech development company. If mathematics is a language, it is time to put an end to overconcentration on its grammar and to study the "literature" that mathematics has created and to interpret that literature. If mathematics is a logico-mechanism of a sort, then just as only a very few of us learn how to construct an automobile carburetor, but all of us take instruction in driving, so we must teach how to "drive" mathematically and to interpret what it means when we have been driven mathematically in a certain manner.

What does it mean when we are asked to create sex-free insurance pools? What are the consequences when people are admitted or excluded from a program on the basis of numerical criteria? How does one assess a statement that procedure A is usually effective in dealing with medical condition B? What does it mean when a mathematical criterion is employed to judge the quality of prose or the comprehensibility of a poem or to create music in a programmatic way? What are the consequences of a computer program whose output is automatic military retaliation? The list of questions that need discussion is endless. Each mathematization-computerization requires explanation and interpretation and assessment. None of these things are now discussed in mathematics courses in the concrete form that confronts the public. If a teacher were inclined to do so, the reaction from his colleagues would probably be "Well, that is not mathematics. That is applied mathematics or that is psychology or economics or social-anthropology or law or whatever." My answer would be: I am trying, little by little, to bring in discussions of this sort into my teaching. It is difficult but important.

If the claim were made, with justice, that these matters cannot be discussed intelligently without deep knowledge both of mathematics and of the particular area of the real world, then I would agree, and point out that this claim forces into the open the conflict between democracy and "expertocracy" (see e.g., Prewitt). This conflict has received considerable attention in areas such as medicine, defense, and technological pollution, but has hardly been discussed at the level of an underlying mathematical language. The tension between the two claims, that of democracy and of expertocracy, could be made more socially productive by an education which enables a wide public to arrive at deeper assessments, moving from daily experience toward the details of the particular mathematizations. While we must keep in mind certain basic mathematical material, we must also learn to develop mathematical 'street smarts' which enable us to form judgments in the absence of technical expertise (cf. Prewitt).

A philosophy of mathematics which is "public" and not "private," lends support to introducing this kind of material into the curriculum. The discussion of such curriculum changes will be assisted by the perception of the mathematical enterprise as a human experience with contractual elements; and by the realization that every civilized person practices and utilizes mathematics at some level, and thereby enters a certain knowledge and belief community

Again, following Kenneth Bruffee's article (with additions and modifications), I would like to suggest several lines of inquiry

(a) Identify and describe the mathematical beliefs, constructions, practices that are now in place. Where and how is mathematics employed in real life?

(b) Describe the mathematical beliefs, constructs, and practices that have been justified by the community What are justifiable and unjustifiable? What are the modes of justification?

(c) Describe the social dimensions of mathematical practice. What constitutes a knowledge community? What does the community of mathematicians think are the best examples that the past has to offer?

As part both of (a) and (b) one should add: describe the nature of the various methods of prediction and the bases upon which prediction can become prescription (i.e., policy).

This type of inquiry is rarely carried out for mathematics. For example, the concrete question of where such and such a piece of university mathematics is used in practical life and how widespread it its use, is seldom answered. Many textbook claims are made in textbooks, but show me the real bottom line. It is important to know. How, in fact, would you define the bottom line?

The technical term for inquiries such as the above is 'hermeneutics'. This word is well established in theology, and in the last generation has been commonly employed in literary criticism. It means the principles or the lines along which explanation and interpretation are carried out. It is time that this word be given a mathematical context. Instruction in mathematics must enter a hermeneutic phase. This is the price that must be paid for the sudden, massive and revolutionary intrusion of mathematizations—computerizations into our daily lives.

CONCLUSION

Mathematics is a social practice. This practice must be made the object of description and interpretation. It is ill-advised to allow the practice to proceed blindly by "mindless market forces" or as the result of the private decisions of a cadre of experts. Mathematical education must find a proper vocabulary of description and interpretation so that we are enabled to live in a mathematized world and to contribute to this world with intelligence.

ACKNOWLEDGEMENT

I wish to thank Professor Reuben Hersh for numerous suggestions.

This paper first appeared in *Zentralblatt für Didaktik der Mathematik International Reviews on Mathematical Education*, Vol. 88, No. 1 (1988) 1015.

REFERENCES

Baruk, S.: *L'âge du capitaine*. Paris: Editions du Seuil, 1985.

Berlin, I.: *Against the current: essays in the history of ideas.* New York: Viking, 1980.

Bruffee, K. A.: Liberal education and the social justification of belief. In: *Liberal Education* 68 (1982). pp. 95–114.

Davis, P. J., & Hersh, R.: *The mathematical experience.* Cambridge: BirkhàuserBoston, 1980.

Davis, P. J., & Hersh, R.: *Descartes' dream: The world according to mathematics.* San Diego: Harcourt Brace Jovanovich, 1986.

Galbraith, J. K.: *The new industrial state,* Boston: Houghton Mifflin, 1967.

Kuhn, T. S.: *The structure of scientific revolutions.* 2nd ed. Chicago: University of Chicago Press.

Iakatos, I.: *Proofs and refutations.* Cambridge: Cambridge University Press, 1976.

Levin, M.: *Dictionary of the history of ideas,* vol. IV. New York: Scribner's, 64, 1973, pp. 251–263.

Morgenthau, H.: *Science: Servant or master?* New York: Norton, 1972

Prewitt, K.: Scientific illiteracy and democratic theory In: *Daedalus* (Spring 1983), pp. 4964.

Roszak, T.: *Where the wasteland ends: Politics and transcendence in post industrial society.* New York: Doubleday, Anchor, 1973.

Spalt, D.: Das Unwahre des Resultatismus. *Seminar für Math. und ihre Didaktik.* Universität zu Köln, Dec. 1986.

Stanley, M.: *The technological conscience: Survival and dignity in an age of expertise.* New York: Free Press, 1978.

Steiner, G.: *After Babel.* Oxford University Press, 1975.

Steiner, G.: Review of Michael Foucault's "On conceptual and semantic coercion." In: *New Yorker Magazine* (March 17, 1986).

Tymoczko, T.: Making room for mathematicians in the philosophy of mathematics. In: *Mathematical Intelligence* (1986).

Tymoczko, T. (Ed.): *New directions in the philosophy of mathematics.* Basel: Birkhàuser, 1985.

Wilder, R. L.: *Evolution of mathematical concepts.* New York: Wiley, 1968.

CHAPTER 22

MATHEMATICS AND CURRICULUM INTEGRATION

Challenging the Hierarchy of School Knowledge

Elizabeth de Freitas
Adelphi University

INTRODUCTION

This paper explores the complex ways that mathematics teachers develop their identity within schools. I discuss a curriculum integration project in which untenured mathematics teachers attempted to cross discipline borders and create an authentic learning experience in a traditional middle school in Canada. I first discuss the procedure for curriculum integration, and explain how such integration is often problematic if too quickly implemented in traditional school contexts. I also discuss the theoretical literature about curriculum integration and explore the role of school mathematics in knowledge integration. I focus my analysis on the conflicts between untenured and veteran teachers in the mathematics department

Critical Issues in Mathematics Education, pages 305–316
Copyright © 2009 by Information Age Publishing
All rights of reproduction in any form reserved.

as the project was implemented, pointing out how tensions between these teachers can be interpreted through the lens of critical theory.

INTEGRATED AND TRANS-DISCIPLINE CURRICULUM

James Beane (1997) argues that integrated curriculum is politically controversial for three important reasons: firstly, integrated curriculum is designed from the bottom up, because content is meant to be responsive to student lives and contexts; secondly, integrated curriculum is collaboratively designed by school teachers and not by curriculum consultants or textbook writers, thus thwarting interest groups who seek to centralize authority over teaching; thirdly, integrated curriculum honours alternative ways of knowing that trouble the borders between traditional disciplines. Since discipline knowledge is a reflection of cultural interests, the dominant culture has a vested interest in maintaining these traditional discipline borders. Disrupting the borders and hierarchies that structure school subjects makes integrated curriculum difficult to implement in traditional schools.

The first stage of integration usually involves finding affinity between disciplines and then collapsing specific course outcomes into more essential cross-curricular outcomes. This stage can be characterized as interdisciplinary, a preliminary stage in the attempt to fully integrate curriculum (Drake, 1999). Interdisciplinary stages tend to sustain discipline borders while searching for common ground, and are usually developed as thematic explorations calling on skills from various subjects. Trans-discipline curriculum, in contrast, disrupts previous knowledge categories and opens up the field of inquiry to a more radical form of integration. At the trans-discipline stage, students play a far more active role in the generating of content outcomes, tapping any and all resources and strategies, without concern for their traditional discipline source. This stage is very difficult to implement within a standards based national curricula, and requires a great deal of institutional independence. Teachers with a certain "liberal" philosophy can often make immediate sense of how to integrate subjects, as they often see themselves as generalists first, and specialists second (Lear, 1992). Despite strong commitment to holistic instruction, many curriculum integration projects have come to naught when teachers return to the classroom, close the door, and teach to the familiar text. Perhaps the most daunting obstacle to curriculum integration (at various stages) is the current emphasis on standardized assessment, which always functions to entrench the traditional borders between disciplines. The strategic and sustained implementation of integration projects often falters, even amongst the most committed (Beane, 1997).

Mathematics as a school discipline is perceived as a highly sequential and linearly structured field of expertise (Stevens, 2000). The linear cumulative acquisition of mathematics knowledge is a well entrenched aspect of school mathematics. But it is precisely this model of linear and sequential knowledge acquisition that is under attack by the advocates of integrated curriculum (Drake, 1999). Integrated curriculum modifies the linear model of knowledge acquisition by suggesting that particular content be introduced according to students' needs in solving authentic problems, instead of introducing content according to a set agenda. Such a differentiated approach to instruction is more democratic in the face-to-face encounters between students and teachers, and seems more organic and authentic in the way it responds to student needs. In practice, integrated curriculum might mean introducing more advanced mathematical concepts earlier to students, depending on the context, if complex "real" world problems demand it. It may also mean not "covering" basic skills at a particular grade, and letting students proceed to the next year without mastering particular algorithms. In many ways, the mathematics teacher has the most to risk in embracing an integrated curriculum. No other subject seems to invest so much in a curriculum arranged in terms of increasing logical complexity. The interruption of sequential linear development threatens the very infrastructure of school mathematics. Indeed, such a radical breaching of protocol around procedural and conceptual development might endanger the discipline by naming the ways in which it reproduces itself as a dominant cultural discourse (Skovsmose, 2005).

In *The Sociology of Mathematics Education: Mathematical Myths/Pedagogic Texts* Paul Dowling (1998) describes a number of "myths" associated with school mathematics, arguing that each myth functions as a way of centering mathematics within our culture. These are not myths in the sense of being untrue or religious, but rather cultural myths that people often take for granted. School mathematics plays a hugely significant role in structuring social life because of the way it functions as a "critical filter." It is a "high status" discipline, impacting the lives of most youth, both negatively and positively (Ahlquist, 2000). Dowling is interested in the socio-cultural framing of school mathematics, offering a way of critically interpreting the role of mathematics within an integrated curriculum. In this case, myths refer to beliefs about mathematics that frequently circulate inside and outside of schools. In this paper, I use two of Dowling's myths, the *myth of reference* and the *myth of utility*. The *myth of reference* refers to mathematics as a "system of exchange-values, a currency"(Dowling, 1998 p. 6) that underwrites "concrete" reality. According to this myth, the mathematician casts a penetrative gaze upon the non-mathematical world and is able to decode or translate the surface appearances into mathematical terms. The *myth of utility*, in contrast, is more utilitarian, constructing mathematics as a "reservoir of use-values" or a tool box (ibid). In this case, mathematics serves other domains and other inquiries, its meaning and value implicated in the ways in which it "works" for that which is outside its borders. Neither of these myths are politically neutral. They participate in the material structuring of society and are themselves consequences of power relations. Both myths, claims Dowling, read culture, community and context as incomplete until inscribed by/within mathematics. In the following case study, the two myths are evidenced in the actions and stated beliefs of teachers involved in a school wide curriculum integration project.

THE CASE STUDY

Our need to revise our middle school curriculum sprang from two parallel problems. The first was the shrinking of a five-year high school program into four, due to a government decision to eliminate grade 13 in Ontario, with the consequent rush to cover content at earlier grades. The second was a recognition that our courses were already overcrowded with too many performance outcomes, and that students were unable to draw connections between subjects. In an attempt to solve the first problem and ease student course loads at the senior school level, we decided to integrate four half credits into the already crowded grade seven and eight curriculum, thereby giving students a two-credit head start on their four-year high school diploma. In order to accommodate these new credits, which focussed on learning strategies, teachers were asked to integrate the new outcomes into their existent units. In an attempt to address the second problem, teachers were asked to create interdisciplinary units that addressed shared expectations. As one of two mathematics teachers working in both the middle and senior schools, I was asked to re-evaluate the arrangement of our mathematics curriculum. We began with a five day planning session attended by all participating teachers.

Teachers were initially concerned that there would now be less time to cover more material. Although duplication of content was occurring across disciplines, teachers were initially unaware of this redundancy. Curriculum had developed over the years to meet changing ministry guidelines, but no dialogue had occurred across discipline borders. Like so many students in all schools, ours were feeling the strain of a fragmented learning experience. Students devised coping mechanisms that involved slotting knowledge into pre-ordained compartments without any feel for the connections between subjects. For instance, they learned bar and line graphing techniques in at least three different courses, but each course framed the knowledge differently, and students failed to comprehend that they were performing the same task in different courses.

The first two planning days focussed on the four half-credit courses, for which government guidelines dictated, in no uncertain terms, the specific skills to be demonstrated. Administrators responsible for collecting the evidence of our having designed the appropriate performance indicators played the role of auditor, through no fault of their own, demanding lesson plans and rubrics as documentation, which were a necessary part of the legal process for obtaining a credit rating. The usual lists of outcomes, all carefully composed using the mandatory active verb tenses, circulated amongst the departments as we tried to accommodate the new content to be embedded in our courses. Major chunks of English and Art courses were then appropriated in the name of the new integrated curriculum. Representatives from these departments felt as though they were sacrificing the most, and indeed they were. At the end of day-two, relations between disciplines were already souring, as the humanities began to realize that mathematics and science were not being asked to "sacrifice" their instructional time. As a mathematics teacher, I altered a few units so as to address the new course expectations, but it was the other courses to which the administration repeatedly turned when asking for additional teaching time. The insidious hegemony that ranked certain kinds of knowledge over other kinds was felt immediately upon launching the project. Tacit assumptions that ranked mathematics and science above the arts were operating at every level.

Once the rather clerical task of documenting how we would address the new material was complete, and the necessary templates were composed, and our administration stamped it with approval, we then moved on to the issue of cross-curricular integration. We began by presenting detailed course outlines to each other which allowed us to identify the areas of overlap and the areas of connection. Many teachers were surprised to learn about course content in other fields and found the exercise exhilarating. We truly began to grasp the connectedness of what we were doing. A more holistic understanding of the student emerged. We were able to get outside of our own subject expertise, and began to comprehend the richness and diversity of student learning. The shear volume of learning outcomes was overwhelming. There were fourteen teachers working in the middle grades, spanning nine subjects, many with their Master's degrees. Most taught higher grades as well as seven and eight, which meant we invariably imposed our higher-level standards of behaviour and learning on the middle school students. As we went around the room, and the course descriptions were shared, the diverse personalities and teaching styles became evident, and I imagined a thirteen year old student trying to juggle all our different expectations, both explicit and implicit–our pedagogic assumptions, our class-management strategies and classroom dynamic, our personal histories, senses of humour, and our own learning styles. This was a potentially powerful affiliation of distinct identities brought together precisely because of our different disciplines. But our collective purpose was to transcend those discipline borders and create a new space for learning. It seemed suddenly very urgent that we recognize the importance of the personal, that we dwell on our diverse teacher identities, and realize that *we teach who we are.*

As the week progressed, and the traditional barriers were undone, teachers shared more of their personal styles. Casting aside some of the conventions of their discipline, teachers began to discuss their own idiosyncratic styles. The luxury of a full week of collaboration allowed us to delve deeply into our assumptions about learning. The other mathematics teacher and I discussed beliefs about education that we felt had been buried beneath the implicitly imposed discipline paradigm, to which we felt we had capitulated. Government guidelines, school expectations, and parental pressure had figured prominently in the shaping of our middle school mathematics curriculum. In discussing our curriculum within this new community, and away from the rest of the mathematics department, we were able to name the ways in which we felt coerced into complying with the dominant departmental practices. Department meetings often focussed on the careful alignment of sequentially organized learning outcomes, ensuring that specific skills and concepts could be traced from the earlier grades to the later grades. In disrupting that affiliation and demanding that we invest in this laterally associated community, where the focus moved from linear development to the co-temporal lateral associations that comprise a saturated learning experience, the integration project seemed to create a space where both students and teachers' diverse ways of knowing might be heard. What emerged during the week were our marginalized beliefs about mathematics and education which had been silenced by the entrenched practices of the status-quo. In this new community we were given a distinct voice and able to question the "common sense" and naturalized practices that were taken for granted by the mathematics department. The "expert" knowledge of veteran mathematics teachers had imposed strict role-playing rules on both of us as beginning mathematics teachers, to such an extent that any possibility of actively and critically interrogating the dominant paradigm was contravened. But as we troubled our discipline subject positions, if only slightly, we remembered these once silenced idiosyncratic beliefs about learners and how they experience mathematics. In breaking open the massive structure of mathematics curriculum, I was able to put the student back at the centre of my planning.

After the initial sharing of outlines, teachers from each discipline met with each other, discussed points where their two curricula might meld into one, and developed lesson plans and a concrete schedule of collaboration. We were ensconced in a large meeting room where lunch and coffee were delivered each day. We circulated about the room, arranging for one-hour brainstorm sessions involving pairs of different discipline teachers. Every discipline teacher sat down with every other discipline teacher, found some affinity between courses, and then collapsed their specific goals into more essential cross-curricular expectations. As a mathematics teacher, I found that my conversations were primarily driven by the other's course content. I chose not to open with my own curriculum objectives. Instead, I asked that they explain in more detail some of their major topics, and I then suggested mathematical tools that might be useful in exploring their topics. Whether it was understanding human anatomy, designing Egyptian art, learning musical forms, or analysing global demographics, there was always a way to insert mathematics. Without intending to, we moved from subject to subject and proffered mathematics as an explanatory foundation of all knowledge. The other discipline teachers accepted the premise that mathematics was a "tool" for exploring other forms of knowledge. In each subject, we suggested that mathematics might be "integrated" by seeing it as an abstraction within the other content, a hidden language of pattern and rule, a means of verifying the "truth" of the phenomena under study. Data management and pattern recognition were the two most often used topics in this regard. Although one might have imagined a scenario where the mathematics was framed by the other disciplines, it often seemed as though teachers began to see the mathematical nature of their own subjects.

When we returned to our classrooms, we implemented some of the integrated lesson plans during the first two weeks of the year, but it quickly became evident that teachers were being drawn back into their discipline borders by their departments. Communication between disciplines broke down, and no new integration emerged. There was no time for planning, and department meetings took priority over cross-discipline meetings. As many of the teachers in the middle grades were untenured, issues of power inhibited the capacity of these teachers to disrupt the traditional practices in the discipline departments. As the year progressed, I found myself more and more concerned with covering the textbook material so as to ensure that my students were prepared for mathematics courses offered the following year by other more veteran teachers. As the days passed, and I discussed issues with other department members, I found it difficult to resist the monumental coercive forces that define "legitimate" mathematics. In the math department meetings, the veteran teachers who taught only senior students expressed disdain and disapproval for an integrated curriculum that would result in eliminating mathematics content. One teacher suggested that the students' future was at risk because of the integrated curriculum, and that disrupting the sequential acquisition model would ruin their chances of obtaining the high grades they needed to enter university. Although such an argument seems initially motivated by concern for student success, it also functions as an argument for maintaining the current discipline borders for the sake of the status quo. Department meetings became sites where normalization of practice occurred. The veteran teachers were very resistant to change. The integrated model was perceived as threatening the unitary and cohesive community of the mathematics department. It was as though an integrated curriculum betrayed the discipline by dispersing the authority of mathematics across other disciplines. My attempt to be partially affiliated with the mathematics department and partially affiliated with the integration project created a conflict of loyalties. Teachers were unable to sustain the multiple affiliations, as departments demanded complete identification with the discipline, and made integration awkward and eventually impossible. In the mathematics department, I often felt coerced into abandoning the integration project, as though I had to choose between mathematics or that "other' project. The implicit refrain might have been "you're either with us or against us."

Senior mathematics teachers were particularly negative about the prospect of an open-ended activity as a form of summative assessment, in lieu of an exam. They were concerned that it would consume precious time without meeting many "concrete" expectations. There was a fear that the culminating activity would be an "organizational nightmare," and that take-home work would inevitably be completed by parents instead of students. Parents' investment in their child's success, within this extremely traditional school, was far-reaching and had huge impact on school culture and teacher identity. Exams had come to function as a way of assessing mathematics teacher accountability. The mathematics department was wedded to exams as a principal form of assessment. Beyond the concern that abandoning the *logical* sequence of mathematical development in favour of a more organic approach would cause a definite loss of content, there was another concern about how to assess simple knowledge and algorithmic skills within an integrated curriculum. Veteran mathematics teachers worried that the students would not have *proven* their understanding if they hadn't done so in isolation from other disciplines. Because this concern was further reinforced by parents' concerns over the prospect of "no objective assessment," the middle-school students were given in-class exams, and their performance in mathematics was documented and recorded separately, as though nothing had been integrated.

CONCLUDING REMARKS

Exams were crucial in repeatedly re-establishing the legitimacy of mathematics as a discipline, validating both the content and the teacher's credibility. The exam, argues Carrie Paechter, "controls the work of the teacher, who is constrained by the pressure for examination success to remain within the boundaries of the examination-defined subject."(Paechter, 2000: 136). She draws from the work of Foucault, suggesting that disciplinary power works through the teachers, "The examiner both carries out the scrutiny and is disciplined by its demands." (151) In our case, assessment and accountability interrupted all attempts at integration because school disciplines were wedded to their specific forms of assessment, and because teachers were ultimately bound by those same forms of assessment.

My attempts at generating cross-discipline connections contributed to what Dowling describes as the "myth of utility" which functions to legitimate mathematics in schools (Dowling, 1998). Again, Dowling does not dispute the utility of mathematics, nor suggest that utility is in itself problematic, but advocates for an increased awareness of how these habits of legitimation are enacted in schools, so as to raise teachers' critical consciousness around the structuring of the institution. By repeatedly declining to define my own needs during the collaborative stage, and inviting the other to specify her own context, I set in motion a power dynamic that inscribed my knowledge into that of all the other disciplines, whether it be geography, art or gym. Although one might see mathematics as a "hand maiden" to the other disciplines in this instance, Dowling argues that the myth of service further centers mathematics within education, establishing itself as foundational to all knowledge. This notion of foundational knowledge presumes the existence of an essential explanatory system and contravenes the more hermeneutic and dynamic epistemology at the heart of integrated curriculum.

The disapproval expressed by the veteran mathematics teachers represents a re-inscribing of the disciplinary power of school mathematics. Their strategic use of "students at risk" in sanctioning sequential curriculum, and their disdain for alternative assessments must be read as acts that reproduce and regulate the status quo. Although their disapproval was prohibitive, it was also a structuring of what was possible, and thereby enacted the disciplinary power that produces and governs subjectivities in schools. Disciplinary power circulates through forms of instruction and assessment, sometimes so tacitly that it is difficult to resist, and always in ways that validate some voices and practices while dismissing others. Speakers often unknowingly participate in the further enculturation of dominant instructional practices, although there is always space for altering and troubling these same practices if we are willing to risk our own comfort.

As a member of the mathematics department, I felt an obligation to share in the emergent collective identity; my resistance to well-entrenched master narratives about curriculum engendered a feeling of alienation which, in turn, made me more vulnerable. It was this vulnerability which became the site for the co-construction of my teacher identity in conformance with the expectations of the department. These moments of identity construction point to the *regimes of experience and accountability* (Wenger, 1998) that regulate membership in communities of practice. Similarly, the government curriculum documents functioned as technologies of surveillance that were all too easily internalized. Despite my original commitment to cross discipline borders, I felt the deep structural impact of an external regulating body. The government curriculum guidelines addressed each teacher or rather each discipline separately, and in doing so, we were "hailed" as subject teachers first, and only obliquely recognized as potentially cross-discipline collaborative educators through the document's reference to "connections." The consequent sense of belonging which emerged was grounded once more within the discipline, where the documents officially (mis)recognized me as a member.

The case study highlights the complex ways in which teachers develop their identities in affiliation with others. We can see that mathematics teacher identity is structured as a singular or unitary social entity with little allowance for diverse or multiple affiliations. This notion of the loyal mathematics teacher whose interest is to serve the discipline by continuing to defend its border and status needs further examination. It may be that such habits of enculturation impact all aspects of instruction and impose a sameness on mathematics departments that might need some disrupting. The case study suggests that there is little tolerance for a multiply affiliated teacher identity that isn't exhaustively defined within the discipline. Further research might examine ways of recognizing multiple affiliations within mathematics teacher identity, and explore the school and department practices that neutralize difference and demand identification. Mathematics departments may rely on a notion of teacher identity that idealizes sameness and consensus while demoting difference and dissent. Homi Bhabha urges us to re-think identity as a process of affiliation rather than autonomy, a site of partial solidarity rather than sovereign mastery, where the complementarity and reciprocity of multiply performed identities might be more ethically recognized (Bhabha, 2003: 6). Re-thinking identity in terms of contingency and provisional commitment suggests that we move away from a paradigm of consensus where contradictions are resolved and difference is sutured over with self-censorship, and towards, possibly, the prospect of a coalition that allows for divergent positions. Amongst mathematics teachers, then, there is a need to confront the notion of identity wholeness or cohesion, to reconfigure identity as an ongoing process that resists closure, so that collaborative or collective enterprises can become sites where mutual alterity is recognized.

REFERENCES

Appelbaum, P.M. (1995). *Popular culture, educational discourse, and mathematics.* Albany, NY: State University Press.

Beane, J. (1997). *Curriculum integration: Designing the core of democratic education.* New York: Teachers College press.

Bernstein, B. (2000). *Pedagogy, Symbolic Control and Identity: Theory, Research, Critique,* Revised edition. Oxford, UK: Rowman & Littlefield Publishers.

Bhabha, H. (2003). A statement for the Critical Inquiry board. *Critical Inquiry* 30(2).

Bhabha, H. (1994). *The location of culture.* New York, NY: Routledge.

Bohl, J. V. & Van Zoest, L. R. (2002). Learning through identity: A new unit of analysis for studying teacher development. Proceedings of the Twenty-sixth of the Annual Meeting of the International Group for the Psychology of Mathematics Education, Norwich, England.

Bourdieu, P & Passeron, C. J. (1977). *Reproduction in Education, Society and Culture.* London, UK: Sage.

Britzman, D. (2003). *Practice makes practice: A critical study of learning to teach, 2nd edition.* New York, NY: State University of New York Press.

Butler, J. (2000). Restaging the universal. In J. Butler, E. Laclau & S. Zizek (Eds.). *Contingency, hegemony, Universality: Contemporary dialogues on the left.* London, UK: Verso Books.

Butler, J. (1993). *Bodies that matter: On the discursive limits of "sex."* New York: Routledge.

Covaleski, J. F. (1993). Power goes to school: Teachers, Students and Discipline. Access at www.ed.uiuc.edu/EPS/PES-Yearbook/93_docs/93contents.html.

Delpit, L. (1995). *Other People's Children: Cultural Conflict in the Classroom.* New York, NY: The New York.

Dowling, P. (1998) *Sociology of Mathematics Education: Mathematical myths/ Pedagogic Texts.* London, UK: The Falmer Press.

Drake, S. M., Bebbington, J., Laksman, S., Mackie, P., Maynes, N., & Wayne, L. (1992). *Developing an integrated curriculum using the story model.* Toronto, Canada: University of Toronto Press

Drake, S. (1998). *Creating integrated curriculum: Proven ways to increase student learning.* Thousand Oaks, CA: Corwin.

Felman, S. (1990). *Jacques Lacan and the adventure of insight: Psychoanalysis in contemporary culture.* Cambridge, MA: Harvard University Press.

Foucault, M. (1977). *Discipline and Punish: The Birth of the Prison.* London, UK: Penguin.

Hargreaves, A., Earl, L. & Ryan, J. (1996). Schooling for Change: Reinventing Education for Early Adolescents. London, UK: Falmer Press.

Lave, J. & Wenger, E. (1991). *Situated learning: Legitimate peripheral participation.* Cambridge, MA: Cambridge University Press.

Paechter, C. (2000). *Changing School Subjects: Power, Gender and Curriculum.* Philadelphia, PA: Open University Press.

Popkewitz, T. (1998). *Struggling for the soul: The politics of schooling and the construction of the teacher.* New York, NY: Teachers College Press.

Sasaki, B. (2002). Toward a pedagogy of coalition. In A. A. Macdonald & S. Sanchez-Casal (Eds.) *Twenty-first-century feminist classrooms: Pedagogies of identity and difference.* New York, NY: Palgrave Macmillan.

Schoenfeld, A.H. (1999). Looking toward the 21st century: Challenges of educational theory and practice. *Educational Researcher* 28(7), 4–14.

Skovsmose, O & Valero, P. (2002). Democratic access to powerful mathematical ideas. In L. D. English (Ed.), *The Handbook of International Research in Mathematics Education*. Mahwah, New Jersey: Lawrence Erlbaum Ass.

Stevens, R. (2000). Who counts what as math? Emergent and Assigned mathematics problems in a problems-based classroom. In J. Boaler (Ed.) *Multiple Perspectives on Mathematics Teaching and Learning*. London, UK: Ablex Publishing, 105–145.

Walshaw, M. (2004). A powerful theory of active engagement. *For the Learning of Mathematics* 24(3). 4–10

Zevenbergen, R. (2003). Teachers' beliefs about teaching mathematics to students from socially disadvantaged backgrounds: Implications for social justice. In L. Burton (Ed.) *Which way social justice and mathematics education?* London, UK: Praeger Publishers.

SECTION 4

SOCIAL JUSTICE IN, AND THROUGH,
MATHEMATICS EDUCATION

CHAPTER 23

SOCIAL JUSTICE AND MATHEMATICS EDUCATION
Issues, Dilemmas, Excellence and Equity

Bharath Sriraman
The University of Montana

Olof Steinthorsdottir
University of North Carolina–Chapel Hill

ABSTRACT

This article explores reasons for educational research and practice in social justice from evolutionary, ideological and philosophical viewpoints. The tension between nihilistic and empathetic tendencies within our history is used to reflexively examine the origins and causes of inequity with emphasis on the works of giants such as Paolo Freire, John Dewey, Karl Marx, and Vivekananda. Finally we address one particular issue in depth, namely the tension between excellence and equity in talent development in schools, east and west.

Critical Issues in Mathematics Education, pages 319–336
Copyright © 2009 by Information Age Publishing
All rights of reproduction in any form reserved.

319

WHY SOCIAL JUSTICE?

It is a basic fact that life around us constantly reveals inequities such as rich versus poor; the educated versus uneducated; those in power versus those without power; wealthy countries versus poor countries; citizens versus guest/transient workers; higher social standing and mobility versus being stuck in abject status quos; affluent neighbourhoods and schools versus ghettos and the remnants of social Darwinism; ad infinitum. While most of the world is caught up in dealing with the excruciating minutiae and the vexing exasperations of day-to-day life simply to survive, we in academia are in the privileged position to ponder over the bigger questions confronting humanity. Why do inequities exist in the first place? What are their origins? Are educators' attempts to address social justice problems in the classrooms simply attempts at "patching up" things that are in essence atomically broken., i.e., an allopathic attempt of getting rid of symptoms so we don't have to deal with the real objective roots of problems. Another analogy is that of surgical procedures done on an ad-hoc basis to remedy defects that arise as opposed to caring for the well being of the whole and getting to the root of problems. Or is social justice research in mathematics education, a well intentioned movement around the world to present arguments for the necessity to address social inequities via mathematics education, i.e., to give a deeper meaning to the purpose of education. A nihilist would choose the allopathic (surgical) answer whereas the empathetic individual would choose the latter. Most of us find ourselves somewhere in between, in perpetual but necessary tension to solve the bigger problems around us. Some positions about the origins of inequity and injustices within educational and societal mechanisms are now presented followed by addressing the issue of the tension between excellence and equity within educational systems, east and west.

The Darwinian explanation suggests that inequity is simply one of the many natural mechanisms that have arisen over the course of our evolution. If we view ourselves as creatures whose sole purpose in life is to survive and to have progeny, then it is evident that the competition for the same natural resources would leave others in the wake. The strictly Darwinian explanation would suggest that certain groups are doomed to perish simply because they are unable to cope with changes occurring in their environment. Unlike other mammals, we tend to hoard natural resources, much more than we can possibly use and at the same time, we also exhibit tendencies towards altruism which are paradoxical and unexplainable in strictly biological terms. In fact, Charles Darwin (1871) in *The Descent of Man*, posed the question whether the phenomenon of moral behaviour in humans could be explained in evolutionary terms, viz., natural selection. The evolution of social systems (religious, ideological, political) of various kinds is not explainable strictly in Darwinian terms. Comte (1972) proposed a stage theory for our social evolution in which humanity moves from a theological stage onto a metaphysical stage onto a "positive" stage. It is too difficult to explain the meaning of the third stage, but simply put, we reject absolutism of all kinds and we strive for knowledge based on rationality.

The present day economic inequity in the world is best illustrated by the fact that many universities in the West have larger budgets than the GNP of many nations in Africa, Asia and South America. Despite the current state of affairs we are also creatures of ideas who over the course of our evolution have moved away from a strictly clannish and genotypic connection to a memetic connection.[1] We conglomerate over common ideas or ideals as evidenced in the spread of the numerous great world religions, which link together people across a spectrum of class, culture, race, socioeconomic status and nationality. These two special issues of PoME journal on social justice are a memetic product. Similarly ideologies such as Marxism connect people from diverse socioeconomic and cultural backgrounds. Even the so-called phenomenon of "globalization" is nothing new from the point of view of history. There is sufficient historical evidence that even in periods when means of transport and communication had not been developed, oriental civilization penetrated into the West. Iran and Greece were in contact with each other, and many South Asians found their way to Greece and vice-versa through this contact (Radhakrishnan, 1964). Asoka's[2] missions to the West, and Alexander's influence on Egypt, Iran, and North West India, produced a cross-fertilization of cultures.

Another big, intensive, but relatively "localized" process, which we may, also call "globalization," occurred in Asia and Europe, in the expansion of Christianity and Islam in the Middle Ages, in the shadows of the Roman and (the emerging) Caliphate Empires. In the late Middle Ages, States began to take shape as components of a new form of Empire. The scenario resulting from this process of European "globalization, prevails until now. In the sort of jig-saw puzzle which characterize the political dynamics present in this process, the idea of a Nation became strong. States and Nations are different concepts, as well as Political Dynamics and Cultural Dynamics. The political dimension of this process prevailed and something vaguely called State/Nation began to take shape as the primary unit of the European scenario. The Empire which emerged in the Late Middle Ages and the Renaissance as the assemblage of such State/Nations, although fragile, mainly due to power struggle, favored the development of the ideological, intellectual and material bases for building up the magnificent structure of Science and Technology, anchored in Mathematics, supporting a capitalistic socio-economic structure. The expanding capitalism, supported by religious ideology and a strong Science and Technology, had, as a consequence, a new form of globalization, now effectively engaging the entire

Globe. The great navigations and the consequent conquest and colonization, completely disclosed the fragility of a possible European Empire. The internal contradictions of State, as a political arrangement, and of Nation, as a cultural arrangement, emerged, in many forms. Religious and linguistic conflicts, even genocide, within a State/Nation became not rare facts. Indeed, they are not over. As a result of all these processes, Education was, probably, the most affected institution. Educational proposals, even curricula, are noticed in this era. The influence of national characteristics interfered with objectives derived from the new World scenario. The development of Science and Technology, obviously related to the educational systems, was unequal. Interchanges intensified. The Industrial Revolution made Science and Technology a determinant of progress. Hence, the enormous competition among European States, which intensified during the 19th century and early 20th century, raised Science and Technology, which became increasingly dependent on Mathematics, to top priority (Sriraman & Törner, 2007). One terrible consequence of this competition between European states was the advent of colonization, the consequences of which the world is still very much experiencing (Sriraman, 2007).

Although many countries in Asia, Africa and South America became "free" from the yoke of colonialism in the last century, this freedom left in its wake uprooted peoples when colonial masters started drawing lines on maps to "equitably" partition land in various regions of the world. Hopefully the reader realizes the irony in my previous statement. There was considerable loss of subsistence lifestyles, loss of indigenous cultures and traditional knowledge. The consequences of colonization were not any different in North America and in Australasia. The outcome of the colonial period of our history was Education as an Institution and a new economic structure being implanted in various regions of the world with the explicit purpose of perpetuating the very structures created to maintain colonialism, namely oppression of the many by a few. Indeed Karl Marx and Friedrich Engels' monumental writings[3] address issues such as exploitation of workers within a capitalistic economic system and the problem of materialism confronting humanity, which would inevitably lead to class struggles and revolutions. Many of the foundational writings of social justice can be traced back to the ideas proposed by Marx and Engels. Today's study of the ecological footprints left by the industrialized nations reveals the obscene differences in resource consumption[4] between rich and poor nations, a natural consequence of materialism run amok as predicted by Marx and Engels.

The question then is: can emancipatory and social justice pedagogies really free individuals from oppression at a societal level? How can this be possible without it occurring at the individual level first? Freire (1998) himself wrote that the central problem was "How can the oppressed, as divided, unauthentic beings, participate in developing the pedagogy of their liberation? Only as they discover themselves to be "hosts" of the oppressor can they contribute to the midwifery of their liberating pedagogy." Clearly Freire is stating that the oppressed adhere to the oppressor and have to break free. If individuals do not subjectively and intrinsically feel free, how can any educational or social mechanism make this happen no matter how good the intention? Cho & Lewis (2005) recently re-emphasized the aforementioned essence of Freire's pedagogy from the point of view of psychology and the problems with the attempts by Marxist theorists to transform Freire's "pedagogy of the oppressed" into a "pedagogy of revolution." They write that "oppression has an existence in the unconscious such that those that are oppressed form passionate attachments to the forms of power that oppress them" (p. 313), and it is necessary for social justice researchers and Marxist theorizers to recognize and address this important issue. Cho & Lewis (2005) formulate several challenges[5] to Marxist theorizers as follows:

> ...part of the discomfort with "revolutionary pedagogy," is that the project of liberation often appears to be presupposing universal notions of what it means to be oppressed, liberated, and how this movement is to be made—often the problem lies in Freire's emphasis on material relations and not on the issue of patriarchy or colonization....[w]ith no clear resolution to the issue of authority, libratory pedagogies can portray particularist notions of oppression and liberation in universal was and to impose these visions of oppression and liberation upon others through a kind of vanguardism, which can ironically replicate relations of oppression other than overcome them thus returning us to the problem with which Freire begins his analysis in the first place. (p. 314)

In India, the problem of individual liberation has been addressed within Hindu philosophy by numerous scholars, especially social reformers in the 19th and 20th centuries. Vivekananda (1863–1902) belonged to a branch of Hindu philosophy called Vedanta (see Sriraman & Benesch, 2005), in particular to a special strand of Vedanta, which holds that no individual can be completely free unless every one else is also free (from oppression). In other words, we as individuals are obliged to act to better society. Vivekananda was able to move beyond the prevalent dogmatic caste system which characterized Indian society and propose a theory of action which necessitated that each of us consciously act towards bettering the lot of our fellow humans, if our goal is to ultimately liberate ourselves and become enlightened. From a Freirean perspective it is not possible to "empower people..."—the best we can do is to create conditions to facilitate, support people empowering themselves, and to work along side in common struggle.

Given this overview of the origins on inequities within society, we now discuss a particular contentious issue especially in the United States, namely gifted/talented education albeit from a social justice viewpoint.

THE CASE OF GIFTED/TALENTED EDUCATION

Gifted education in the United States has been subject to much criticism due to the perception that it is either elitist, or caters to students who are socio-economically privileged (Clark, 1997). Some theorists have also reduced this problem to that of nature versus nurture, i.e., students who are labeled academically promising or gifted are more likely to have received the benefits of a socio-cultural upbringing where reading, music and other creative activities are encouraged with parents providing additional support for children to pursue intellectually stimulating activities. This is in contrast to students that have not had the privilege of growing up in an intellectually stimulating social environment. Gifted education is also construed as being elitist because it caters to the needs of a small proportion of students based on identification measures which may be biased. There is also considerable debate over the definitions of giftedness with the word "gifted" lending itself to negative connotations. On the other hand, public schools in the U.S do provide a lot of resources towards students in need of remediation in reading and mathematics at the elementary levels. Kent & Lawrence (2002) pointed to the fact that in the U.S, on an average $30 billion is spent on special education programs, whereas funding for gifted education is less than 1% of this amount.

Sriraman & Steinthorsdottir (2007a) argue that if we view catering/nurturing talent of students that are academically promising as being elitist, then are we not being "unjust" towards the abilities of these students and squandering the opportunity to develop their talent. So, addressing this issue becomes a catch-22 situation unless we try to construct a completely different theoretical perspective which moves away from the regressive and dogmatic spirals of various arguments. According to Sriraman (2005) it is a basic fact that the comforts and security of today's technologically evolved society is due to the innovative spirit and the toil of scientists, inventors, investors, artists and leaders who have made the present comfortable state possible. In a similar vein, Martinsen (2003) in his introduction to a special issue of the *Scandinavian Journal of Education* focused on creativity wrote:

> Despite differences of opinion as regards the definition of this construct it can be argued that creativity is fundamental to all individual and societal development. Societies need inventors, creative artists, managers, teachers, authors, philosophers, entrepreneurs, therapists and more. Moreover, most people will need to restructure their understanding, find new solutions, new

challenges and new ideas frequently during their lifetime. One may also suspect that the capacity to create and to solve complex or novel problems will become more and more important in an increasingly regulated, technology-oriented and complex world. (p. 227)

The Special Place of Mathematics Education

The field of mathematics has been criticized for its academic elitism. There is a growing canon of studies which indicates that the institution of mathematics tends to marginalize women and minorities (Burton, 2004; Herzig, 2004). Moreover several studies have shown that the knowledge produced by the institution of mathematics is based on a patriarchal structure and a male-centered epistemology. There is also adequate empirical evidence in the U.S that academic fields related to mathematics continues to be predominantly male (Seymour, 1995). Further, in the U.S, the representation of minorities (African America, Native American) at the post-graduate level is still miniscule (Seymour & Hewitt, 1997; Sriraman & Steinthorsdottir 2007b, 2007c). Mathematics has also historically served as the gatekeeper to numerous other areas of study. For instance the hard sciences, schools of engineering and business typically rely on the Calculus sequence as a way to filter out students unable to fulfill program pre-requisites.

In numerous countries around the world, particularly in Asia, entry to government subsidized programs in engineering and the sciences is highly competitive and require students to score in the top 1 percentile in entrance exams in which mathematics is a major component. The situation is not so different in North America as evidenced in the importance of standardized tests like SAT or ACT to gain entry into college programs. It is not uncommon to hear politicians use schools' performance on mathematics assessments as a reference point to criticize public school programs and teachers (e.g., the passing of the No Child Left Behind Act in the U.S).

MATHEMATICS EDUCATION: DEMOCRATIZATION AND GLOBALIZATION

Based on the previous paragraphs, we can say that mathematics education has everything to do with today's socio-cultural, political and economic scenario. In particular, mathematics education has much more to do with politics, in its broad sense, than with mathematics, in its inner sense (D'Ambrosio, 1990, 1994a, 1994b, 1998, 1999, 2007; Moreno & Trigo, 2007; Sriraman & Törner, 2007). Mathematics seen in its entirety can be viewed as a means of empowerment as well as a means to oppress at the other end of the spectrum. For instance Schoenfeld (2004) in his survey of the state of mathematics education in the U.S wrote "Is mathematics for the elite or for the masses? Are there tensions between "excellence" and "equity"? Should mathematics be seen as a democratizing force or as a vehicle for maintaining the status quo?" (p.253). Skovsmose (2004) poses the questions: Is it true that mathematics has no social significance? Or does also mathematics provide a crucial resource for social change? in other words: How may mathematics and power be interrelated? We further ask what does this have to do with current educational structures and pedagogy. Skovsmose (2005) further discusses critically the relations between mathematics, society and citizenship. Skovsmose's program of critical mathematics education give challenges connected to issues of globalization, content and applications of mathematics, mathematics as basis for actions in society, and on empowerment and mathematical literacy (mathemacy).

These questions are more generally addressed by Spring (2006), who summarizes the relationship between pedagogies and the economic needs of nation/states. His thesis is that the present need for nation/states to prepare workers for the global economy has resulted in the creation of an "educational security state" where an elaborate accountability-based system of testing is used to control teachers and students. Spring correctly points out that:

> Both teachers and students become subservient to an industrial–consumer paradigm that integrates education and economic planning. This educational model has prevailed over classical forms of education such as Confucianism, Islam, and Christianity and their concerns with creating a just and ethical society through the analysis and discussion of sacred and classical texts. It has also prevailed over progressive pedagogy designed to prepare students to reconstruct society. In the 21st century, national school systems have similar grades and promotion plans, instructional methods, curriculum organization, and linkages between secondary and higher education. Most national school systems are organized to serve an industrial–consumer state. As later explained, the industrial–consumer state is premised on the idea that a good society involves economic growth resulting from increased production and consumption of goods. In the industrial–consumer state, education is organized to serve the goal of economic growth. (p. 105)

Therefore, in order to counter this organized push for eliminating progressive education, it is important that educators be open to alternative models of pedagogies which attempt to move beyond the current dominant "industrial consumer state" model of education. In order to do so, it is imperative that we first understand the dichotomy between excellence and equity in mathematics education. The general questions we raise in this article are:

1. Is there a way in which one can resolve talent development, particularly in mathematics education so that that the curriculum and/or instruction is equitable to all the students in the classroom?
2. Can excellence and equity co-exist or does attending to one compromise the other, i.e., excellence at the sacrifice of equity; equity at the sacrifice of excellence?

REFORMULATING THE PROBLEM
WITHIN A HEGELIAN DIALECTIC

In western philosophy, the use of dialectics is seen in ancient Greek philosophy (e.g., Socrates, Plato etc). The dialectic consists of theses and anti-theses which are in opposition to each other. The tension between these two opposing forces eventually leads to a synthesis which in turn becomes the thesis of a new dialectic. It should be noted though that the idea or concept of a dialectic is much older and found in Hindu and Buddhist philosophy. It was however, Hegel (1770–1831), who applied the dialectic method to create a model of a direction in which history unfolds. Although Hegel was in very abstract terms speaking of this evolution towards an absolute idea (a kind of philosophical idealism), his dialectic was also applied by Marx towards material conditions to formulate what is known as the Marxist dialectic or dialectical materialism. Figure 23.1 shows a visual representation of Hegel's dialectical model.[6] The Hegelian dialectic is applicable to the problem of resolving/understanding the tension between the opposing forces of excellence and equity within the framework of education. The tension between these opposing forces is seen in educational systems in numerous parts of the world. We present some examples (from east and west) and urge the reader to attempt to apply the dialectic to resolving this issue.

EAST VERSUS WEST

Educational systems are heavily influenced by the social and cultural ideologies that characterize the particular society (Clark, 1997; Kim, 2005; Spring, 2006). Kim (2005) characterizes "western" systems of education as fostering creativity and entrepreneurship when compared to "eastern" systems where more emphasis is laid on compliance, memorization and repetitive work. However East Asian countries stress the values of effort, hard work, perseverance and a general high regard for education and teachers from society with adequate funding for public schools and family support. Again in comparison, in the U.S., public schools are poorly funded, teachers are in general not adequately compensated nor supported by parents, and there is a

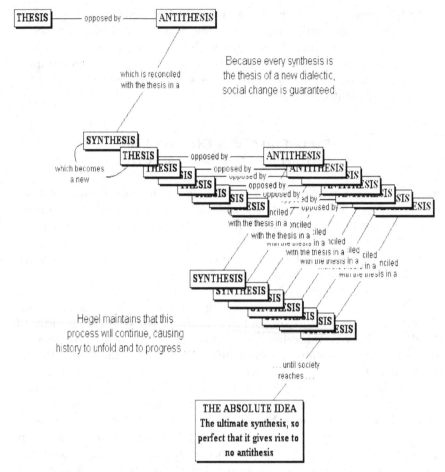

Figure 23.1   The Hegelian Dialectic.

decline in the number of students who graduate from high school (Haynes & Chalker, 1998; Hodgkinson, 1991). Among the western developed democratic nations, the U.S has the highest prison population proportion, 30% of whom are high school dropouts (Hodgkinson, 1991).

In China, Japan and Korea, the writings of Confucius (551–479 BCE), namely a system of morals and ethics, influenced the educational systems. The purpose of studying Confucian texts was to create a citizenry that were moral and worked toward the general good of society. Competitive exams formed a cornerstone of this system, in order to select the best people for positions in the government. The modern day legacy of this system is the obsession of students in these societies to perform well on the highly competitive college entrance exams for the limited number of seats in the science and engineering tracks. The tension and contradiction within this system is apparent in the fact that although these societies value education, the examination system is highly constrictive, inhibits creativity and used to stratify society in general. Late bloomers do not have a chance to succeed within such an educational system. In the U.S., despite the problems within the educational system and the general lack of enthusiasm from society to fund academic programs that benefit students, the system in general allows for second-chances, for individuals to pursue college later in life in spite of earlier setbacks. On the other hand, for many students, particularly from poorer school districts, socio-economic circumstances may not allow for such second chances. The U.S model of an industrial-consumer state based on the capitalistic ideal of producing and consuming goods, forces students into circumstances which make it economically unfeasible particularly for students from poor socio-economic backgrounds to veer vocations and pursue higher education. Clearly both systems, based on different ideologies have strengths and weaknesses that are a function of their particular historical and cultural roots. Social change is possible within and across both systems but requires changes within cultural and socio-political ideals of eastern and western societies. Both systems have intrinsic flaws that undermine developing the talents of students. There are however solutions proposed by numerous educational philosophers and activists which reveal a synthesis of eastern and western ideas and provide for the possibility of systemic change for society as presented in the Hegelian dialectic (Sriraman & Steinthorsdottir, 2007a).

PROGRESSIVE EDUCATION AND CRITICAL PEDAGOGIES

The principles of progressive education as outlined by John Dewey inspired educational structures in the former Soviet Union and China in the early part of the 20th century and ironically have been forgotten by policy makers in the U.S and today viewed as a "dissenting" tradition (Spring, 2006). The goal of the progressive movement was to empower students by attending to student strengths and interests; stressing constructivist approaches to learning; and integrating the curriculum in order to improve society. Interdisciplinary approaches to the teaching and learning of mathematics create natural differentiation and enrichment opportunities for all students including the talented students (Sriraman & Dahl, 2007). In the age of globalization, societies have realized that there is an urgent need to move beyond the ego-centric needs of their particular society towards the shared needs of the planet. Numerous educational theorists have stressed the need for educational structures to stress co-operation, social and environmental justice and wisdom through schooling (Atkinson, 1994; Kurth-Schai, 1992; Sternberg, 1998, 2001).

The Greek philosopher, Socrates, said that our ultimate purpose is to will Good for humankind. If one accepts this premise and connects this to the purpose of contemporary education (and not only the discipline of mathematics), then it becomes clear that two broad goals of education must be:

1. to produce citizenry who are capable of thinking critically and willing to engage in such thought; and
2. to develop an awareness for the value of making reasoned choices that seek to will Good for humanity.

Critical thinking is often not associated with the teaching and learning of mathematics, however the two disciplines share many common traits. Historically, training in critical thinking makes explicit use of formal logic in order to draw inferences and/or make comparisons. Mathematics can be presented as being structured and rigid in the same way, but this need not be so. Plato in *Apology* suggested that one should not blindly accept a persuasive argument without being aware of the reasons why the argument is persuasive (Plato, 1999). In other words, a critical thinker must be able to examine the validity of the logic used in an otherwise eloquent and persuasive argument, as well as to verify the facts and assumptions that are involved. Likewise, students of mathematics can be taught to question the didactic claims of their teachers and can be taught to validate mathematical propositions based on their own emerging skills and frames of reference. For the ancient Greeks, critical thinking not only involved an examination of the eloquent words and actions of other people but also an examination of one's own thoughts and capabilities. The traditional constructs of critical thinking have been criticized as being "a narrow way of thinking, excessively centered on reasoning and argumentation" (Smith, 2001, p.349) which do not take into account imagination or intuition, and do not nurture the creative (generative) side of thinking (Walters, 1994). If we believe well-stated arguments based on the position of the author (ad hominem) or the authority of the teacher, then we can easily be misled by charismatic voices and thus fail to question and think critically for ourselves. Another criticism about the traditional view of critical thinking is that the excessive focus on formal logic, rhetorical ploys, fallacies and argument construction encourages students to view critical thinking as merely an arduous mental exercise without any wide-ranging applicability (Adler, 1991; Baron, 1988; McPeck, 1984). Students may lack the confidence to challenge propaganda or advertising because they might feel the rigorous tools required to think critically are beyond their abilities. Likewise when mathematics is taught as formal algorithms, with learning restricted to successful computation without any requirement to apply this mathematics to the real world, then it weakens the growth of knowledge for students (Sriraman & Adrian, 2004). Students must learn not only to perform in the context of their own world, but to explain what they are doing and why it is important (Sriraman, Knott & Adrian, in press).

Given these criticisms of the traditional definitions of critical thinking, we adopt a modified view of critical thinking that is compatible with the expectations we generally hold for beginning students. We define critical thinking as "reasonable reflective thinking that is focused on deciding what to believe or do" (Ennis, 1991, p.6) with the added requirement that it be connected to real life. This pragmatic view enables students at very early stages to understand the cultural and instructional influences that ought to influence accepted thought (Bacon, 1902). The rationale for choosing this definition is that it requires that critical thinking skills apply to real world problems, brings to the forefront the issue of bias in critical thinking (Paul, 1990), and makes use of appropriate questioning to stimulate students' reflections on problems (Simpson, 1996).

Paolo Freire (1921–1997), the Brazilian educator and social reformist, came of humble backgrounds. His book *Pedagogy of the Oppressed* (Freire, 1998) is perhaps the most frequently cited Marxist-influenced work in educational literature. Freire (1998) addressed the power dynamics between the oppressed and the oppressors (including the dynamic between teacher and student), and that the way toward liberation is through political movements and political struggle, of which literacy is but one part. Thus his emphasis on *writing* the world, is beyond literacy. Clearly, literacy (i.e., reading the world) is also an integral and necessary part of this process. Freire's banking concept holds that students are knowledgeable beings with the intrinsic capacity of creating knowledge *with the teacher*, as opposed to being empty buckets of ignorance or simply "files" or automatons dependent on the teacher's absolute authority to learn and construct new knowledge. It is also important to note that Freire emphasized critical literacy as opposed to functional literacy. The Organization for Economic Co-Operation and Development (OECD, 2004) has attempted to promote mathematical literacy in numerous countries through international tests like the Program for International Student Assessment (PISA).

Freire (1998) suggested that pedagogical practices should support education for liberation and emphasized problem-posing pedagogies that strive "for the emergence of consciousness and critical intervention in reality" (p.62). Problem posing pedagogies are necessary if the goal of education is to challenge inequities. Freire's writing suggests a pedagogy which promotes greater social awareness or a social consciousness appropriate for initiating major shifts in thinking. An outstanding example of this pedagogy in practice is Gutstein's (2006) work *Reading and Writing the World with Mathematics.*

Gutstein's work also points out the obstacles to such a pedagogy within a school system, particularly institutional resistance from administration and other stake holders within a school district (Sriraman, 2007; Sriraman & Steinthorsdottir, 2007c).

CONCLUDING NOTE

The present day tension between equity and excellence in the school system in the U.S is symptomatic of the deeper political problems polarizing politics and people in this society. While the Hamiltonian tradition stresses elitism and division of classes based on the "cognitive" capital possessed by people, the Jacksonian tradition suggests everyone is equal no matter what. Sternberg (1996) points out that there exists a forgotten third alternative to the polarizing positions characterizing education today, namely the Jeffersonian tradition whose essence is that "people are indeed all equal in terms of political and social rights and should have equal opportunities; but they do not necessarily avail themselves equally of these opportunities and hence do not get rewarded for what they accomplish, *given equal opportunity*, rather than what might have, should have or could have accomplished" (pp. 262–263). The challenge facing society today (in the U.S and elsewhere) is to first create this equality in educational opportunity. This is a necessary first step in resolving the tension between equity and excellence.

ACKNOWLEDGMENT

A portion of this article is based on the first monograph of the Montana Mathematics Enthusiast entitled International Perspectives on Social Justice in Mathematics Education. http://www.math.umt.edu/TMME/Monograph1/

The section on the tension between equity and excellence has been adapted from an article appearing in a special issue of the *Mediterranean Journal for Research in Mathematics Education* on talent development. Both these books were edited by the first author.

REFERENCES

Adler, J.E. (1991). Critical thinking, a deflated defense. A critical study of John. E. McPeck's Teaching Critical Thinking: Dialogue and dialectic. *Informal Logic,* 13(2), 61–78.

Atkinson, D. (1994). A Tao for schools. *Interchange,* 25(2), 145–155.

Bacon, F. (1902). *Novum Organum* (1620). P.F. Collier & Son, New York.

Baron, J. (1988). *Thinking and deciding.* Cambridge, UK: Cambridge University Press.

Burton.L. (2004). *Mathematicians as Enquirers: Learning about Learning Mathematics.* Kluwer Academic Publishers.

Cho, D., & Lewis, T. (2005). The persistent life of oppression: The unconscious, power and subjectivity. Interchange: A Quarterly Review of Education, 36(3), 313–329.

Comte, A. (1972). Das Drei-Stadien-Gesetz. In H.P. Dreitzel (Ed). Sozialer Wandel (pp. 95–111).Neuwied and Berlin.

Clark, B. (1997). Social ideologies and gifted education in today's schools. *Peabody Journal of Education,* 72 (3&4), 81–100.

D'Ambrosio, U. (1990). The role of mathematics education in building a democratic and just society. *For the Learning of Mathematics,* 10 (3), 20–23.

D'Ambrosio, U. (1994a). Cultural framing of mathematics teaching and learning. In R.Biehler et al. (Eds.) *Didactics of Mathematics as a Scientific Discipline* (pp. 443–455), Dordrecht, Kluwer Academic Publishers.

D'Ambrosio, U. (1994b). On environmental mathematics education. *Zentralblatt für Didaktik der Mathematik,* 94(6), 171–174.

D'Ambrosio, U. (1998). Mathematics and peace: Our responsibilities. *Zentralblatt für Didaktik der Mathematik.* 98 (3), 67–73.

D'Ambrosio, U. (1999). Literacy, Matheracy, and Technoracy: A Trivium for Today. *Mathematical Thinking and Learning,* 1 (2), 131–153.

D'Ambrosio, U. (2007). Peace, social justice and ethnomathematics. In B. Sriraman (Ed). *International Perspectives on Social Justice in Mathematics Education. The Montana Mathematics Enthusiast,* Monograph 1, pp. 25–34.

Darwin, C. (1871). *The Descent of Man.* London: John Murray.

Dawkins, R. (1976). *The Selfish Gene.* Oxford: Oxford University Press.

Ellsworth, E. (1989). Why doesn't this feel empowering? Working through the repressive myths of critical pedagogy. *Harvard Educational Review,* 59(3), 297–324.

Ennis, R. (1991). Critical thinking: A streamlined conception. *Teaching Philosophy,* 14(1), 5–24.

Freire, P. (1998). *Pedagogy of freedom: ethics, democracy, and civic courage.* Lanham: Rowman & Littlefield Publishers.

Gore, J. (1990). What can we do for you! What can "we" do for "you"? Struggling over empowerment in critical and feminist pedagogy. *Educational Foundations,* 4(3), 5–26.

Gutstein, E. (2006). *Reading and Writing the World with Mathematics: Toward a Pedagogy for Social Justice.* New York, Routledge.

Haynes, R.M., & Chalker, D.M. (1998). The making of a world-class elementary school. *Principal,* 77, 5–6, 8–9.

Herzig, A. H. (2002). Where have all the students gone? Participation of doctoral students in authentic mathematical activity as a necessary condition for persistence toward the Ph.D. *Educational Studies in Mathematics,* 50, 177–212.

Hodginkson, H. (1991). Reform versus reality. *Phi Delta Kappan,* 73, 8–16.

Kent, G.S., & Lawrence, B. (2002). Celebrating mediocrity? How schools shortchange gifted Students. *Roeper Review,* 25(1), 11–13.

Kim, K.E. (2005). Learning from each other: creativity in East Asian and American Education. *Creativity Research Journal*, 17(4), 337–347.

Kurth-Schai, R. (1992). Ecology and equity: Toward a rational re-enhancement of schools and society. *Educational Theory*, 42(1), 147–163.

Martinsen, O. (2003). Introduction: Special Issue on creativity. *Scandinavian Journal of Educational Research* ,47(3) 227–233.

Marx, K., & Engels, F. (1879–1882). *Gesamtausgabe*, Edited by the Institut für Marxismus-Leninismus.

McPeck, J. E. (1984). Stalking beasts, but swatting flies: The teaching of critical thinking. *Canadian Journal of Education*, 9(1), 28–44.

Moreno, L & Trigo, M.S. (2007, in press). Democratic access to powerful mathematics in a developing country. To appear in L. English (Ed). *Handbook of International Research in Mathematics Education (2nd Edition)*. Lawrence Erlbaum and Associates.

Organization for Economic Co-Operation and Development [OECD] (2004). *Problem Solving for Tomorrow's World—First Measures of Cross Curricular Competencies from PISA 2003*, http://www.pisa.oecd.org/dataoecd/25/12/34009000.pdf. Retrieved 29.09.2005.

Paul, R. (1990). *Critical Thinking: What Every Person Needs to Survive in a Rapidly Changing World*. Rohnert Park, CA: Center for Critical Thinking and Moral Critique.

Plato (1999). *Great Dialogues of Plato: Complete Texts of the Republic, Apology, Crito Phaido, Ion and Meno*, Vol. 1. Translated by W.H. Rouse. Mass Market Paperback.

Radhakrishnan, S. (1964). The Dignity of Man and the Brotherhood of Peoples. Foreign Affairs Record of the Government of India, (January).

Schoenfeld, A. (2004). The Math Wars. *Educational Policy*, 18(1), 253–286 .

Seymour, E. (1995). The loss of women from science, mathematics and engineering undergraduate majors: An explanatory account. *Science Education*, 79(4), 437–473.

Seymour, E., and Hewitt, N.M. (1997). *Talking about leaving: Why Undergraduates leave the Sciences*. Boulder, CO: Westview Press.

Skovsmose,O. (1997). Critical mathematics education: Some philosophical remarks. The Royal Danish School of Educational Studies, Copenhagen (Denmark). Dept. of Mathematics, Physics, Chemistry and Informatics).12 p.

Skovsmose,O. (2004) Mathematics: Insignificant? *Philosophy of Mathematics Education Journal*, no.18, 19 p.

Skovsmose, O. (2005). Kritisk matematikkundervisning–for fremtiden Tangenten. *Tidsskrift for Matematikkundervisning*,16(3) p. 4–11.

Simpson, A. (1996). Critical questions: Whose questions? *The Reading Teacher*, 50, 118–126.[EJ 540 595] ED436007 .

Smith, G.F. (2001). Towards a comprehensive account of effective thinking. *Interchange*, 32(4), 349–374.

Spring, J. (2006). Pedagogies of globalization. *Pedagogies: An International Journal*, 1(2), 105–122.

Sriraman, B. (2005). Major Book Review of L.V. Shavinina& M. Ferrari (Ed.) (2004). Beyond Knowledge: Extra Cognitive aspects of developing high ability. Lawrence Erlbaum & Associates. *Interchange: A Quarterly Review of Education*, 35(6), 455–460.

Sriraman, B. (2007). On the origins of social justice: Darwin, Freire, Marx and Vivekananda. In B. Sriraman (Ed). *International Perspectives on Social Justice in Mathematics Education. The Montana Mathematics Enthusiast,* Monograph 1, pp. 1–6.

Sriraman, B., & Adrian, H. (2004). The pedagogical value and the interdisciplinary nature of inductive processes in forming generalizations. *Interchange: A Quarterly Review of Education,*35(4) 407–422. Offprint available at http://www.math.umt.edu/sriraman/20_Interchange2004.pdf

Sriraman, B., & Benesch, W. (2005). Consciousness and Science: An Advaita-Vedantic perspective on the theology-science dialogue. *Theology and Science,* 3(1), 39–54.

Sriraman, B., & Dahl, B., (2007). Interdisciplinary Ideas in Gifted Education. To appear in L.V. Shavinina (Editor). *The International Handbook of Giftedness.* Springer Science. Pre-print available at http://www.umt.edu/math/reports/sriraman/abstract_14.html

Sriraman,B., Knott,L., & Adrian, H. The Mathematics of Estimation: Possibilities for Interdisciplinary Pedagogy and Social Consciousness. Accepted for publication in *Interchange: A Quarterly Review of Education.* Pre-print available at http://www.umt.edu/math/reports/sriraman/abstract_16.html

Sriraman,B., & Steinthorsdottir, O. (2007a). Excellence and Equity in education and talent development-Components of a Hegelian dialectic. (in press) in *Mediterranean Journal for Research in Mathematics Education,* 6 (1&2).

Sriraman, B., & Steinthorsdottir, O. (2007b). Research into Practice: Implications of research on mathematics gifted education for the secondary curriculum. (In press) in C. Callahan & J. Plucker (Editors) *What the Research Says: Encyclopedia on Research in Gifted Education.* Prufrock Press.

Sriraman, B., & Steinthorsdottir, O. (2007c). Emancipatory and Social Justice Perspectives in Mathematics Education. *Interchange: A Quarterly Review of Education,* 38(2), 195–202.

Sriraman, B & Törner, G. (2007) Political Union/Mathematics Education Disunion: Building Bridges in European Didactic Traditions. To appear in L. English (Editor) *Handbook of International Research in Mathematics Education (2nd Edition).* Taylor and Francis. Pre-print available at http://www.umt.edu/math/reports/sriraman/abstract_13.html

Sternberg, R. (1996). Neither elitism nor egalitarianism: Gifted education as a third force in American Education, *Roeper Review,* 18(4), 261–263.

Sternberg,R.J.(1998).A balance theory of wisdom. *Review of General Psychology,*2,347–365.

Sternberg, R. J. (2001). Why schools should teach for wisdom: The balance theory of wisdom in educational settings. *Educational Psychologist,* 36(4), 227–245.

Walters, K.S. (1994). *Re-thinking reason: New perspectives in critical thinking.* Albany: State University of New York Press.

Weiler, K. (1991). Freire and a feminist pedagogy of difference. *Harvard Educational Review,* 61(4), 449–474.

NOTES

1. See Richard Dawkins (1964) *The Selfish Gene*
2. Asoka (c. 299–237 BCE) is credited with the establishment of the so-called "first" Indian empire, accomplished through decades of bloody conquests. His deep remorse over the carnage at Kalinga led him to embrace the peaceful doctrines of Buddhism. Under his protection, Buddhism flourished and numerous Buddhist texts were written. Asoka also sent numerous emissaries of Buddhism to places like South East Asia, Egypt, Libya, and Macedonia, which resulted in the "golden" age for Buddhism.
3. Karl Marx, Friedrich Engels, Gesamtausgabe, Edited by the Institut für Marxismus-Leninismus
4. *World Resources 2000–2001: People and ecosystems: The fraying web of life. United Nations Development Programme, United Nations Environment Programme, World Bank, World Resources Institute.*
5. Here Cho & Lewis are synthesizing the writings of Ellsworth (1989), Gore (1990) and Weiler (1991). These particular writings convey a completely different conception of the complexities of empowerment from the point of view of feminist pedagogy. See references.
6. Figure reprinted with permission from Daniel Waldspurger http://www.calvertonschool.org/Waldspurger/pages/hegelian_dialectic.htm.

CHAPTER 24

SOCIAL JUSTICE AND MATHEMATICS

Rethinking the Nature and Purposes of School Mathematics

Kurt Stemhagen
University of Mary Washington

INTRODUCTION: THE NEED FOR AN ETHICS OF MATHEMATICS EDUCATION

It is interesting to see how different subject area teachers view their role in the wider community. Nearly all teachers I have spoken with acknowledge that schools can and should play a part in helping our society work toward social justice. However, it has been my experience that when mathematics teachers are pressed on this point they often explain that, because of the nature of mathematics, there is not much that they can do in this regard. They explain that the content of their subject matter reduces their obligations when it comes to teaching for social justice, that is, it is out of their hands. Furthermore, I have spoken with social justice-oriented teacher educators who succeed in exciting most of their students about the enterprise

Critical Issues in Mathematics Education, pages 337–349

of teaching for social justice yet struggle with how to help future mathematics teachers link their curricula and classroom practices to the movement toward a more just, equitable, and democratic society.

Conceding that many important ways in which teachers can work toward social justice have little to do with their specific subject areas (e.g., the physical setup of the classroom, the types of assignments given, and the way students and other teachers are treated), I am nonetheless troubled by the conception of mathematics as devoid of social implications. In this article, I consider what needs to change if mathematics class is to become a place where its aims stretch beyond the narrow transmission of mathematical skills and knowledge. A central facet of my argument is that for mathematics to have broad social implications, the way that mathematics is typically thought about needs to change. I will offer a sketch of one such reconceptualization. It is important to note that a larger argument, that mathematics has ethical implications, does not require the adoption of this particular philosophy of mathematics. I have included this account because I believe offering one such possibility will add specificity to, and illuminate the possibilities of, an ethics of mathematics education.

There will be expectations that connecting social justice to mathematics education ought to involve the inclusion of marginalized groups in curriculum materials, improving access to higher mathematics courses of study,[1] tracking and mathematics, using mathematical skills to analyze social injustices, and other issues more traditionally thought of as relating to equity in education. While each of the preceding ideas is certainly important and will need to be analyzed and confronted if mathematics education is to become an arena in the battle for civil rights, I am arguing that such practical changes are necessary but not sufficient and that a change in the way we conceptualize the subject matter of mathematics is also critical in any effort to include mathematics education in the mission of social betterment. In addition to seeking to equalize educational structures and methods, we must also find ways for the very subject matter to be able to take part in aiding this reconstruction. I submit that school mathematics needs to help students recognize their ability to, and the value of, creating and evaluating mathematical knowledge as a means to improve the world around them.

I must admit that my belief that increasing students' *mathematical agency* (a term that I will define more clearly later) will aid in the push toward social justice is at least somewhat intuitive. However, there is precedent for arguing that conceptual shifts can have social implications. One reasonably closely related example can be found in situated theory. Jean Lave and others have argued that rethinking the nature of our thought-action dichotomy is a way to recognize the rationality of situated, non-academic cognitive activity in an effort to increase social equality. Likewise, Jo Boaler has studied various ways in which situativity might help lessen the gender gap in school mathematical performance as well as to increase participation in mathematics in general.[2] Although a philosophical reconceptualization of mathematics might seem a strange way to promote social justice, I hope that readers will consider it an avenue worthy of exploration.

AN ALTERNATIVE PHILOSOPHY OF MATHEMATICS

Historically, mathematics has been held up as a bastion of certainty. The popular, common sense version of mathematics as objective, logical, neutral and extra-human has fostered resistance to pedagogical shifts similar to the other subject areas.[3] Recent reform efforts that have sought to introduce psychosocial components to mathematics education have had to work against particularly entrenched understandings of mathematics. Consequently, reformers have postured against this absolutism by offering constructivist versions of mathematics that are thoroughly subjective, relative, and fallible. The "math wars" have been raging for several years and show no signs of letting up, pitting traditionalists—those calling for more rigor and a "back to basics" approach to mathematics education—against reformers—those advocating a child-centered, applied approach to mathematics education.

The "math wars" are about more than just teaching methods and curriculum decisions. Undergirding this split are pronounced differences as to how the nature of mathematics is conceived. Absolutists tend to view mathematics as certain, permanent, and independent of human activity. Constructivists, on the other hand, focus on the ways in which humans actually create mathematical understandings and knowledge. A simple yet powerful way to characterize this split is to borrow from philosopher Rorty's distinction between those who view phenomena as *found* versus those who view it as *made* (1999, p. xvii).

I contend that absolutism and constructivism, while having much to offer, ultimately fail as philosophies of mathematics. Absolutism suggests an understanding of mathematics that captures its unique stability but that does not acknowledge its human dimensions. Conversely, constructivism tends to encourage understandings of mathematics that feature human involvement but, in doing so, seem to lose the ability to explain the remarkable stability and universality of mathematical knowledge.

Elsewhere, I have worked to develop a useful and different philosophy of mathematics education given this stalemate (both practically and philosophically speaking) existent within the context of contemporary mathematics education.[4] I use the work of several thinkers and schools of thought to develop an evolutionary philosophy of mathematics education. This perspective acknowledges how the empirical world, that is the world of experience, contributes to mathematics. As Philip Kitcher and others have noted, the very origins of mathematics were probably empirical, most likely originating in Mesopotamia, arising out of the practical experiences of farmers and others.[5] Whereas some mathematical empiricists (particularly Kitcher) have had trouble explaining how mathematics has gone from an empirical to a highly rational and abstract enterprise, this evolutionary account, through recognition of the development of mathematics as a series of individual-environment interactions, emphasizes the ways in which simple, applied, and directly empirical mathematics can be quite rational. Conversely, the evolutionary account also develops the empirical and pragmatic dimensions of contemporary mathematics. Furthermore, the origins of mathematics are not conceived of as crudely empirical, but rather as arising out of pragmatic endeavors that possessed both physical and mental aspects, as human organisms developed and used mathematics as a means to interact with their environments.

A functional account of the nature of mathematics is suggested by this presentation of ideas. Whereas past philosophies of mathematics tended to advance structural approaches to explaining mathematical knowledge, this functional approach posits mathematics as a series of evolving, humanly-constructed tools that are created in order to solve genuine problems. Additionally, whether a mathematical activity functions well in its role as a solution to the particular problem it was employed to contend with presents an opportunity to judge its "correctness." The educational implications of the evolutionary perspective's functional account are potentially quite broad and powerful. However, the scope of this article is limited to a consideration of how a reconceptualization of mathematics might encourage the development of mathematical agency and ultimately work toward social justice.

EMPOWERMENT AND AGENCY AS AIMS
OF MATHEMATICS EDUCATION

While I would not argue that most teachers view mathematics as a way to teach powerlessness, I do believe that, unfortunately, mathematics class frequently has such an effect. My claim here is that if empowering students is an aim of mathematics education (and I argue that it ought to be if increased social equity and democratic participation are more general aims of education), then rethinking the nature of mathematics is called for. A necessary step toward social justice is helping children recognize that their voice matters. Real and lasting social change cannot come about until individuals realize the power that they possess. The mathematics class version of this is that they must develop *mathematical agency*.

In "Empowerment in Mathematics Education," (2002), Paul Ernest identifies three different but overlapping domains within which mathematics can be personally empowering for students: mathematical, social, and epistemological. Mathematical empowerment refers to becoming fluent in the ways and language of school mathematics. Social empowerment involves using mathematics to: "better one's life chances" (2002). Ernest explains that the world in which we live is highly quantified and that knowledge of and the ability to use mathematics is critical to being able to negotiate it:

> Our understanding is framed by the clock, calendar, work timetables, travel planning and timetables, finances and currencies, insurance, pensions, tax, measurements of weight, length, area and volume, graphical and geometric representations, etc. Much of our experience of life is already mathematised. Unless schooling helps learners to develop the knowledge and understanding to identify these mathematisations of our world, and the confidence to question and critique them, they cannot be in full control of their own lives, nor can they become properly informed and participating citizens. (Ernest, 2002)

The third type of mathematical empowerment is epistemological. It is concerned with the ways in which individuals come to view their role in the creation and evaluation of knowledge, both mathematical and in general. Ernest rightly claims that epistemological empowerment: "is perhaps the most neglected in discussions of the aims of teaching and learning of mathematics" (2002). It is a critical component of what I earlier referred to as mathematical agency. Epistemological empowerment refers to the degree to which children recognize that they can construct new knowledge and that they have the power to determine the value of their constructions.

Those who possess primarily absolutist or constructivist outlooks face severe problems fostering genuine *mathematical agency* in mathematics classrooms. Mathematical agency is some combination of Ernest's social and epistemological empowerments. For mathematical agency to be addressed in mathematics classrooms, teachers must commit to helping students learn how to deal with their already mathematized existences and also to recognize that they are agents capable of altering such mathematizations and also to create new ones when they see fit. Finally, teaching mathematical agency requires that teachers help students develop the means to judge the merit of different forms of mathematics.

In viewing mathematics as a static body of preexistent truths, absolutists have a problem in placing the student in a position to become a mathematical agent in any robust sense. As Ernest explains:

> Many students and other individuals, including mathematics teachers (Cooper, 1989), are persuaded by the prevailing ideology that the source of knowledge is outside themselves, and that it is both created and sanctioned solely

by external authorities. They are led to believe that only such authorities are legitimate epistemological agents, and that their own role as individuals is merely to receive knowledge, with the subsequent aim of reproducing or transmitting it as accurately as possible. (2002)

Constructivists face a different, yet equally daunting set of challenges. Elsewhere, I have analyzed constructivist mathematics education through scrutiny of a textbook for mathematics educators and an account of constructivism in practice (Stemhagen, 2004). I found that in constructivist classrooms, students are certainly empowered in the sense that their individual ideas, methods, and findings are given value. The problem is that following through with this way of thinking tends to foster the view that all mathematical constructions are valuable, regardless of their power to solve "real world" or even theoretical problems. That is, the primary means by which a mathematical idea can be evaluated is whether and how it matches a child's existing mental structures. Math educator John Van De Walle demonstrates the constructivist's tendency to evaluate the worth of mathematical constructions according to internal criteria: "Children (and adults) do not learn mathematics by remembering rules or mastering mechanical skills. They use the ideas they have to invent new ones or modify the old. The challenge is to create clear inner logic, not master mindless rules" (1990, p. vii).

Jere Confrey is more explicit and succinct: "...reflection is the bootstrap for the construction of mathematical ideas" (p. 116). The result is that although children learn to create mathematical constructions, they are discouraged from developing understandings of how mathematics can help outside of mathematics class (or even beyond their own minds), as according to this way of thinking mathematics is connected not primarily to the physical world so much as it is to the prior mental structures of each student. This "bootstrap" theory provides no explanation of how engaging with the physical world can foster new mathematical constructions and also help students to judge the merits of what they have constructed. Consequently, in an effort to empower students, constructivist teachers run the risk of encouraging students who are emboldened to create mathematical constructions that may or may not be truly empowering in the sense that they can help children live in and negotiate a mathematized world and lead to a recognition of how they can be epistemological agents, creating and evaluating their own functional mathematical constructions.

THE NON-NEUTRALITY OF MATHEMATICS: WINNER'S MAKING-USE DISTINCTION

The content of mathematics is typically thought of as neutral. That is, to most, mathematics is considered a domain that is devoid of ethical-moral implications. One can use mathematics for whatever purposes one wishes, but the mathematics itself is not good or bad, it just is. If I am right and mathematics classrooms frequently teach powerlessness, then the notion that mathematics is essentially neutral needs to be revisited. Furthermore, if the content of mathematics class fosters a particular way of looking at the world (a mathematical one), then it seems reasonable that this way of looking at the world, to the extent that it is different from non-mathematical perspectives, can be conceived of as more or less valuable. Thus, it is not devoid of ethical implications.

Langdon Winner argues similarly about technology. In *The Whale and the Reactor* (1986), Winner writes that technology is often viewed as a neutral tool that can be used for good or ill purposes. His argument is that this common conception is mistaken and that technology is not neutral, in that its very invention and employment alter our social arrangements. Winner's ideas about technology can be helpful in thinking about mathematics.[6] Perhaps most relevant, is his idea that much of the reason why technology is mistakenly thought of as neutral is that there is a sharp distinction between its creation and its use:

> The deceptively reasonable notion that we have inherited from much earlier and less complicated times divides the range of possible concerns about technology into two basic categories: making and use. In the first of these our attention is drawn to the matter of 'how things work' and of 'making things work.' We tend to think that this is a fascination of certain people in certain occupations, but not for anyone else. 'How things work' is the domain of inventors, technicians, engineers, repairmen, and the like who prepare artificial aids to human activity and keep them in good working order. Those not directly involved in the various spheres of 'making' are thought to have little interest in or need to know about the materials, principles, or procedures founding those spheres." (Winner, p. 5)

Winner goes on to explain how, to most, it is only the use of the tools that matters. Our interactions with these tools are instrumental, and take place to achieve certain desired outcomes: "One picks up a tool, uses it, and puts it down. One picks up a telephone, talks on it, and then does not use it for a time. A person gets on an airplane, flies from point A to point B, and then gets off" (Winner, p. 6). According to this view of technology our interactions with the tools in question are: "occasional, limited, and nonproblematic" (p. 6).

Not surprisingly, Winner finds the making-use dichotomy unacceptable and damaging. He contends that if people were aware of what went into the making of some forms of technology that there would be greater awareness of how use is not so simple. In a section titled, "Return to Making," Winner eloquently makes this point with a question: "As we 'make things work,' what kind of *world* are we making?" (p. 17).

As far as the relevance of Winner's making-use distinction to mathematics and mathematics education, there seems to be a similar divide between those who do or use mathematics professionally and those who do not. Furthermore, there is a similar lack of consideration as to how mathematics can alter our lives in ways more radical than simply existing as an instrumental aid to be intermittently used in a non-complicated manner. Winner writes of the small group interested in the making of technology. With mathematics, the group is even smaller, perhaps consisting of professional mathematicians, those who employ much mathematics in their work (scientists and engineers, for example), and teachers of mathematics. It seems that to most, mathematics is some sort of language or system that supports or silently undergirds the technologies that sustain their lifestyle.

THE REINTEGRATION OF *MAKING* AND *USE*: HERSH'S METAPHOR

A first step in countering the making-use distinction in mathematics education might be to help students experience some of what goes on in the world of those who are involved in the making of mathematics. In *What is Mathematics, Really?*, Reuben Hersh explains that mathematics can be thought of as divided into two areas, front and back. The idea, an application of sociologist Erving Goffman's work (1973), is that the finished product of mathematicians belongs in the well-ordered and more-or-less highly polished front of mathematics while the back is the area where mathematicians are busy engaging in the messy but often practically fruitful activities of mathematicians. He uses the analogy of a restaurant. The front of a restaurant is the dining room and the back is the kitchen. In the dining room everything is to appear orderly and under control. Those in the front are not privy to all that goes on behind the scenes (in the back) in order to create the seamless experience of dining in the front. Hersh explains math in these terms:

> The front and back of mathematics aren't physical locations like dining room and kitchen. They're its public and private aspects. The front is open to outsiders; the back is restricted to insiders. The front is mathematics in finished form—lectures, textbooks, journals. The back is mathematics among working mathematicians, told in offices or at café tables...Front mathematics is formal, precise, ordered, and abstract. It's broken into definitions, theorems, and remarks. Every question either is answered or is labeled: "open question." At the beginning of each chapter a goal is stated. At the end of each chapter,

it's attained. Mathematics in back is fragmentary, informal, intuitive, tentative. We try this or that. We say "maybe," or "it looks like." (1997, p. 36)

It is a common belief that the front part of mathematics is all that exists. The application of the Hersh/Goffman metaphor is an invitation for all students to leave the well-ordered dining room and to see what's cooking in the kitchen (as well as *how* things are cooking and most importantly, to do some cooking themselves!).

Hersh stresses that we do mathematics first and philosophize about it later. He does not deny the seeming banality of declaring that mathematics is a human activity that takes place in the context of a society. He goes on to assert that failing to recognize the importance of mathematics' socio-historical context is the source of the intractability of many of the problems of the philosophy of mathematics, and by extension, mathematics education.

MATHEMATICS AND THE FABRIC OF OUR LIVES

Hersh's metaphor is offered as a means to suggest the possibility of mathematics classrooms where the *making* and not just the *use* of mathematics is taught and learned. If Winner's work with technology has any relevance to mathematics education, this could be a very important development. Recall Winner's notion that people tend to think of our relationship to technology as a simple instrumental one. We employ something for a time for a given purpose, then we put it down; we use the telephone or airplane for briefly, then we stop using them, etc. When thought about in this way, certain technologies tend to change the way we live much more broadly than by simply altering our immediate mode of communication or transportation. The question, as I see it, is: Do the development and acquisition of mathematical skills and techniques alter our existence in profound, not immediately clear ways? If so, it is interesting and even disturbing that mathematics class rarely, if ever, confronts such issues.

One interesting explanation about the power of mathematics to alter our social arrangements is detailed by Lewis Mumford. In *Technics and Civilization*, Mumford considers the impact of certain machines and technologies on human life. He details the development of the clock in European monasteries in the fourteenth century and he considers its effect on the monks and also its eventual impact on humanity as a whole. It should be noted that, in talking about the mechanical clock, Mumford bundles technological development with a more general and perhaps underlying mathematization of human experience: "The application of quantitative methods of thought to the study of nature had its first manifestation in the regular measurement of time..." (Mumford, p. 12). Later, Mumford furthers an interesting clock-mathematics/science connection: "The clock, moreover, is a piece of power-machinery whose product is seconds and minutes; by its essential nature it dissociated time from human events and helped create the belief in an independent world of mathematically measurable sequences; the special world of science" (p. 15).

It is not too much of a stretch, I should think, to consider ways in which other mathematizations might alter the way we experience our lives. We tend to see most things as able to be mathematically modeled. Some, such as Jean Baudrillard, argue that our simulations (mathematical and otherwise) are taking the place of our real experiences.[7] This might sound preposterous on the surface, but models are, virtually by definition, "cleaner" and more understandable than the "real thing" so it seems understandable that models might creep in as substitutes for portions of our lives. In fact, Baudrillard points out that on some level, this is necessary. *Simulations* starts with a retelling of a fable about how a map became so detailed that it eventually covered exactly the same area as the territory it represented. Maps are models of reality that can be quite useful. The trouble, according to Baudrillard, is when we cannot distinguish between the model and the reality and perhaps, when we can but when we prefer to experience our lives through models and not reality.

CONCLUSION

What does all of this have to do with differing philosophies of mathematics education, the argument that mathematics is non-neutral (that is, that it has moral/ethical implications), and the notion that social justice should be on the minds of mathematics educators? If meaningful change in mathematics education is going to take place, a reconceptualization of the nature of mathematics and what we want to accomplish in mathematics classrooms is needed. That is, Hersh's back part of mathematics only becomes important as anything more than a teaching technique if the traditionalists' absolutism is abandoned. Likewise, the hyper-empowerment of many constructivist accounts does not lead to genuine social and epistemological empowerment, or what I am calling mathematical agency. The adoption of some alternative conception of mathematics is needed if we are to recognize that mathematics is non-neutral and that it has very real yet not always obvious effects on our actual experience. In other words, absolutist philosophies of mathematics tend not to acknowledge the non-neutrality of mathematics and constructivist accounts tend to under-emphasize the ways in which mathematics matters to anything outside of the mind of the individual learner or beyond the confines of groups of professional practitioners.

The links from pondering the nature of mathematics to teaching mathematics as a means of changing the world for the better might seem hidden and even tenuous, yet they are both present and important. While I agree that getting more females and students of color into advanced mathematics class is important, my concern with this project is with what gets taught and learned in such classes. A more meaningful and genuinely agency-producing mathematics education would, to paraphrase Neil Postman, teach us not only how to use mathematics, but also how mathematics uses us.[8] For this to happen, two gulfs will need to be bridged. First, the making and use of mathematics need to be reintegrated. Second, the absolutist understanding of mathematics as stable, universal, and inert and the competing constructivist version that emphasizes its contingency and uniqueness needs to be reconciled. Whether the evolutionary philosophy of mathematics that I have sketched in this paper is the particular philosophical bridge is not as important as the fact that one gets constructed.

Is the recognition of an ethics of mathematics education possible? I certainly hope so, as a latent premise of this paper is that it can not be and never was neutral. Furthermore, if we do not actively consider and attempt to shape the ethical meta-messages of mathematics, we might not be pleased with that ones that will nonetheless emerge. Thinking of mathematics class as a forum for students to learn to analyze, understand, and improve their world is a radical shift from both traditional and contemporary notions. I concede that affecting this shift will not be easy but that taking the time to think deeply about the dual enterprises of teaching and learning mathematics is an important first step. Hopefully this can pave the way for recognition of the possibilities of mathematics class becoming a legitimate arena in the battle for increased democratic participation and social justice.

REFERENCES

Baudrillard, J. (1983). *Simulations.* New York: Semiotext, Inc.

Boaler, J. (1994). When do girls prefer football to fashion? An analysis of female underachievement in relation to 'realistic' mathematic contexts. British Educational Research Journal, v 20; 5, 551–562.

Boaler, J. (1999). Participation, knowledge and beliefs: A community perspective on mathematics learning. *Educational Studies in Mathematics*, 40, 259–281.

Confrey, J. (1990). What constructivism implies for teaching. In R. Davis, C. Maher & N. Noddings (Eds.), *Constructivist Views on the Teaching and Learning of Mathematics.* Reston, VA: National Council of Teachers of Mathematics.

Ernest, P. (1998). *Social constructivism as a philosophy of mathematics.* Albany, New York: State University of New York Press.

Ernest, P. (2002). Empowerment in mathematics education. *Philosophy of Mathematics Journal, 15.* Retrieved April 18, 2003, from http://www.ex.ac.uk/~PErnest/pome15/contents.htm.

Goffman, E. (1973). *The presentation of self in everyday life.* Woodstock, NY: The Overlook Press.

Hersh, R. (1997). *What is mathematics, really?.* New York: Oxford University Press.

Kitcher, P. (1983). *The nature of mathematical knowledge.* New York: Oxford University Press.

Moses, R., & Cobb, C. (20010). *Radical equations: Math literacy and civil rights.* Boston: Beacon Press.

Mumford, L. (1963). *Technics and civilization.* New York: Harcourt, Brace & World, Inc.

Postman, N. (1995). The end of education: Redefining the value of school. New York: Alfred A. Knopf.

Rogoff, B. & Lave, J. (1984). *Everyday cognition: Its development in social context.* Cambridge, MA: Harvard University Press.

Rorty, R. (1999). *Philosophy and social hope.* New York: Penguin Books.

Stemhagen, K. (2004). *Beyond absolutism and constructivism: The case for an evolutionary philosophy of mathematics.* (Doctoral dissertation, University of Virginia, 2004).

Stemhagen, K. (2003). Toward a pragmatic/contextual philosophy of mathematics: Recovering Dewey's 'Psychology of Number'. In *2003 Philosophy of Education Yearbook.* Urbana, IL: Philosophy of Education Society.

Van de Walle, J. (1990). *Elementary school mathematics: Teaching developmentally.* White Plains, NY: Longman.

Von Glasersfeld, E. (1991). *Radical constructivism in mathematics education.* Norwall, MA: Kluwer Academic Publishers.

Winner, L. (1986). The whale and the reactor: A search for limits in an age of high technology. Chicago: The University of Chicago Press.

NOTES

1. Moses and Cobb's *Radical Equations* (2001) is an important recent work in this vein. They look at how Algebra I is used as a means to separate the college-bound from the non-college bound. The book details a project in the rural south designed to increase minority participation and success in Algebra I in an effort to work toward increased equity.

2. See Rogoff and Lave's *Everyday Cognition* (1984) for more on the link between the mathematical and the social. Boaler's "When Do Girls Prefer Football to Fashion? An Analysis of Female Underachievement in Relation to 'Realistic' Mathematic Contexts" (1994) and "Participation, Knowledge and Beliefs: A Community Perspective on Mathematics Learning" (1999) address gender and wider participation, respectively.

3. In science, philosophical work (including Kuhn's and Popper's), coupled with a post-Sputnik concern for relevant, applicable science education has led to a shift in the way science is taught and learned in school, as science class has become a place where students often play the role of fledgling scientists.

History class, with the advent of new forms of technology, has undergone a similar metamorphosis; from emphasis on the memorization of names, dates, and places to a dynamic forum for students to act as mini-historians, using the newfound ease of access to primary source data to discover and interpret material.

4. See Stemhagen (2003, 2004).

5. *The Nature of Mathematical Knowledge* (1983) is Kitcher's fullest account of mathematics as an empirical enterprise.

6. Bryan Warnick and I are in the process of writing a paper on the educational implications of conceiving of mathematics as a non-neutral technology. We are making a case that mathematics can actually be thought of as a form of technology. For the sake of this article, I am merely suggesting that some work in the philosophy of technology can be useful in thinking about mathematics.

7. Much of Baudrillard's *Simulations* (1983) deals explicitly with what he sees as the replacement of reality with this new "virtual reality" in which models are substituted for genuine experience.

8. In *The End of Education* (1995), Postman speaks this way of technology.

CHAPTER 25

POSSIBILITIES AND CHALLENGES IN TEACHING MATHEMATICS FOR SOCIAL JUSTICE

Eric Gutstein
University of Illinois–Chicago

INTRODUCTION

This article is about some of the possibilities and challenges in teaching mathematics for social justice. As such, it reflects ongoing, in-process work. It is based on my collaboration with others over the past thirteen years (1994–2007) in Chicago public schools. I first studied culturally relevant mathematics teaching in a Mexican American context (Gutstein, Lipman, Hernández, & de los Reyes, 1997), then taught my own 7th and 8th grade classes at *Rivera* (a pseudonym) school from 1997–2003 (Gutstein, 2003, 2006a). More recently (2003–present), as an outgrowth of my work at Rivera, I have been working at the Greater Lawndale/Little Village School for Social Justice (GLLVSSJ—not a pseudonym; Gutstein, 2007-b, in-press b). At GLLVSSJ, I was a member of the design team that helped start the school, which opened in Fall 2005. Since then (two plus years), I have been

Critical Issues in Mathematics Education, pages 351–373
Copyright © 2009 by Information Age Publishing
351

supporting the school's mathematics teachers, co-teaching, working with students, and continuing to teach and study mathematics for social justice. This piece focuses more on the research, practice, and current issues at the GLLVSSJ, as I explained my work at Rivera in depth in Gutstein (2006a).

Where we come from—our histories—and who we are in this hierarchical, racialized, gendered, and class-based world—our locations—matter in what we say and do. For that reason, I share a little bit about myself. Without belaboring the details, I am white, anti-racist, political activist involved in social movements in the U.S. from my youth through the present. I was fortunate enough to come of age during the radical social period of the 1960s, grew up in inner-city New York, and was strongly influenced by the Black Liberation Movement, the anti-Vietnam War movement, and other social movements at the time. I was a politically active high school student, graduated in 1970, and have been involved in political struggles ever since. My path took me eventually to academia (late), and since 1993, I have been a mathematics educator as part of my continuing efforts to effecting social change in the U.S.

OVERVIEW OF TEACHING MATHEMATICS FOR SOCIAL JUSTICE

Teaching mathematics for social justice, or *critical mathematics* (Frankenstein, 1987; Skovsmose, 1994), has many variations and meanings. It is safe to make some general observations about it, such as it is a critical pedagogy (Freire, 1970/1998; Giroux, 1988); builds on culture and experiences; and attempts to engage students to use mathematics to think about, and act on, the world. A good deal has been written on it outside of the U.S., from South Africa (e.g., Julie, 1993, 1998, 2004; Vithal, 2002a, 2002b; Volnick, 1994) to South America (Knijnik, 1997; Valero, 1999), Europe (Skovsmose, 1994, 2004, 2005), and Australia (Atweh & Clarkson, 2001; Zevenbergen, 2000). In the U.S., Frankenstein (1987, 1990, 1995, 1997, 1998) was the principal person writing about critical mathematics in academic journals for years, and there are now several others.

The goals of teaching and learning (mathematics) for social justice include that students learn important competencies in mathematics (or whatever subject they study). For example, students should develop mathematical power,[1] as well as be able to surmount the various hurdles preventing them from accessing advanced mathematical and educational opportunities and full participation in civil society (Moses & Cobb, 2001). In addition, students ideally will change their orientations towards mathematics, away from viewing it is a series of random rules to be rotely memorized and regurgitated, to seeing mathematics as a way to create meaning and make sense of human and social experiences. However, just as important, and intimately related, students also need to develop a critical comprehension of those experiences, using mathematics as a key analytical tool. That is, students, through using mathematics and in mathematics classes, can develop sociopolitical consciousness of their immediate and broader contexts and can also develop a sense of *social agency,* or an understand of themselves as actors capable of working with others to effect change towards social justice. The two sets of goals—mathematical and social justice—dialectically interact with each other. Thus, although there are specific times at which one set may take precedence over the other, they are inextricably connected in complicated ways that need to be resolved in the practice of the moment and over the long term.

TEACHING AND LEARNING FOR SOCIAL JUSTICE— HISTORY AND FRAMEWORKS

Each of us draws on multiple sources to create our evolving theoretical perspectives. In my case, the two primary ones informing my educational research and practice are the work of Paulo Freire and the historical traditions of African American education for freedom. I turned to Freire while teaching at Rivera in the late 1990s. Freire's contributions to liberatory, critical education are many and include, among others, his writing on the role of *conscientização* (critical sociopolitical consciousness), his assertion that education was always political and never "neutral," his advocacy of problem-*posing* (as opposed to problem-*solving*) pedagogies, his contention that the starting point of liberatory education be learners' *generative themes* (the dialectical relationship of key social contradictions in people's lives and how they understand them), and the unmasking of *banking* education in which teachers "deposit" dead morsels of pre-digested "knowledge" into the open mouths (minds) of "passive recipients" (students). Finally, Freire consistently wrote about the need for teachers and students to join together in partnerships in the struggle to make a better world, and that teachers be learners and learners be teachers; this point is particularly relevant to my discussion below.

Predating Freire, and eventually influencing him, is the tradition of education for freedom that was a central component of African American liberation struggles from the time of slavery in the Americas to the present (Anderson, 1988; Bond, 1934/1966; DuBois, 1935; Foster, 1994; Marable & Mullings, 2000; Payne, 1995; Perry, 2003; Siddle Walker, 1996; Watkins, 2001; Woodson, 1933/1990). A common theme running through this history is the collective efforts of communities so that youth could have educational opportunities in order to lead their people (Anderson, 1988). Humanization was also a fundamental precept. Echoing Freire's philosophical tenets, Perry (2003) paraphrased the exhortations of Malcolm X:

> Read and write yourself into freedom! Read and write to assert your identity as a human! Read and write yourself into history! Read and write as an act of resistance, as a political act, for racial uplift, so you can lead your people well in the struggle for liberation! (p. 9).

A key idea central to both African American liberatory education history and Freire's work is that teachers and students are (and need to be) partners in the joint struggle for freedom and humanization. This was the foundation of many of the sacrifices adults made for their children's education. For example, Anderson (1988) wrote about African American adults in the South who were told in 1866 (just one year after emancipation) that their children's schools would be closed by the Freedman's Bureau (the agency responsible for funding southern schools during Reconstruction). The adults requested that they themselves be additionally taxed to pay for the schools, and they submitted a 30-foot petition of 10,000 names (many of them signed with an "X") demanding that the Bureau reverse the decision. Similarly, Freire wrote often about the need to reconcile what he called the basic contradiction between teachers and students and referred to their common struggle, "...we cannot say that in the process of revolution someone liberates someone else, nor yet that someone liberates himself [sic], but rather that human beings in communion liberate each other" (1970/1998, p. 114). Along the same lines, he also wrote, "No one can, however, unveil the world for another" (p. 150). Rather, this "naming" (coming to consciousness) had to be done in genuine collaboration: "So it is that the leaders [including teachers] cannot say their word alone; they must say it with the people [including students]" (p. 159).

In general, despite these histories and traditions, understandings and/or definitions of social justice pedagogies may be somewhat, or sometimes, limited. That is, there are sources of liberatory educational practice that the progressive education research community at times overlooks. For example, generations of African American teachers have taught mathematics in their communities and to their people. The goal of many Black teachers historically, especially in the segregated schools of the south before the 1954 U.S. Supreme Court decision that stated that such schools were inherently unequal, was the advancement, education, survival, and freedom from oppression of African American communities (Anderson, 1988; Bond, 1934/1966; Perry, 2003; Siddle Walker, 1996). In mathematics education, the work of Bob Moses, the *Algebra Project* (AP), and the *Young People's Project* (YPP) (Moses & Cobb, 2001); *Project Seed* (Phillips & Ebrahimi, 1993); that of renowned African American mathematician Abdulalim Shabazz (Hilliard, 1991); and other mathematics teachers would be similarly described. However, these educators have generally not used explicitly critical contexts as learning sites. That is, neither the AP, YPP, nor Project Seed usually have students use mathematics to investigate injustice, study racism, or examine institutional discrimination. The situations from which students learn mathematics may be sports, games, other real-world settings familiar to students, or abstract mathematical structures, but they tend not to be critical analyses of social relations and institutional arrangements. Nonetheless, the framework of these programs is firmly grounded in self-knowledge, social justice, equity, and self- and community-empowerment. It is important to recognize and value the multiple paths towards social justice, and to appreciate that the form and content, and meanings and practices, of social justice pedagogy will differ in specific historical circumstances and localities.

SITUATING MATHEMATICS EDUCATION IN A GLOBAL CONTEXT

Discussions in mathematics education, especially in the United States, do not often involve broader analyses of world affairs, contending geopolitical forces, and larger economic contradictions and reverberations as they relate to teaching and learning. But in the present context of the neoliberal drive to privatize, commodify, and marketize all forms of life, it is relevant to ask the question of what role could, and should, mathematics education play in resisting the domination of the world and its peoples by U.S. empire, and contributing to the creation of a counter-hegemonic trend.

In the U.S., the current agenda promoted by the Bush administration includes the American Competitiveness Initiative (ACI), a relatively far-reaching plan with multiple components. An important part of the ACI is the National Mathematics Panel that was formed to recommend policy initiatives to the Bush administration and Department of Education. The ACI, developed in part from several influential reports (Kirsch, Braun, Yamamoto, & Sum, 2007; National Academy of Sciences, 2006; National Association of Manufacturers, 2005; National Center on Education and the Economy, 2007), is based on a series of interlocking premises—none of which by themselves are new—that have significant implications on a global scale as well as on mathematics education in the U.S. (Gutstein, in preparation). The first of these basic contentions is that the U.S. is becoming (or is already) a second-rate global economic power, despite its clear military dominance. The second major assumption underlying the ACI is that education is a major way (if not the primary way) for the U.S. to counter the economic efforts of other nations such as China, India, Brazil, South Korea, Russia, Japan, and the European Union. The final argument, embedded in the series of reports and the ACI itself, is that mathematics, science, and technology education are the real motive forces needed to overcome the economic threats and potential woes to the U.S. position posed by these global competitors.

The ACI includes, besides the *National Mathematics Panel,* the *Math Now* programs (in elementary and middle schools), a corollary of the *Reading First* initiative which will only promote "scientifically based research"; the recruitment and rapid preparation of 30,000 mathematics and science professionals to be high school math/science teachers (a program that emphasizes content knowledge rather than pedagogical content knowledge, knowledge of students and their communities, or sociopolitical context); relatively small grants for either high-achieving high school and college students, or those in technological fields (for example, grants of only $750 a year for first-year college students); a proposed large increase of advanced placement courses and test-takers in urban high schools; as well as a huge infusion of research dollars in technological fields. However, what the ACI fails to do, in a major way, is to support students who are neither high-achieving in school, nor in mathematics/science fields—that is, the majority of low-income/working-class students and students of color. There are no plans to sink massive resources into neighborhood public high schools, such as those in Chicago, where the ACT mathematics mean is 15 in non-selective schools, and where three-quarters of the African American male students fail to graduate (Greater West Town Community Development Project, 2003) and where this abysmal failure of the schools to educate Black students can be traced to the long history of racism within the U.S. including disinvestment and deindustrialization in urban communities, segregated and ghettoized substandard schooling, and a profound lack of resources compared to more affluent areas (Demissie, 2006; Kozol, 1992; Lipman, 2004; Boger & Orfield, 2005). There are no scholarships for struggling students at the margins, with C averages, only for high-achieving ones. Nor are there plans to explicitly reach students who have been so mis-educated (Woodson, 1933/1990) that they are on the brink of being forced out or no longer even in school. The ACI's goals are to improve the U.S.'s standing and position vis-à-vis global economic competitors, and nothing in it directly aids the mass of low-income youth of color—only those who can be selected (or "creamed," as U.S. social activists in the 1960s referred to this) and adequately prepared to alleviate the perceived national calamity of being second rate economically. Its documents do not even give lip service to economically and socially marginalized and disempowered urban and rural communities *except* where their needs coincide with those of capital and the corporate elite; this latter is the *interest convergence* principal about which Derrick Bell (1992) wrote. In other words, equity is not part of the real agenda, and the idea of increasing access for those who lack it, such as it is in the proposals, exists only to defeat the presumed competition. This is a step backwards even from the NCTM's (1989) infamous statement that "We cannot *afford* [emphasis added] to have the majority of our population mathematically illiterate. Equity has become an economic necessity" (p. 4). These words, then—and now—shamelessly strip any moral imperative and genuine commitment to rectify historic injustice from social and educational reforms in general, and those within mathematics education in particular. In this, the ACI is thoroughly complicit.

Thus, to consider the role of mathematics education, and the life possibilities for low-income or working-class students and communities of color in the U.S. today without contextualizing them within this broader situation and corporate/governmental programs prevents mathematics educators, researchers, and teachers from more fully comprehending and responding to various initiatives. Furthermore, the lack of the broader framing also potentially hampers our vision in conceptualizing, envisioning, and actualizing alternative mathematics education programs. One does not often view mathematics education as a vehicle through which students may study their social surroundings and learn to be active change agents for social justice. Mathematics education is rarely referred to as a "weapon in the struggle" for social justice and equity, but nothing in principle prevents us from enacting such a vision. And, there are increasing numbers of teachers and mathematics educators engaged in such practices (see, for example, Gutstein & Peterson, 2005). I turn now to the potential of how mathematics education might resist the global designs of empire and instead support the efforts of people of the world to be free from oppressive forces. As a practitioner involved in this work, I am well aware of the challenges we face in actualizing these possibilities; these also are part of my discussion.

POSSIBILITIES AND CHALLENGES IN TEACHING MATHEMATICS FOR SOCIAL JUSTICE

In Gutstein (2006a), I argued for the following reconceptualization of mathematics education:

> Students need to be prepared through their mathematics education to investigate and critique injustice, and to challenge, in words and actions, oppressive structures and acts—that is, to "read and write the world" with mathematics. (p. 4)

The *reading of the world*, to paraphrase Paulo Freire (Freire & Macedo, 1987), refers to the process of developing sociopolitical consciousness of immediate and broader contexts through studying the life conditions of one's community, being involved in social movements, reflecting on experiences, and various other means. Freire developed, and put into practice, many of his ideas in Latin America and Africa (e.g., his native Brazil, Chile, Nicaragua, Guinea Bissau, and other countries) in which he led, provided support to, or participated in literacy campaigns (as well as post-literacy, political development work). In Freire's (Freire & Macedo, 1987) endeavors, learning to *read the word* (i.e., develop text literacy) always followed a critical reading of one's world through which learners examined their lives using political, social, economical, historical, and cultural lenses in an attempt to distance themselves from their immediate circumstances and begin to investigate the root causes of oppression and injustice. A variety of mathematics education projects have attempted to "reinvent" (Freire's word) his principles, theories, and practices and have tried to provide students the opportunities to read the world with mathematics (Brantlinger, 2006, Frankenstein, 1987, 1998; Gutstein, 2006a; Peterson, 1995; Turner, 2003).

But reading the world is only part of the dialectical process of transforming society, and needs always to be linked to *writing* the world (Freire & Macedo, 1987). This refers to the process of acting in the world as a conscious human agent to remake reality. Freire linked reflection to action (a unity that he called *praxis*), and this unity is captured in his description of the dialectically connected processes of reading and writing the world:

> Reading the world always precedes reading the word, and reading the word implies continually reading the world.... In a way, however, we can go further and say that reading the word is not preceded merely by reading the world, but by a certain form of *writing* it, or *rewriting* it, that is, of transforming it by means of conscious, practical work. (Freire & Macedo, 1987, p. 35)

Some reading and writing the world with mathematics work has occurred in Chicago (and other cities') public schools, often under the conditions of high-stakes testing and draconian accountability regimes (see Lipman, 2004; Lipman & Haines, 2007). An instance of these practices is the work at the GLLVSSJ in Chicago (Blunt, Buenrostro, González, Gutstein, Hill, Rivera, & Sia, 2007; Gutstein, 2007b, in press-b). The school grew out of a hunger strike initiated mainly by Mexican immigrant mothers for a new school in an overcrowded neighborhood (Russo, 2003; Stovall, 2005). At the school, students study a NCTM-reform-based mathematics curriculum (the *Interactive Mathematics Program*, Fendel, Resek, Alper, & Fraser, 1998) and also use mathematics to investigate aspects of injustice through what we call "real-world projects." A small core group of students has emerged who voluntarily spend extra time after school and weekends and who function as a coresearch team. They study mathematics, code and analyze data from previous real-world projects, prepare presentations, and develop their own capacity to lead and teach others how to read and write the world through social justice mathematics projects. In the 2006–2007 school year, tenth-grade students from the school presented a social justice mathematics project, or their own research on how students in the school as a whole have been learning social justice mathematics, at six national or regional academic and educational activist conferences (in addition to graduate-level university classes). Part of their role is to help conceptualize and plan how to spread their high level of commitment and engagement in social justice mathematics to the rest of the school, as well as to the broader public. In these ways, they are inserting themselves in the teaching and learning process as conscious agents of social change, and they represent the seeds and possibilities of a potentially powerful counter-hegemonic trend.

The reality is that what is doable is always constrained by objective conditions and power relations, but teacher agency and spaces to work against the grain do exist, despite the obstacles (Carlson, 2002). There is clear evidence of students beginning to read and write the world in classrooms, in a variety of settings, in which they participated as co-constructors of a social justice learning environment (e.g., Morrell, 2004). In these classrooms, teachers have begun the process of developing *political relationships* (Gutstein, in press-a) with students that incorporate, and go beyond, the caring relationships many competent, caring teachers build with their students and families; I say more about this below. While it is always difficult and problematic to assign responsibility for students' growth to any specific experiences (perhaps particularly to those in school), both epistemologically and methodologically, and also hard in terms of the processes through which people develop critical awareness, there is evidence to suggest that urban youth of color in public school mathematics classes can begin to develop sociopolitical consciousness and a sense of social agency (Blunt et al., 2007; Gutstein, 2003, 2006a, 2007a; Turner, 2003, 2005).

This is not to suggest that using mathematics to read and write the world in urban U.S. schools is simple—the effort is filled with complexities. Some are well known and familiar to teachers, such as the accountability constraints I mention above that can drive out culturally relevant pedagogies, critical literacies, and teachers' use of students' home languages to develop bi- and multi-lingualism (Lipman, 2004). Student resistance to new modes of mathematics pedagogy and curriculum are commonly acknowledged. Other challenges include that students sometimes initially exhibit a sense of powerlessness when they are confronted with sober realities beyond their immediate experiences, although they can move beyond this state (Bigelow, 2002). In addition, schools in this country not only do poor work in teaching students how to think, but in addition, they do an excellent job in teaching students how to *not* think; Macedo (1994) referred to this as "literacy for stupification."

Of the various challenges in teaching mathematics for social justice, the two on which I concentrate in the rest of this article are one, the complexity of using students' generative themes (Freire, 1970/1998) from which to develop social justice curricula, and two, teachers not seeing their role as *political militants* (Freire, 1998) and consequently not building the political relationships with students that can support their development of sociopolitical consciousness and sense of social agency.

CURRICULUM BASED ON GENERATIVE THEMES

As I mention above, Freire (1970/1998) described generative themes to mean the dialectical interaction between key social contradictions in people's lives (objective reality) and how people understand and act in relation to them (subjective interpretation). He called them "generative" because, like fans, they may be unfolded to reveal deeper or overarching, other themes. For example, in our research group with students in the social justice high school, we raised the (mathematical) issue that there are two males for every three females in one of the communities in which students live. When we asked students to explain the data, they said things like, "because the men are dead or locked up." Missing Black men from the community is itself a generative theme leading to the theme of Black men in prison, and that leads to other themes when we ask students to further explain the phenomenon, such as uncovering the reasons behind the over proportion of African Americans in the U.S. prison system (yet another concept with important mathematical components).

Freire's (1970/1998) concept of a *problem-posing pedagogy* is relevant here. He meant by this that teachers and students together needed to ask questions about significant generative themes and to dig beneath surface-level phenomena, that is, to make a "problem" out of what one sees, and often normalizes, such as the dearth (and death) of Black males in the neighborhood. Objective phenomena in students' lives have complex sociopolitical geneses. When educational settings (including, but not limited to, schools) are structured and afford the opportunities so that participants (teachers and students) can deeply investigate, study, reflect on, and interrogate the roots of injustices, with the ultimate aim of rectifying the causes that spawn them, then they are enacting a problem-posing pedagogy. A core component of the intricate (and non-linear) process is the uncovering of generative themes that lie in the community's knowledge and experience so that teachers (and other partners including parents and students) can create curriculum based on the study of the themes as aspects of complex social reality (Freire, 1978). The process does not end there. One has to also provide a framework for the investigations so that students will develop the academic competencies, and cultural capital, that not only has society demanded that they need to achieve "success"—but equally or more importantly, so that students themselves are more fully prepared to read and write the world. Curriculum development, then, is a necessary aspect of social justice teaching and learning, from a Freirean problem-posing framework, but definitely not sufficient by itself. As educators well know, quality curriculum (in the narrow sense of curriculum), of whatever form, is only one of many components. This may be especially true in social justice-oriented classrooms.

Creating problem-posing pedagogies and developing curriculum based on students' generative themes is difficult work. I briefly discuss some of the challenges in our work at the GLLVSSJ in Gutstein (2007b, in-press b). In addition, there is international experience that we may draw upon. In Porto Alegre, Brazil, the *Citizen Schools Project* has been developing such a praxis for the past 15 years (Gandin, 2002). Gandin described the process in Porto Alegre, where educators refer (with similar meaning) to generative themes as *thematic complexes*:

> The starting point for the construction of curricular knowledge is the culture(s) of the communities themselves, not only in terms of content, but in terms of perspective as well. The whole educational process is aimed at inverting previous priorities and instead serving the historically oppressed and excluded groups. The starting point for this new process of knowledge construction is the idea of Thematic Complexes. This organization of the curriculum is a way of having the whole school working on a central generative theme, from which the disciplines and areas of knowledge, in an interdisciplinary effort, will structure the focus of their content. (p. 140)

In Porto Alegre, the work has been challenging enough (I. Martins de Martins, personal communication, 3/22/05), even though the city administration was supportive of the Project and worked with its participants to rearticulate the notion of citizen involvement to counter neoliberal trends in Brazil (Gandin & Apple, 2003). Developing curriculum based on generative themes in Porto Alegre required a detailed, not entirely linear 10-step process, including interviews of community residents, large open forums with community members, and much interdisciplinary and collaborative work among teachers (Gandin, 2002). To do so in a U.S. or similar context, with high-stakes testing and accountability regimes may be substantially more difficult. Nonetheless, the possibilities exist. I do not have space here, but refer readers to the earlier references I cite about developing curriculum based on generative themes (and what we refer to as *community knowledge*) at the GLLVSSJ (Gutstein, 2007b, in-press b).

TEACHERS AS "POLITICAL MILITANTS"

Freire's (1998) use (above) of the term "militant" refers, according to Carmen St. John Hunter, one of his translators, to "persons actively committed to justice and liberation—political activists" (Freire, 1978, p. 73). Freire (1998), in a letter to teachers, wrote:

> We are political militants because we are teachers. Our job is not exhausted in the teaching of math, geography, syntax, history. Our job implies that we teach these subjects with sobriety and competence, but it also requires our involvement in and dedication to overcoming social injustice. (p. 58)

For many teachers in U.S. schools, the idea and practice of being a "political militant" are not second nature. The demands of teachers' work are difficult enough, even without accountability mandates. To ask them to be activists, above and beyond their already over-taxed work-lives, borders on the extreme. However, the literature demonstrates (and my personal experience corroborates) that teachers who are engaged as activists themselves (in a myriad of ways) also provide their students to have opportunities to become agents of change (Christensen, 2000; Gutstein, in-press a; Ladson-Billings, 1994; Peterson, 1991).

In Gutstein (2006a), I described how my students and I co-constructed a classroom oriented toward social justice when I taught mathematics in a Chicago middle school. I outlined three features that were necessary in my context and in no way assumed nor suggested that these fit all situations. They were what I called "normalizing politically taboo topics" (e.g., making discussions about racism and injustice part of ongoing classroom discourse), creating a "pedagogy of questioning" (Gutstein, 2006b), and developing "political relationships" with students that

> ...subsume the personal, supportive relationships with students that some teachers see as essential to their pedagogy. Many teachers build quality relationships with students both in and out of class, and they spend time with students and families when appropriate; share stories from their own lives; and talk, listen, and respond to students about any concerns they have. However, political relationships go further. They include taking active political stands in solidarity with students and their communities about issues that matter. Political relationships also entail teachers sharing political analyses with students as much as possible. Finally, they include talking with students about social movements, involving students themselves in studying injustice, and providing opportunities for them to join in struggles to change the unjust conditions. (Gutstein, 2006a, pp. 132–3)

These relationships may not be necessary in all settings to teach for social justice nor to help students develop sociopolitical consciousness and a sense of social agency. However, evidence in the literature on culturally relevant teaching, critical pedagogy, and teaching for social justice suggests that these are important components. For example, teachers may work with students to examine the politics of language, as Delpit (1988) described; directly confront racism in dominant narratives (Ladson-Billings, 1997); critique the political nature of knowledge (Bigelow, 1998; Christensen, 2000); explicitly engage in anti-colonial, liberatory educational practices (Camangian, 2006; Yang, 2006); and support students in learning and using mathematics for social justice, to examine (and ultimately try to change) injustice while explicitly naming racism and other forms of discrimination. I am not prescribing a formula or delineating how these relationships might occur. They are partially based on who teachers are and their own strengths, weaknesses, experiences, knowledge, and orientations/dispositions toward knowledge, social movements, and political struggle. Clearly, like any social process, this is complicated. But educators are involved in this work, and we need to learn the lessons (Gutstein, in press-a).

An important aspect is to support the development of others to become "political militants" and to use and further develop social justice curriculum and pedagogy. This is a challenge for those who have some of the knowledge and experience needed to teach for social justice, much of it gained through participation in social struggles. Traditional teacher education programs are not oriented toward, or, in my view, able to prepare social justice educators. Some have proposed ideas on social justice teacher education (e.g., Darling-Hammond, French, & Garcia-López, 2002; Gau-Bartell, 2005). While there is no room in this article to fully examine the question of initial or in-service teacher education, I offer some thoughts that apply to mathematics teaching and learning and hopefully beyond, to all subject areas.

First, the experiences that contribute to teachers' sociopolitical knowledge mostly occur out of school and in social interaction. If one subscribes to "learning by doing," one would accept that people learn about social activism by being involved in social movements, as Freire (1998) suggested. Anecdotally, this is the case for myself and for all the teachers I know who are politically active and who have experience in teaching for social justice. If a teacher wants to teach for social justice, then it is important for that person to concretely express solidarity with her of his students, their communities, and their struggles, in both words *and* deeds. This may be particularly true for teachers who are outsiders to their students' communities and to those who teach "other people's children" (Delpit, 1988).

Second, teachers whose lives have not yet taken them towards political action may need to seize the opportunity to take such steps. In Chicago, for example, there are ongoing struggles in students' communities, such as the Little Village and North Lawndale neighborhoods in which the GLLVSSJ is located—for justice and against gentrification, anti-immigrant discrimination, and police violence. These struggles need allies from without, and there are real possibilities for teachers to participate in this role.

Third, collective study within teacher inquiry groups, for example, of Freire's work, can be important. At the GLLVSSJ, we have started a study group with the mathematics teachers in which we are reading about neoliberalism to understand its impact on the city and the school communities. In particular, we are investigating the relationship of transnational capital and the financialization of the economy to gentrification in North Lawndale; we are also studying how free trade agreements, agricultural subsidies, and other neoliberal policies and international entities (such as the World Trade Organization, International Monetary Fund, and the World Bank) contribute to displacing Mexicans from their lands and creating an impetus for their immigration to the U.S. The teachers have expressed that a deeper understanding of these issues, as part of the larger sociopolitical context enveloping us all, will be useful to develop social justice mathematics projects about generative themes in students' lives.

Fourth, collaborating with students, as I describe above, to design, teach, study, assess, and discuss social justice mathematics is an important component in the learning of all participants. Working with students' as co-researchers and paying deep attention to their knowledge, views, analyses, and perspectives, not only on their own lives (and generative themes) but also on the broader questions involved in teaching and learning about social justice mathematics, is yet another way to support teachers (and researchers) in deepening their understanding of sociopolitical contexts. Those who are open to learning from and with students can significantly enrich their collective understanding of the issues. As well, it can strengthen their political commitment to liberatory education and contribute to the development of political relationships between teachers and students. As an aside, at the GLLVSSJ, we have yet to begin working with parents and other community adults whom we are confident will share many of our goals (Gutstein, 2006c), but whom we believe will challenge and expand our political and pedagogical thinking and direction as well.

Finally, we need to continue to theorize, as well as to capture, the process of creating social justice mathematics programs and the particulars of the mutual teaching and learning that occurs. It is through the documentation and analysis, again, collaboratively with students, teachers, and university faculty as co-researchers, that we accumulate knowledge about critical education. A responsibility of those engaged in this practice is to find ways to describe and share the lessons, appropriately theorized, and to support the development of others.

CONCLUSION

Despite the myriad of challenges of critical mathematics teaching and learning, the possibilities are many. They include students developing sociopolitical consciousness, a sense of social agency, and positive social and cultural identities (Frankenstein, 1998; Gutstein, 2006a; Turner, 2003). There are several central issues involved in the processes which I do not discuss here for lack of space. These include (and are not limited to) a) the dialectical relationship of teaching mathematics for social justice and teaching to develop mathematical power; b) the thorny dilemma of explicitly politicizing teaching and learning while maintaining space for students to develop and advocate their own views (Freire, 1994; Freire & Faundez, 1992); c) the difficulties of teaching across differences such as race and ethnicity, culture and language, gender, and social class (that is, in teaching "other people's children," Delpit, 1988); and d) the constraints of teaching against the grain while in oppressively hierarchical accountability systems (such as Chicago Public Schools).

I conclude this article with a short description of a social justice mathematics project that Joyce Sia (the GLLVSSJ 11th grade mathematics teacher) and I just created and co-taught in September 2007 to the five 11th grade classes. I present this to give readers a more concrete feel for what social justice mathematics can actually look like, at least at a surface level. We have not yet co-analyzed students' work and responses with the student co-research team, but there are some initial observations that relate to the main themes in this article. The project was on the *Jena 6*, six African American high school students in the small town of Jena, Louisiana, located in the U.S. South. Jena has about 3000 adults and was about 85.6% white according to the 2000 census. In September 2006, an African American student at Jena High School asked his principal for permission to sit under a large shade tree on the school grounds known as the "white tree" because only whites sat under it. The principal told him "yes," but the next day, three nooses were found hanging from the tree painted in school colors. While some (e.g., the school superintendent) wrote it off as a "prank," for an African American community in the South, the historical record of thousands of their people lynched from such nooses, made it anything but. In addition, the three nooses were inferred by many to represent the three letters "KKK" for the Ku Klux Klan, a home-grown U.S. terror organization. The racial tensions mounted with a number of fights, and a white man pulled a gun pulled on some Black students. In December 2006, an African American student was beaten up by whites at a party, and two days later, a white student taunted the student in school, at which point some African American students beat him up. They were initially charged with attempted murder with potential prison terms of 100 years. The first of these students, Mychal Bell, was tried in adult court and found guilty by an all-white jury (his conviction was overturned because he was improperly tried as an adult), and the story continues as of this writing.

On the second day of class this fall, we began an intense two-week project with the framing question based on Mychal Bell's conviction: If a jury (12 people in the U.S.) was randomly selected in Jena (with 2154 adults, 85.6% white) what is the probability of choosing an all-white jury? (We urge readers to guesstimate the answer before going further.) Briefly, we started out by showing students a short video clip from alternate media summarizing the situation, read some interview transcripts from involved individuals, and had students figure out what the jury would have looked like *if* it matched Jena's demographics (i.e., 10 whites, 2 people of color). We did the latter to both politically contextualize the issue (since the expected did not occur, then what are the odds, or chances, that the jury turned out all white?) and to situate the mathematics. One way to answer the principal question is to use combinatorial analysis and compute the quotient of the possible number of all-white juries in Jena divided by the total possible number of Jena juries. These numbers are on the order of 10^{30} and 10^{31} respectively; dividing them yields a probability of ~15.4%. Most people (ourselves included) thought the chances would be substantially higher. The specifics of how students worked on this and developed the generalizations of n C r, (necessary because we did not tell students formulas, but rather provided them opportunities to "reinvent significant mathematics, Freudenthal) are in Sia & Gutstein (in preparation).

In our short writeup of the curriculum we created for a teacher activist group based in the U.S., we outlined the social justice and mathematics objectives (see Table 25.1). Although this project was not based on a generative theme directly related to students' lives, since they have no connection to Jena and our students' neighborhoods are almost entirely people of color, it nonetheless resonated. While we have not yet fully unpacked students' reasons for their engagement in the project, we can summarize from our in-class observations and knowledge of their lives that they related because of their personal knowledge and experiences of racism which leads to a deep sense of justice that they bring with them to school. The African American students live in a community in which an astounding 57% of the adults are in the prison system, either in prison, or probation/parole, or awaiting sentencing (McKean & Raphael, 2002); this means that roughly two thirds of the males are involved. Anecdotally, all the Black students in the school I have asked have reported that they have male family members involved (present or past) with the prison system. The Latino/a students, overwhelmingly Mexican American, have similar knowledge and experience, even if their community experiences less incarceration. Furthermore, they know, directly or indirectly, of undocumented people in

TABLE 25.1   Social Justice and Mathematics Goals for Jena 6 Project

Social Justice objectives:

- Grow in being able to "read the world with mathematics," that is, develop deeper sociopolitical consciousness of reality using mathematics.
- Provide some concrete support to the Jena 6, i.e., take some action;
- Raise awareness about the Jena 6;
- Inform students about how juries are selected;
- Have students answer: "Was the jury for Mychal Bell selected randomly and without bias?"
- Connect the Jena 6 situation to students' own lives and communities.

Mathematics objectives:

- Determine the probability of randomly selecting a 12-person, all-white jury from a town that is 85.6% white, 14.4% people of color (mainly African American), of 2,154 adults (2000 census).
- Generalize the formula for combinatorics, that is, n C r.
- Develop a better understanding of the mathematical concept of "randomness."
- Gain experience in "thinking like a mathematician."
- Understand the role mathematics has in understanding a key social justice issue—and that without relatively sophisticated mathematics, one cannot know the answer.

the area who are forced to live "underground" (e.g., they cannot obtain a drivers license), and the community recently was raided by immigration authorities which outraged (and terrorized) many neighborhood residents (Little Village is the area's largest Mexican immigrant community, and the Chicago region is estimated to have up to 400,000 of the nations 12,000,000 undocumented immigrants).

It is mathematics/social justice projects like this that provide students context, history, and opportunities to learn about, and be engaged in, aspects of social justice as well as social movements—using and learning mathematics at the same time. Although we are well aware that all our goals are never realizable through one particular project, we understand the deepening of sociopolitical consciousness and a sense of social agency to be a dialectical process taking place over a period of years. On September 20, 2007, a major demonstration demanding all charges be dropped on the six took place in Jena (which is 900 miles from Chicago). Many of our students were so moved that they organized a protest at the school and walked out of school (during the school day) to make signs and posters and held their own impromptu rally at the nearest major intersection near the building. This was led by African American eleventh graders, supported by Latino/a students, which had a unifying impact on the school. Furthermore, the tenth- and ninth-grade mathematics teachers in the school also decided to do the project (although they did not emphasize the process of mathematical generalization as much). Thus, the whole school, through mathematics classes, became involved in the Jena 6 struggle, opening up avenues for continual investigation, links to Chicago issues, and it further develops commitment and awareness on the part of students, as well as of their teachers. It is precisely opportunities such as these that embody the possibilities of teaching mathematics for social justice.

ACKNOWLEDGMENT

Correspondence concerning this article should be addressed to the author at 1040 W. Harrison St., Chicago, IL 60607 or to gutstein(at)uic.edu.

I would like to acknowledge Patricia Buenrostro, Phi Pham, Joyce Sia, and Jon Reitzel, my math team teacher colleagues in 2005–07, and Nikki Blunt, Veronica Gonzalez, Darnisha Hill, and Rogelio Rivera, my student colleagues, at the social justice high school in Chicago. The work I describe here regarding the school was collaborative and informed by all.

REFERENCES

Anderson, J. (1988). *The education of Blacks in the south, 1860–1935.* Chapel Hill, NC: University of North Carolina Press.

Atweh, B. & Clarkson, P. (2001). Internationalization and globalization of mathematics education: Towards an agenda for research/action. In B. Atweh, H. Forgasz, & B. Nebres (Eds.), *Sociocultural research on mathematics education: An international perspective* (pp. 77–94). New York: Erlbaum.

Bigelow, B. (1998). Discovering Columbus: Re-reading the past. In B. Bigelow & Peterson, B. (Eds.), *Rethinking Columbus* (pp.17–22). Milwaukee, WI: Rethinking Schools, Ltd.

Bigelow, B. (2002). Defeating despair. In B. Bigelow & B. Peterson (Eds.), *Rethinking globalization: Teaching for justice in an unjust world* (pp. 329–334). Milwaukee, WI: Rethinking Schools, Ltd.

Bell, D. (1992) *Race, racism and American law.* Boston: Little & Brown.

Blunt, N., Buenrostro, P., González, V., Gutstein, E., Hill, D., Rivera, R., & Sia, J. (2007, April). *Developing social justice mathematics curriculum in a Chicago public school.* Paper presented at the annual meeting of the American Educational Research Association, Chicago.

Boger, J. C., & Orfield, G. (2005). *School resegregation: Must the South turn back?* Chapel Hill, NC: University of North Carolina Press.

Bond (1934/1966). *The education of the Negro in the American social order.* New York: Octagon Books.

Brantlinger, A. (2006). *Geometries of inequality: Teaching and researching critical mathematics in a low-income urban high school.* Unpublished doctoral dissertation. Evanston, IL: Northwestern University.

Camangian, P. (2006, March 30). *Transformative teaching and youth resistance.* Talk given at DePaul University, Chicago, IL.

Carlson, D. L. (2002, April). *Small victories: Narratives of hope in a neo-conservative age.* Paper presented at the Annual Meeting of the American Educational Research Association, New Orleans.

Christensen, L. (2000). *Reading, writing, and rising up: Teaching about social justice and the power of the written word.* Milwaukee, WI: Rethinking Schools, Ltd.

Darling-Hammond, L. French, J., & Garcia-López, S. P. (Eds.). (2002). *Learning to teach for social justice.* New York: Teachers College Press.

Delpit, L. (1988). The silenced dialogue: Power and pedagogy in educating other people's children. *Harvard Educational Review, 58,* 280–298.

Demissie, F. (2006). Globalization and the remaking of Chicago. In J. P. Koval, L, Bennett, M. I. J. Bennett, F. Demissie, R. Garner, & K. Kim (Eds.), *The new Chicago: A social and cultural analysis* (pp. 19–31). Philadelphia, PA: Temple University Press.

Du Bois, W. E. B. (1935). Does the Negro need separate schools? *Journal of Negro Education, 4,* 328–335.

Fendel, D., Resek, D., Alper, L., & Fraser, S. (1998). *Interactive mathematics program.* Berkeley, CA: Key Curriculum Press.

Foster, M. (1994). Educating for competence in community and culture: Exploring the views of exemplary African American teachers. In M. J. Shujaa (Ed.), *Too much schooling, too little education: A paradox of Black life in white societies* (pp. 221–244). Trenton, NJ: African World Press, Inc.

Frankenstein, M. (1987). Critical mathematics education: An application of Paulo Freire's epistemology. In (I. Shor, Ed.), *Freire for the classroom: A sourcebook for liberatory teaching* (pp. 180–210). Portsmouth, NH: Boyton/Cook.

Frankenstein, M. (1990). Incorporating race, gender, and class issues into a critical mathematical literacy curriculum. *Journal of Negro Education, 59,* 336–359.

Frankenstein, M. (1995). Equity in mathematics education: Class in the world outside the class. In W. G. Secada, E. Fennema, & L. B. Adajian (Eds.), *New directions for equity in mathematics education* (pp. 165–190). Cambridge: Cambridge University Press.

Frankenstein, M. (1997). In addition to the mathematics: Including equity issues in the curriculum. In J. Trentacosta & M. Kenney (Eds.), *Multicultural and gender equity in the mathematics classroom* (pp. 10–22). Reston, VA: National Council of Teachers of Mathematics.

Frankenstein, M. (1998). Reading the world with math: Goals for a criticalmathematical literacy curriculum. In E. Lee, D. Menkart, & M. Okazawa-Rey (Eds.), *Beyond heroes and holidays: A practical guide to K–12 anti-racist, multicultural education and staff development* (pp. 306–313). Washington D.C.: Network of Educators on the Americas.

Freire, P. (1970/1998). *Pedagogy of the oppressed.* (M. B. Ramos, Trans.). New York: Continuum.

Freire, P. (1978). *Pedagogy in process: The letters to Guinea-Bissau* (C. St. John Hunter, Trans.). New York: Continuum.

Freire, P. (1970/1998). *Pedagogy of the oppressed* (M. B. Ramos, Trans.). New York: Continuum.

Freire, P. (1994). *Pedagogy of hope: Reliving* Pedagogy of the Oppressed. (R. R. Barr, Trans.). New York: Continuum.

Freire, P. (1998). *Teachers as cultural workers: Letters to those who dare teach.* (D. Macedo, D. Koike, & A. Oliveira, Trans.). Boulder, CO: Westview Press.

Freire, P., & Faundez, A. (1992). *Learning to question: A pedagogy of liberation.* New York: Continuum.

Freire, P., & Macedo, D. (1987). *Literacy: Reading the word and the world.* Westport, CT: Bergin & Garvey.

Gandin, L. A. (2002). *Democratizing access, governance, and knowledge: The struggle for educational alternatives in Porto Alegre, Brazil.* Unpublished doctoral dissertation, University of Wisconsin, Madison.

Gandin, L. A., & Apple, M. W. (2003). Educating the state, democratizing knowledge: The Citizen School Project in Porto Alegre, Brazil. In *The state and the politics of knowledge* (pp. 193–219). New York, RoutledgeFalmer.

Gau-Bartell, T. (2005). *Learning to teach mathematics for social justice.* Unpublished dissertation. University of Wisconsin, Madison, WI.

Giroux, H. (1988). *Teachers as intellectuals: Toward a critical pedagogy of learning.* Westport, CN: Bergin & Garvey.

Greater West Town Community Development Project (2003). *Chicago's dropout crisis: Continuing analysis of the dropout dilemma by gender and ethnicity.* Chicago: Author.

Gutstein, E. (2003). Teaching and learning mathematics for social justice in an urban, Latino school. *Journal for Research in Mathematics Education, 34,* 37–73.

Gutstein, E. (2006a). *Reading and writing the world with mathematics: Toward a pedagogy for social justice.* New York: Routledge.

Gutstein, E. (2006b). "So one question leads to another": Using mathematics to develop a pedagogy of questioning. In N. S. Nasir & P. Cobb (Eds.), *Increasing access to mathematics: Diversity and equity in the classroom* (pp. 51–68). New York: Teachers College Press.

Gutstein, E. (2006c). "The real world as we have seen it": Latino/a parents' voices on teaching mathematics for social justice. *Mathematical Thinking and Learning, 8,* 331–358.

Gutstein, E. (2007a). "And that's just how it starts": Teaching mathematics and developing student agency. *Teachers College Record, 109,* 420–448.

Gutstein, E. (2007b). Connecting *community, critical,* and *classical* knowledge in teaching mathematics for social justice. *The Montana Mathematics Enthusiast, Monograph 1,* 109–118.

Gutstein, E. (in press-a). Building political relationships with students: What social justice mathematics pedagogy requires of teachers. In (E. de Freitas, & K. Nolan, Eds.), *Opening the research text: Critical insights and in(ter)ventions into mathematics education.* New York: Springer.

Gutstein, E. (in press-b). Developing social justice mathematics curriculum from students' realities: A case of a Chicago public school. In (W. Ayers, T. Quinn, & D. Stovall, eds.), *The handbook of social justice in education.*

Gutstein, E, (in preparation). *Mathematics education, neoliberalism, and U.S. global competitiveness.*

Gutstein, E., Lipman, P., Hernández, P., & de los Reyes, R. (1997). Culturally relevant mathematics teaching in a Mexican American context. *Journal for Research in Mathematics Education, 28*, 709–737.

Gutstein, E., & Peterson, B. (Eds.). (2005). *Rethinking mathematics: Teaching social justice by the numbers.* Milwaukee, WI: Rethinking Schools, Ltd.

Hilliard, Asa, III. (1991). Do we have the *will* to educate all children? *Educational Leadership, 31*(1), 31–36.

Julie, C. (1993). People's mathematics and the applications of mathematics. In J. de Lange, C. Keitel, I. Huntley, & M. Niss (Eds.), *Innovation in maths education by modelling and applications* (pp. 31–40). Chichester: Ellis Horwood.

Julie, C. (1998). Ideal and reality: Cross-curriculum work in school mathematics in South Africa. *ZDM, 30*, 110–115.

Julie, C. (2004). Can the ideal of the development of democratic competence be realized within realistic mathematics education? The case of South Africa. *The Mathematics Educator, (14)*, 2, 34–37.

Kirsch, I., Braun, H., Yamamoto, K., & Sum, A. (2007). *America's perfect storm: Three forces changing our nation's future.* Princeton, NJ: Educational Testing Service.

Kozol, J. (1992). *Savage inequalities: Children in America's schools.* New York: Harper-Collins.

Knijnik, G. (1997). Popular knowledge and academic knowledge in the Brasilian peasants' struggle for land. *Educational Action Research, 5*, 501–511.

Ladson-Billings, G. (1994). *The dreamkeepers: Successful teachers of African American Children.* San Francisco: Jossey-Bass.

Ladson-Billings, G. (1997). I know why this doesn't feel empowering: A critical *race* analysis of critical pedagogy. In P. Freire (Ed.), *Mentoring the mentor: A critical dialogue with Paulo Freire* (pp. 127–141). New York: Peter Lang Publishing.

Lipman, P. (2004). *High stakes education: Inequality, globalization, and urban school reform.* New York: Routledge.

Lipman, P., & Haines, N. (2007). From accountability to privatization and African American exclusion: Chicago's "Renaissance 2010." *Educational Policy, 21*, 471–502.

Macedo, D. (1994). *Literacies of power: What Americans are not allowed to know.* Boulder, CO: Westview Press.

Marable, M., & Mullings, L. (Eds.). (2000). *Let nobody turn us around: Voices of resistance, reform, and renewal.* Lanham: MD: Rowman & Littlefield Publishers.

McKean, L., & Raphael, J. (2002). *Drugs, crime, and consequences: Arrests and incarceration in North Lawndale.* Chicago: North Lawndale Employment Network.

Morrell, E. (2004). *Becoming critical researchers: Literacy and empowerment for urban youth.* New York: Peter Lang Publishing.

Moses, R. P., & Cobb, C. E. Jr. (2001). *Radical equations: Math literacy and civil rights.* Boston: Beacon Press.

National Academy of Sciences (2006). *Rising above the gathering storm: Energizing and employing America for a brighter economic future.* Washington, DC: Author.

National Association of Manufacturers (2005). *The looming workforce crisis: Preparing American workers for 21st century competition.* Washington, DC: Author.

National Center on Education and the Economy (2007). *Tough times or tough choices: The report of the* new *commission on the skills of the American workforce.* Washington, DC: Author.

National Council of Teachers of Mathematics (1989). *Curriculum and evaluation standards for school mathematics.* Reston, VA: Author.

National Council of Teachers of Mathematics (2000). *Principles and standards for school mathematics.* Restón, VA: Author.

Payne, C. M. (1995). *I've got the light of freedom: The organizing tradition and the Mississippi freedom struggle.* Berkeley, CA: University of California Press.

Perry, T. (2003). Up from the parched earth: Toward a theory of African-American achievement. In *Young, gifted, and black: Promoting high achievement among African-American students* (pp. 1–108). Boston: Beacon Press.

Peterson, B. (1991). Teaching how to read the world and change it: Critical pedagogy in the intermediate grades. In C. Walsh (Ed.), *Literacy as praxis: Culture, language, and pedagogy* (pp. 156–182). Westport, CT: Ablex.

Peterson, B. (1995). Teaching math across the curriculum: A 5th grade teacher battles "number numbness." *Rethinking Schools, 10*(1), 1 & 4–5.

Peterson, B. (2003). Understanding large numbers. *Rethinking Schools, 18*(1), 33–34.

Phillips, S., & Ebrahimi, H. (1993). Equation for success: Project SEED. In G. Cuevas & M. Driscoll (Eds.), *Reaching all students with mathematics.* Reston, VA: NCTM.

Russo, A. (2003, June). Constructing a new school. *Catalyst.* Retrieved March 3, 2004 from http://www.catalyst-chicago.org/06-03/0603littlevillage.htm.

Sia, I. J., & Gutstein, E. (in preparation). *Mathematizing racism: Free the Jena 6!*

Siddle Walker, V. (1996). *Their highest potential: An African American school community in the segregated south.* Chapel Hill, NC: University of North Carolina Press.

Skovsmose, O. (1994). *Towards a philosophy of critical mathematical education.* Boston: Kluwer Academic Publishers.

Skovsmose, O. (2004). *Critical mathematics education for the future.* Aalborg, Denmark: Aalborg University, Department of Education and Learning.

Skovsmose, O. (2005). *Traveling through education: Uncertainty, mathematics, responsibility.* Rotterdam, Netherlands: Sense Publishers.

Stovall, D. (2005). Communities struggle to make small serve all. *Rethinking Schools, 19*(4). Retrieved April 30, 2007 from http://www.rethinkingschools.org/archive/19_04/stru194.shtml

Turner, E. (2003). Critical mathematical agency: Urban middle school students engage in mathematics to investigate, critique, and act upon their world. Unpublished doctoral dissertation. Austin, TX: University of Texas-Austin.

Turner, E. E., & Font Strawhun, B. T. (2005). With math, itís like you have more defense. Rethinking Schools, 19(2), 38–42.

Valero, P. (1999). Deliberative mathematics education for social democratization in Latin America. *ZDM, 99*(1), 20–26.

Vithal, R. (2002a). A Pedagogy of conflict and dialogue for a mathematics education from a critical perspective. *For the Learning of Mathematics, 22*(1).

Vithal, R. (2002b). Methodological challenges for mathematics education research from a critical perspective. In P. Valero & R. Zevenbergen (Eds.) *Researching the Socio-political Dimensions of Mathematics Education: Issues of Power in theory and*

Methodology (pp. 227–248). Dordrecht, The Netherlands: Kluwer Academic Publishers.

Volnick, J. (1994). Mathematics by all. In S. Lerman (Ed.), *Cultural perspectives on the mathematics classroom* (pp. 51–67). Dordrecht, The Netherlands: Kluwer Academic Publishers.

Watkins, W. (2001). *The white architects of black education: Ideology and power in America, 1865–1954.* New York: Teachers College Press.

Woodson, C. G. (1933/1990). *The mis-education of the Negro.* Trenton, NJ: Africa World Press, Inc.

Yang, K. W. (2006, March 30). *Transformative teaching and youth resistance.* Talk given at DePaul University, Chicago, IL.

Zevenbergen, R. (2000). ìCracking the codeî of mathematics classrooms: School success as a function of linguistic, social and cultural background. In J. Boaler (Ed.) Multiple Perspectives on Mathematics Teaching and Learning (pp. 201–223). Westport, CT: Ablex Publishing.

NOTE

1. I use the following as a definition of mathematical power, "Students confidently engage in complex mathematical tasks . . . draw on knowledge from a wide variety of mathematical topics, sometimes approaching the same problem from different mathematical perspectives or representing the mathematics in different ways until they find methods that enable them to make progress . . . are flexible and resourceful problem solvers . . . work productively and reflectively . . . communicate their ideas and results effectively . . . value mathematics and engage actively in learning it." (National Council of Teachers of Mathematics, 2000, p. 3)

CHAPTER 26

MATH EDUCATION AND SOCIAL JUSTICE

Gatekeepers, Politics and Teacher Agency

Peter Appelbaum and Erica Davila
Arcadia University

INTRODUCTION

This article addresses several urgent mathematics education conversations through the lens of critical educators. As teacher educators, we see and hear the experiences pre-service and in-service teachers undertake while teaching math that is grounded in social justice, or when thinking through teaching math using pedagogies rooted in emancipatory frameworks. We deconstruct the resistance to teaching math that is grounded in social justice education. The two overarching conversations we address are:

(1) The gatekeepers that surface for teachers who teach mathematics with an emphasis on social inequity. Gatekeepers are those individuals or groups of people who are perceived as potentially expressing dissatisfaction with the teachers' choice of social justice pedagogies, e.g., a principal or parents. For many pre-service and practicing teachers, the conversation regarding gatekeepers is more about perceptions they have regarding the

Critical Issues in Mathematics Education, pages 375–394
Copyright © 2009 by Information Age Publishing
All rights of reproduction in any form reserved.

struggles they will face when faced with gatekeepers within the structure of the school system than with actual realities.

(2) The curriculum politics that determine who decides what is taught in K–12 mathematics, and how these political forces connect to the implementation of socially just curricula and pedagogy, specifically for mathematics. We highlight the experiences of early childhood educators because of the recurring question posed to us as teacher educators: "How can I teach about social justice to the younger children?" The deconstruction of conversations that pre-service and practicing teachers pursue can empower the participants to be intellectuals and social agents in their classrooms and schools, and to make critical decisions around curriculum and pedagogy in mathematics.

THEORETICAL FRAMEWORK

Critical race theory offers a way to understand how ostensibly race-neutral structures in education—knowledge, merit, objectivity, and "good education"—in fact help form and police the boundaries of white supremacy and racism (Parker, Deyhle & Villenas 1999; Ladson-Billings & Tate 1995). Specifically for this research project, the perception of math as a neutral subject is analyzed through the experiences of K–12 educators. In particular, critical race theory is used to deconstruct the meaning of "educational achievement," to recognize that the classroom is a central site for the construction of social and racial power. There are reoccurring conversations unfolding in our university classrooms about social justice math versus "real" math education; this juxtaposition speaks to the meaning our students, future and current teachers, are making through the use of the phrase, "educational achievement." As we analyze this dichotomy between social justice math and "real" math, the lens of critical race theory helps us to interpret this concern. When our students begin to understand the integral role teachers must take to empower *their* students, they shift to discourses in which teaching math for social justice *is* teaching "real" math.

We also reference the Standards of the National Council of Teachers of Mathematics (NCTM 2000), local State Standards (PDE undated), and the *Manifesto* of the International Commission for the Study and Improvement of Mathematics Education (CIEAEM 2000), in order to place our analysis in the context of current curriculum reform efforts. We theorize teachers' participation as change agents in the reform process as central to the success of social justice curriculum development. Leu (2005), Remillard (1999), and Drake and Sherin (2006) indicate that teacher participation in the theorizing of mathematics education reform in is critical to our understanding of the reform process. Teacher interpretations of curriculum materials, the philosophy of education supported by specific curriculum materials, and the meanings and purposes of classroom activities, are the nodes of multiple networks of social and cultural discourses in educational reform.

MODES OF INQUIRY

Analyses of university classroom discussions, online discussions, class assignments and presentations serve as our modes of inquiry. We analyze the processes students encounter while exploring two assignments: Each of us shares an example of an assignment we currently implement in our teacher education course work at the graduate level. Both of these projects help authorize teachers to become *empowered teachers* who can teach social justice and "real" math at the same time. One project is part of a course on mathematics and the curriculum; this project requires students to develop a piece of curriculum centered in social justice themes, and to demonstrate that their lesson/unit idea can cover enough of their current curriculum to justify using it as a "replacement unit." The second project is part of a foundations course on culture and education and requires students to review and critique a lesson they have taught through a sociological perspective; students must subsequently design a modified lesson. Both projects also include a reflection paper in which students identify two potential gatekeepers; they are required to explain how they would convince each gatekeeper to use this new curriculum. The ideas students develop through these two learning activities highlight the experiences teachers are sharing in terms of teaching math for social justice, or thinking math through social justice. The analysis of these experiences deconstructs the struggle to transform curriculum and construct hope.

POINT OF VIEW

As teacher educators who are experiencing the intersection of teaching and researching, we have gathered data from our classroom dialogues for this article. The conversations unfolding in our classrooms are the catalyst for this research project. The perceptions and practices that pre-service and in service K–12 teachers share in class need to be deconstructed if we are to carve new spaces of inquiry in their teaching. The students in our courses teach one another. Their lived experiences in their classrooms and field work sites provide many opportunities for them to reflect and analyze. More specifically, in the area of math education, many of these teachers are struggling with the same dissonances between learning how to implement "best practices" and actually implementing them.

The rationale for the dissonances varies from student to student. Many fear that teaching math for social justice does not fit with the mandated scope and sequence of their school/district's curriculum; they perceive various gatekeepers that serve as barriers. Others fear that they will bring issues to a group of children who are innocent and naïve about social inequity. This latter fear has been shared by many teachers and future teachers who teach in the early grades. Another common reaction among our students is the perspective that teaching math for social justice is "too political." We have also experienced students who don't share the dissonance because they hold the perception that they are implementing "best practices" and do not see the urgency of integrating social justice in their curriculum. Students who do not see the urgency often offer explanations about their curriculum being "multicultural" (and therefore they are teaching for social justice). However, most of the teachers who practice teaching math for social justice still feel the urgency and are usually eager for resources and support.

A PEAK AT THE IMPLICATIONS OF OUR STUDY

The notions of urgency, dissonance, and "best practices" can be interpreted as the ongoing social construction of ethics. In this sense, teaching is an ethical stance one takes with the world (Block 2003). Through one's social actions, a person continually invents, or reinvents, an ethical self, in each and every moment. According to Jim Neyland (2004), the ethical self is prior to all codification, and instead founded on the direct, face-to-face ethical encounter of responsibility between persons. "In contrast to currently dominant approaches, mathematics education should ensure that legislative protocols do not override the ethical primacy of the direct encounter." (p.55) This would imply that teachers must understand their practices in ways that take heed of Standards and other protocols, but only insofar as these protocols do not displace direct encounters between people. However, we see our work as contributing to a postmodern sensibility that avoids a reproduction of "reform versus authentic" dualities: in our work, teachers create reform projects and critique existing curriculum in ways that do not set up Standards and Innovation in opposition, and instead see both as mutually supporting the other. Standards are used to justify the social justice mathematics activities; and the social justice mathematics activities are used to accomplish the standards as well as other goals.

The various perceptions shared in our teacher education courses are critical to explore in order to make effective changes in curriculum reform. We need to understand the resistance to transformative math education. In order for social justice math to be valued, teachers must first become aware of the social injustices around them. Teacher educators should encourage K–12 teachers to infuse socially just curriculum and resist the perspective that schools and teachers are apolitical. Teachers should be empowered and be conscious of their beliefs and values and reflect on who they are and what they believe in and most importantly how this impacts their teaching. During this reflection process many teachers can begin to see the moral responsibility of teaching and thinking math through social justice. Altogether this research project has served to highlight the experiences teacher educators and teachers are having around the need for infusing our classroom with socially just curricula. The two specific learning activities that empower teachers to rethink their practice in teaching mathematics in K–12 classrooms are discussed to bridge our theory and practice.

CREATING A REPLACEMENT UNIT

The *Mathematics and the Curriculum* course included eleven students: one first-grade teacher, one fifth-grade teacher, four middle school mathematics teachers, three secondary school mathematics teachers, one alternative high school special education teacher, and one pre-service secondary mathematics teacher. We began the course with an overview of contemporary State, National, and International Standards documents, and compared the content with the curriculum innovations in Gutstein and Peterson's (2005) *Rethinking Mathematics.* Student reactions were varied, but each spoke in one way or another of perceived barriers to implementing the ideas presented in *Rethinking Mathematics.* Examples of activities appeared initially to focus on content that is not central to the given mathematics curriculum found in Standards documents and textbooks that are provided by the schools in which they teach. Organization of classroom learning experiences also appeared to require very different forms of teacher and student behaviors from what these teachers usually encountered themselves in their own work. It became clear to us that this perception of social justice mathematics as alien to common practice needed to be confronted head-on. What was originally required in the course as a curriculum design project became a mutually negotiated "trajectory paper," in which students outlined the trajectory of their mathematics curriculum for an extended period of time, incorporating social justice ideas from the *Rethinking Mathematics* book in ways that supported the achievement of expected outcomes and Standards.

One particular feature of the assignment that presented a unique challenge was to transform the teachers' fears and anxieties into a productive task. We discussed the underlying reasons why we were not able to directly implement the social justice mathematics ideas in our current work, and determined that we unconsciously constructed "gatekeepers" whom we imagined as standing in the way of innovative curriculum development and mathematics education practices. These "gatekeepers" seemed very real as the teachers attempted to describe concrete reasons for why they would not be able to implement social justice mathematics. However, the gatekeepers were in some sense imaginary symbolic tools of resistance to change. It may very well be the case that other teachers and/or the school principal would worry about what is going on in a teacher's classroom if the activity is very different from the usual practice in the school. But the fear that such a reaction would occur is no more real than the possibility that an innovative teacher would receive accolades or simply be ignored by others. It may very well be that parents of students might worry that their children are not learning what should be learned, or that the Standards are not being met; yet this too is no more than a fiction created to concretize fears of the unknown. One such interesting fiction that arose in our collaborative work was the teacher herself as gatekeeper: a teacher may worry herself that trying something new without previous practice, or approaching mathematics from alternative perspectives, could be unsuccessful when judged from the position of whether or not students have mastered expected skills and developed particular conceptual understandings.

One clear requirement for introducing social justice mathematics into the school curriculum was thus made evident: we needed to confront potential gatekeepers directly and proactively. Whether fictional tools of resistance to change or actual constituents in the school community, gatekeepers should be recruited as supporters of innovation before they can begin to formulate concerns. With the notion of gatekeepers in mind, the group decided that the best proactive stance would be to demonstrate convincingly that the planned social justice activities not only contributed to successful mastery of Standards-based curriculum, but that the social justice curriculum would seem to promise better achievement of these objectives.

In an ancillary examination of particular kinds of experiences within a mathematics curriculum, we also identified the location of a classroom activity at the beginning, middle, or end of a unit of study as an important planning decision for a teacher and/or curriculum developer. Placement at the beginning of a unit works well for initial pre-assessment of skills and concepts that students already know and would therefore be available for elaboration or as the basis for the construction of new knowledge. Placement in the middle of a unit is useful for a teacher's ongoing assessment of what has been learned so far, and for making decisions about whether any materials needs to be re-taught or supported with supplementary topics. An activity in the middle can also introduce new perspectives on content, or redirect students' attention to critical ideas. Sometimes such an activity can allow for reflection on new concepts or skill practice. Placement at the end of a unit can provide a culminating experience, can serve as a more definitive assessment tool for the teacher, and can serve as the context for concept integration; such an activity might also enable students to use skills in new contexts, or to provide a transition to the following unit. Appelbaum (2008) describes this curriculum planning analysis as the placement of the activity within a "trajectory."

Through such considerations, we created the trajectory assignment, which involved the following criteria for materials and networking tools:

Incorporate one or several ideas from the Rethinking Mathematics book in order to plan an extended unit or multi-day lesson.

1. Demonstrate that this lesson/unit idea can cover enough of "my" current curriculum to justify using it as a "replacement unit."

2. Describe how the larger curricular context would change if this lesson/ unit were at the beginning, middle, or end of the design.

3. Demonstrate working knowledge of NCTM, CIEAEM, and PA Standards/goals.

4. Identify 2 gatekeepers who will need to be convinced that it would be OK for you to use this and describe how you will convince them.

Create any materials that are needed in order for this lesson/unit to be immediately implementable (for example, create any student handouts that would be needed, schedule appointments with colleagues and community members, compose and deliver letters and memos that must be written, and so on).

The teacher-designed units and lessons that were developed in this project were highly successful in the sense that each design incorporated social justice mathematics ideas from Gutstein and Peterson while also establishing explicit ways that these units and lessons were carefully constructed to meet more traditional content standards. A common form of resistance is to label the social justice content as other than mathematics, so that one might distance oneself from social justice mathematics by declaring it more relevant to social studies or other curriculum areas. By carefully outlining the traditional content that is addressed by the social justice curriculum, these teachers were not only able to make a reasonable argument for the social justice curriculum as a replacement unit, but were moreover better prepared to make the replacement unit more successful than the commonplace curriculum at achieving the expected content objectives. More powerfully, we note how the teachers' relationships with social justice mathematics was transformed through the design experience. Each teacher was convinced that their unit or extended lesson would in fact easily meet the Standards and curriculum objectives that they believed were assigned to them by higher authorities. They were comfortable explaining how and why their unit would do a better job than the already existing set of daily lessons based on their textbooks. Furthermore, because the group supported each other throughout the design process through in-class collaborative workshops, they were convinced of the efficacy of each of the other designs as well as their own. Because each participant had given feedback and offered ideas to the others in small working groups over the course of four class meetings, they were each personally committed to each others' designs, seeing something of their own response to a concern in the others' final materials.

The gatekeeper materials served a unique psychoanalytic function. When a teacher imagines a potential gatekeeper—e.g., principal, parent, colleague, community member, student, themselves—they are projecting their own anxieties fears onto an imaginary construct. As noted above, it is possible that any one of these real people may in fact display certain anticipated reactions. However, in the proactive process of working *with* gatekeepers, the teacher has an opportunity to interact with their own projections and concerns. Whether a teacher presents herself or her principal as worried about the lack of attention to the assigned content for the next month of the school year, the worry about not meeting obligations is the most important concern that the teacher must address. Whether explicitly addressing their own concern or not, the proactive composition of a letter to parents, or the preparation of a proposal for a meeting with an administrator, allows the teacher to work through any possible fears and worries that they may even be having trouble consciously articulating. For example, a letter that lists twenty standards that are being addressed by the forthcoming unit, along with specific suggestions for ways that skills can be practiced at home simultaneously helps a teacher recognize that the unit indeed addresses the twenty standards and indeed provides adequate coverage of particular skills. Furthermore, the composition of such a letter prepares a teacher to better accomplish these goals now that they are explicit, while also making clear to the teacher what must be assessed on an ongoing basis in order to meet the goals.

The trajectory notion serves to highlight both the importance of the social justice aspects of the unit/lesson and the ways that the social justice mathematics and the traditional Standards-based mathematics mutually support each other. Our trajectory designs were really three different unit/lesson designs, and the juxtaposition of each with the others helped to clarify our goals for the designs themselves. The main social justice activities from the *Rethinking Mathematics* book were placed at the beginning, middle, or end of the plans. The consideration of how this placement would change the skills and concepts that would be developed through the designed curriculum helped the participants to determine what skills and concepts were possible in the first place, and in the process, helped the teachers to better understand the mathematics of the social justice experiences more deeply.

Teachers rarely have the opportunity to reflect on the kind of mathematical thinking that they can spark for their students through the introduction of a specific experience. In this project, each teacher pursued the subtle differences that would be enacted by the choice that they needed to make about when and where to place a particular experience within the trajectory of the unit or lesson. They also were participants in the discussion of nine other unit/lesson designs that were also carefully analyzing these issues.

A maxim of psychoanalytic thought is that resistance is essential for critical learning to occur (Appelbaum 2008). In this assignment, the teachers' resistance to new curriculum ideas became the explicit focus of the task. By labeling the fears and anxieties as external "gatekeepers," we were able to distance ourselves from our own resistance, objectify it, analyze it, and move through it to a new understanding of a social justice mathematics curriculum. Social justice mathematics was no longer set up as an alternative to "real" mathematics, and no longer positioned as oppositional to accepted practices. What was initially "othered" became an ally, a tool for accomplishing our own and others' goals for school mathematics.

CULTURAL FOUNDATIONS AND CURRICULUM

The Cultural Foundations of Education course is also a graduate level foundations course. In this course students are empowered to be introspective, reflective and action oriented teachers. Many students are practicing teachers working towards their master's degree or teacher certification. Several are pre-service teachers, some straight out of undergraduate teacher preparation programs; others are career changers who spent significant parts of their lives working in various career tracks. The course is taught from a theoretical stance that is rooted in an emancipatory framework that empowers teachers and future teachers to be activists and advocates for their students. The students work towards developing "conscientization" as defined by Freire (1972), which is a level of consciousness that evokes the power to transform reality.

The course begins with introspection and reflection, and then moves to the lived experience of privilege and oppression; the final portion of the course is designed to explore the implications of the previous conversations within the context of K–12 schools. The students read and explore ideas on the interconnectedness of multicultural education and social justice. The experience the students have while developing their culminating project in this course is what has peaked this inquiry in mathematics education and social justice.

Review & Critique of Curriculum or *Lesson Plan*

Individually or in pairs, students will find either a textbook chapter/unit or a lesson plan to:

1. Evaluate and critique from a multicultural lens (e.g., What is missing from the material that would make it more multicultural? What is in the material that is stereotyped, Eurocentric and/or biased in specific ways?); and

2. Create an alternative text that articulates your ideal multicultural lesson and reflects issues raised in the course (e.g., What could be changed or added to enhance the material? To make it more inclusive and reflective of multicultural ideals?).

Presentations can be structured in a variety of ways. You are encouraged to use this as an opportunity to conceptualize and practice critiquing educational materials from a multicultural perspective as well as developing creative and critical material to use in your classroom. You will be evaluated based on the thoroughness of your critique and your ability to develop practical materials/methods that are inclusive and culturally responsive. Curricula and lesson plans as well as any additional material for presentations should be copied so that all class members have a copy on the day of your presentations.

For this article, we focus on several students who are teaching or hope to teach mathematics. One of the students is a kindergarten teacher in the Philadelphia public schools who struggled to see the bridge from theory to practice when reading Skilton-Sylvester's (1994) *Elementary School Curricula and Urban Transformation* and Gutstein and Peterson's (2005) *Rethinking Mathematics*. She kept saying, "but they're only in kindergarten." It was not until she chose a specific lesson to review that she had taught for several years on currency in the United States (US), specifically pennies and dimes, that she not only saw the bridge from theory to practice, but built her own. The Pennsylvania state standard she was aiming to meet was:

PA—*Pennsylvania Standards for Kindergarten*
 Key Learning Area: Mathematics
   Standard: 2.1 Numbers, Number Systems and Number Relationships
     Content: K. Count pennies and dimes

Her initial idea was to include international coins in her lesson, and to have the kids compare and contrast US coins to other coins. However, after one of her peers asked a simple question: "Will you tell them the US is the wealthiest country?" She was empowered to teach these very young children about the power of US currency, and to begin a critical conversation on global inequities. As she presented this lesson to her peers she shared the experience she had with her students, during which she came to a realization: she held the power to subsequently empower her students.

She knew that presenting them with a chance to think of global inequity in kindergarten was a critical component in developing their sense of agency. Furthermore, she described this new lesson on coins and power to be the catalyst in her transformation. She described social justice as a moral issue; she shared her sense of urgency for teaching and thinking about math and social justice. This experience was rooted in a shared dialogue between a teacher and her peers; together they wrestled with heavy questions regarding teaching kindergarteners about social inequity.

During the same semester (spring 2007), another student preparing to become an elementary school teacher decided to create an interdisciplinary lesson for 4–5th grade that combined teaching perimeter and geography. The state math standards she was aiming to meet were:

PA—*Pennsylvania Academic Standards*
    Subject: Mathematics
        Area 2.3: Measurement and Estimation
            Grade 2.3.5: Grade 5
            Standard A.: Select and use appropriate instruments and units for measuring quantities (e.g., perimeter, volume, area, weight, time, temperature).
            Standard E.: Add and subtract measurements

This future math teacher decided to review a lesson on perimeter she found in a widely-used American textbook series, *Everyday Mathematics* (UCSMP 2004). She expanded the concept of perimeter to include man-made, natural and political perimeters in addition to the mathematical concept of perimeter offered in the text. She critiqued the real world application provided in the text, which discussed kitchen layouts. As her own replacement context of study, she explored the designated borders within the United States Native American Reservations, focusing on the 21 federally recognized tribes in Arizona. This future teacher was helping her students with making meaning of political borders while teaching them math skills in measuring perimeters. Similar to the kindergarten teacher, this student's project was rooted in class dialogue. Although she came to the class knowing that she wanted to teach math for social justice, and knew she must given her teaching philosophy, she struggled with the "how" as well. Her peers supported her; it was clear through discussions in class and on an online discussion board that the project grew and evolved as the teachers and future teachers worked together to rethink teaching perimeter.

The final example of a curriculum project in mathematics was developed by a secondary math teacher who teaches algebra in the suburbs of Philadelphia. In rethinking a lesson on linear programming, she had her students think about price discrimination. The lesson facilitated the application of linear programming within this context. The state standards she was aiming to meet are:

PA—Pennsylvania Academic Standards
  Subject: Mathematics
    Area 2.8: Algebra and Functions
      Grade 2.8.11: Grade 11
        Standard D.: Formulate expressions, equations, inequalities, systems
          of equations, systems of inequalities and matrices to model routine
          and non-routine problem situations.
        Standard S.: Analyze properties and relationships of functions
          (e.g., linear, polynomial, rational, trigonometric, exponential,
          logarithmic).

She shared with the other teachers and future teachers her experience re-teaching this lesson, and how she used the airline industry as an example of price discrimination. First, she grabbed the attention of her students with an opening activity. She posed some critical questions, such as. "Have you ever been the victim of discrimination?" After a discussion on discrimination, they moved to connecting those lived experiences to an application of linear programming to think about profit and cost. When she shared this lesson with her peers, an empowering conversation developed regarding the digital divide as well as the difference in people's experiences around having the flexibility to make travel plans in advance.

The three examples of curriculum critique projects exemplify the process teachers and pre-service teachers may undergo when thinking about transforming mathematics curriculum. Most of the students in this course were struggling with rethinking their own pedagogy. The curriculum critique assignment helped them to reach a level of "conscientization" (Freire 1988) in two ways: through dialoging with their peers, and through rethinking a piece of curriculum they have themselves already taught.

INTERPRETATION

The experiences we discuss above highlight the critical component of dialogic teaching while offering a glimpse of the current condition of mathematics education within the framework of struggle and hope in teacher preparation. As we prepare teachers to be the most effective educators for *their* students, we work to model teaching practices that are rooted in emancipatory frameworks. This in turn provides the real application which the teachers are craving. They can see what these practices look like and they can build bridges between their theories and practices regarding the process of teaching and learning. The power of classroom dialogue must be instrumental within the teaching for social justice. The sharing of lived experiences both inside and outside classrooms is an authentic way to think about one's own values, beliefs and perceptions of teaching, and more importantly, of their students. Nieto (2004) offers an insightful thought about the role of teachers, asserting that teachers are students of their students. Without dialogic teaching and the perception of students being integral to the design and delivery of the course, we cannot fully learn about our students, and they cannot learn about one another. This is relevant for the university course as well as the K–12 teaching contexts that our students are grappling with. Through the sharing and subsequent analysis of personal experience and observation, the teachers and future teachers in our classrooms are constructing meaning. This is not a smooth or linear process. Our students struggle through this. For many, they are questioning their own reality, and that takes work. As teacher educators we see the role of teachers as transformative intellectuals. On the role of teaching, Giroux (1988) states

> If we believe that the role of teaching cannot be reduced to merely training in the practical skills, but involves, instead, the education of a class of intellectuals vital to the development of a free society, then the category of intellectual becomes a way of linking the purpose of teacher education, public schooling and in-service training to the very principles necessary for developing a democratic order and society. (1988, p. 126)

Giroux (2002) pushes for student empowerment, and for human agency rooted in the belief of teachers as transformative intellectuals. He offers insight on educators contemplating the role that public schools might play in facilitating an alternative discourse grounded in a critique of militarism, consumerism, and racism.

Our experiences with teachers and pre-service teachers highlight the importance of working with and through sites of resistance rather than attempting to weave an alternative discourse or struggling to circumvent anxieties. Most professional development resources offer alternative "pictures" of what is possible in mathematics education. They are often accompanied by philosophical arguments that claim ethical or moral grounding. While these "pictures" and ethical arguments are important, they do not do the work of empowerment for teachers. The work of empowerment is a form of learning that is first indicated as potential by the statements and actions of resistance that teachers exhibit in their own learning contexts. Directly working to understand the meaning and sources of these resistances instigates dialogue with peers and forms of self-reflection that can lead to new forms of understanding.

IMPLICATIONS FOR FURTHER RESEARCH AND
FOR FURTHER INNOVATIONS IN TEACHER EDUCATION

Schools function as sites of struggle where teachers can critically explore their emancipatory potential as educators and public intellectuals. The political nature of classrooms entrenched in traditional notions of power, knowledge and truth that reproduce social categories of inequality, evokes the need for transformative intellectuals in schools at every level, from early childhood through higher education. Thus, critical educators at all levels of schooling must create/re-create and question the discourse around public intellectuals by reflecting on ideas that can rupture the anti-intellectualism that is woven in the current structure of public schools. When a teacher hears her- or himself questioning the appropriateness of social justice mathematics for his or her students, or when he or she must collaborate with colleagues who cannot yet see the mathematics in the social justice curriculum, these are moments of opportunity for teacher leadership. Teachers can provide gatekeepers with professional articulations of how and why the standards and tested objectives can be placed in symbiotic relationships with curricula that exceed the minimum goals that such bureaucratic expectations hold for our youth. Furthermore, Freire's thorough discussion of the political nature of classrooms entrenched in traditional notions of power, knowledge and truth that reproduce social categories of inequality actualizes the context for the call for transformative intellectuals in schools. As teachers participate in collaborative curriculum design, they are no longer subjects of power but are active constructors of knowledge, working within and across regimes of truth and power. Such active design may be taken as an articulation of the ethical stance that teaching embodies.

In our work with teachers, we find a focus on assessment helps to raise issues of critical race theory. Given the ways that power is implicated in the determination of the "good" student, notions of whether or not a student is displaying evidence of conceptual growth or understanding, for example, are grounded in cultural and historical legacies of discourse and power. Since a learner's trajectory through the differentiated tracks that constitute 'the school' is determined partially by his or her assessed performance at key branching points, any sociology of learning, broadly conceived, will need to address assessment (Cooper 2007). When the teachers and pre-service teachers created dialogues around the efficacy of an activity's placement at the beginning, middle or end of a lesson/unit trajectory in our classes and on-line, they were able to confront the complexities of this seemingly commonplace and undertheorized teacher role. Because assessment is so central to decision-making and to the judgment of "success"—both of the teacher and the students, assessment is a critical site of power/knowledge and thus of the important work of learning through resistance. Much of the gatekeeper work is focused on assuring the gatekeepers (whether this be the teacher him- or herself and/or supervisors of their teaching) that students will perform well on standardized paper and pencil achievement tests. As teachers broaden their comprehension of assessment within the critique and design of curriculum, they are better able to place such assessment within a broader field of possibilities, each of which defines mathematical understanding and ability in different ways. Because learning is currently defined in this pragmatic way, as based on the performance of the learning by the student, the ability to imagine possible forms of performance and their potential interpretation enables teachers to explore the ways that varying assessments can mutually support as well as potentially conflict with one another.

CONCLUSION

The call for transformative intellectuals speaks to the struggles that have been manifested in a society that leaves families living in poverty, people of color, linguistic minorities, women and children on the margins of full citizenship and denies us equal educational opportunities. However, the rays of hope and victory pierce the struggle from multiple angles and burn through it. Unfortunately, there are many people who are still left to struggle in the shadows missed by the rays of hope. Schools continue to be organized and structured to perpetuate inequality. More importantly, our children continue to be the products of this inequality. It is time for schools to change these inequalities through providing spaces of resistance, coupling the discourse of critique with that of possibility and helping teachers play their role as transformative intellectuals who witness the urgency in teaching for social justice.

The call for transformative intellectuals is not a well packaged recommendation that can be achieved with a workshop or even a course; instead, it is an ideological shift that is occurring within the way we think about learning and teaching within the institution of schools. Thus, if whole communities tap into their standpoint, thoughts, perspectives and ideas of teaching and learning, we can begin conversations with more than just scholars at the table. We can build collective conversations with the people that are in the lives of children every day: family members, child care providers, neighbors, teachers, counselors, secretaries, cafeteria workers, coaches, tutors, and bus drivers.

Critical Mathematics Education demands a critical perspective on both mathematics and the teaching/learning of mathematics. In doing so, it takes one step further in questioning our assumptions about what critical thinking could mean and what democratic participation should mean. As

Ole Skovsmose (1994) describes a critical mathematics classroom, the students (and teachers) are attributed a "critical competence." A century ago (see, e.g., Fawcett 1938), we moved from teaching critical thinking skills to using the skills that students bring with them. We accepted that students, as human beings, *are* critical thinkers, and would display these skills if the classroom allowed such behavior. It seemed that we were not seeing critical thinking simply because we were preventing it from happening; through years of school, students were unwittingly "trained" *not* to think critically in order to succeed in school mathematics. So we found ways to lessen this "dumbing down of thinking through school experiences." Now we understand human beings more richly as exhibiting a *critical competence*, and because of this realization, we recognize that decisive and prescribing roles must be abandoned in favor of all participants having control of the educational process. In this process, instead of merely forming a classroom community for discussion, Skovsmose suggests that the students and teachers together must establish a "critical distance." What he means with this term is that seemingly objective and value-free principles for the structure of the curriculum are put into a new perspective, in which such principles are revealed as value-loaded, necessitating critical consideration of contents and other subject-matter aspects as part of the educational process itself. (See also Skovsmose 2005)

Keitel, Klotzman and Skovsmose (1993) together offer a new way for teachers to think about the mathematics that is being taught. New ideas for lessons and units emerge when teachers describe mathematics as a technology with the potential to work for democratic goals, and when they make a distinction between different types of knowledge based on the object of the knowledge. The first level of mathematical work, they write, presumes a true-false ideology and corresponds to much of what we witness in current school curricula. The second level directs students and teachers to ask about right method: are there other algorithms? Which are valued for our need? The third level emphasizes the appropriateness and reliability of the mathematics for its context. This level raises the particularly technological aspect of mathematics by investigating specifically the relationship between means and ends. The fourth level requires participants to interrogate the appropriateness of formalizing the problem for solution; a mathematical/technological approach is not always wise and participants would consider this issue as a form of reflective mathematics. On the fifth level, a critical mathematics education studies the implications of pursuing special formal means; it asks how particular algorithms affect our perceptions of (a part of) reality, and how we conceive mathematical tools when we use them universally. Thus the role of mathematics in society becomes a component of reflective mathematical knowledge. Finally, the sixth level examines reflective thinking itself as an evaluative process, comparing levels 1 and 2 as essential mathematical tools, levels 3 and 4 as the relationship between means and ends, and level 5 as the global impact of using formal techniques. On this final level, reflective evaluation as a process is noted as a tool itself and as such becomes an object of reflection. When teachers and students plan their classroom experiences by making sure that all of these levels are represented in the group's activities, it is more likely that students, and teachers, can be attributed the critical competence that we envision as a more general goal of mathematics education.

In formulating a democratic, critical mathematics education, it is also essential that teachers grapple with the serious multicultural indictments of mathematics as a tool of post-colonial and imperial authority. What we once accepted as pure, wholesome truth is now understood as culturally specific and tied to particular interests. Philip Davis and Reuben Hersh (1987) and David Berliner (2000), for instance, have described some aspects of mathematics as a tool in accomplishing a fantasy of control over human experience. They use the examples of math-military connections, math-business connections, and others.

Critical mathematics educators ask why students, in general, do *not* see mathematics as helping them to interpret events in their lives, or gain control over human experience. They search for ways to help students appreciate the marvelous qualities of mathematics without adopting its historic roots in militarism and other fantasies of control over human experience. Arthur Powell and Marilyn Frankenstein (1997) have collected valuable essays in *ethnomathematics* and the ethnomathematical responses that educators can make to contemporary mathematics curricula. Ethnomathematics makes it clear that mathematics and mathematical reasoning are cultural constructions. This raises the challenge to embrace the global variety of cultures of mathematical activity and to confront the politics that would be unleashed by such attention in a typical North American school. That is, ethnomathematics demands most clearly that critical thinking in a mathematics classroom is a seriously political act.

One important direction for critical mathematics education is in the examination of the authority to phrase the questions for discussion. Who sets the agenda in a critical thinking classroom? Stephen Brown and Marion Walter (1999) lay out a variety of powerful ways to rethink mathematics investigations through *The Art of Problem Posing*, and in doing so they give us a number of ideas for enabling students both to "talk back" to mathematics and to use their problem solving and problem posing experiences to learn about themselves as problem solvers and posers. In the process, they help us to frame yet another dilemma for future research in mathematics education: Is it always more democratic if students pose the problem? The kinds of questions that are possible, and the ways that we expect to phrase them, are to be examined by a critical thinker. Susan Gerofsky (2001) has recently noted that the questions themselves reveal more about our fantasies and desires than about the mathematics involved. Critical mathematics education has much to gain from her analysis of mathematics problems as examples of literary genre. And Mark Boylan (2007) has offered insights on the micropolitics of teacher questioning as constitutive of the larger political context for mathematics teaching and learning. His work, too, must be integrated into discussions of innovative social justice mathematics curricula.

And finally, it becomes crucial to examine the discourses of mathematics and mathematics education in and out of school and popular culture (Appelbaum 1995). Critical thinking in mathematics education asks how and why the split between popular culture and school mathematics is evident in mathematical discourse, and why such a strange dichotomy must be resolved between mathematics as a "commodity" and as a "cultural resource." Mathematics is a commodity in our consumer culture because it has been turned into "stuff" that people collect (knowledge) in order to spend later (on the job market, to get into college, etc.). But it is also a cultural resource in that it is a world of metaphors and ways of making meaning through which people can interpret their world and describe it in new ways. Critical mathematics educators recognize the role of mathematics as a commodity in our society; but they search for ways to effectively emphasize the meaning-making aspects of mathematics as part of the variety of cultures. In doing so, they make it possible for mathematics to be a resource for political action.

REFERENCES

(CIEAEM) International Commission for the Study and Improvement of Mathematics Education (2000). *Manifesto 2000 for the Year of Mathematics.* http://www.cieaem.net/50_years_of_c_i_e_a_e_m.htm.

(NCTM) National Council of Teachers of Mathematics (2000). *Principles and Standards.* http://standards.nctm.org/.

(PDE) Pennsylvania Department of Education. Undated. *Academic Standards for Mathematics.* http://www.pafamilyliteracy.org/k12/lib/k12/MathStan.doc

(UCSMP) University of Chicago School Mathematics Project (2004; 2007). *Everyday Mathematics.* Second edition 2004; third edition 2007. DeSoto, TX: McGraw-Hill/Wright.

Appelbaum, Peter (1995). *Popular culture, educational discourse, and mathematics.* Albany, NY: State University of New York Press.

Appelbaum, Peter. (2008, in press). *Embracing Mathematics: On Becoming a teacher and Changing with* Mathematics. NY: Routledge.

Berlin, David (2000) *The advent of the algorithm: The idea that rules the world.* New York: Harcourt Brace.

Block, Alan. (2003). They Sound the Alarm Immediately: Anti-intellectualism in Teacher Education. *Journal of Curriculum Theorizing,* 19 (1): 33–46.

Boylan, Mark (2007). Teacher Questioning in Communities of Political Practice. *Philosophy of Mathematics Education Journal* 20. http://www.people.ex.ac.uk/PErnest/pome20/index.htm.

Brown, Stephen and Walter, Marion (1999). *The art of problem posing*. Mahwah, NJ: Erlbaum.

Clandinin, D. J. & Connelly, F. M. (2002). Personal Experience Methods. In Denzin, N. K & Lincoln, Y. S. (Eds.). *Collecting and Interpreting Qualitative Materials*: 150–178. London: Sage Publications.

Cooper, Barry (2007). Dilemmas in Designing Problems in 'Realistic' School Mathematics: A Sociological Overview and some Research Findings. *Philosophy of Mathematics Education Journal* 20. http://www.people.ex.ac.uk/PErnest/pome20/index.htm.

Davis, Philip, and Hersh, Reuben (1986). *Descartes' dream: The world according to mathematics*. San Diego: Harcourt, Brace, Jovanovich.

Denzin, N. K& Lincoln, Y.S. (Eds.) (1998).*The Landscape of Qualitative Research: Theories and Issues*. California: Sage Publications.

Drake, Corey and Miriam Gamoran Sherin. 2006. Practicing Change: Curriculum Adaptation and Teacher Narrative in the Context of Mathematics Education Reform. *Curriculum Inquiry* 36, no. 2: 153–18.

Fawcett, Harold (1938/1995). *The nature of proof* (NCTM 1938 Yearbook). Reston, VA: National Council of Teachers of Mathematics.

Freire, P. (1988). Introduction: Teachers as Intellectuals: Critical Educational theory and the Language of Critique. In Giroux *Teachers as Intellectuals: Toward a Critical Pedagogy of Learning*. Massachusetts: Bergin & Garvey Publishers, Inc.

Gerofsky, Susan (2001). Genre analysis as a way of understanding pedagogy n mathematics education. In John Weaver, Peter Appelbaum, and Marla Morris (eds.) *(Post) modern science (education): propositions and alternative paths*: 147–176. New York: Peter Lang.

Giroux, Henry (1988). *Teachers as Intellectuals: Toward a Critical Pedagogy for the Opposition* South Hadley, MA: Bergin & Garvey.

Gudmundsdottir, S. (2001) Narrative Research on School Practices. In Richardson, V. (Ed.) *Handbook of Research on Teaching* (4th ed.). Washington D.C.: American Educational Research Association.

Gutstein, Eric, and Bob Peterson. 2005. Rethinking Mathematics: Teaching Social Justice by the Numbers. Milwaukee, WI: Rethinking Schools.

Gutstein, Eric. 2007. *Reading and Writing the World with Mathematics: Toward Pedagogy for Social Justice*. NY: Routledge.

Hones, D. F. (1998). Known in part: the transformational power of narrative inquiry. *Qualitative Inquiry* 4(2), pp.225–249.

Keitel, Christine, Klotzmann, Ernst, and Skovsmose, Ole (1993). Beyond the tunnel vision: Analyzing the relationship between mathematics, society and technology. In Christine Keitel and Kenneth Ruthven (eds.), *Learning from computers: mathematics education and technology*, 243–279. NY: Springer-Verlag.

Ladson-Billings, G. (1995). Just what is critical race theory and what's it doing in a nice field like education? *International Journal of Qualitative Studies in Education*, 11(1), 7–24.

Ladson-Billings, Gloria and William F. Tate. 1995. Toward a Critical Race Theory of Education. *Teachers College Record* 97:1 (Fall 1995): 47–68.

Leu, Yuh-Chyn. 2005. The Enactment and Perception of Mathematics Pedagogical Values in an Elementary Classroom: Buddhism, Confucianism, and Curriculum Reform. *International Journal of Science and Mathematics Education* 3, no. 2 (2005): 175–212.

Neyland, Jim. 2004. Toward a Postmodern Ethics of Mathematics Education. In *Mathematics Education within the Postmodern*, edited by Margaret Walshaw: 55–73. Greenwich, CT: Information Age Publishing.

O'Daffer, Phares G. and Bruce Thomquist (1993). Critical thinking, mathematical reasoning, and proof. In Patricia S. Wilson (ed.), *Research Ideas for the Classroom: High School Mathematics*, NY: Macmillan/NCTM.

Parker, Laurence, Donna Deyhle, and Sophia Villenas (eds). 1999. *Race Is...Race Isn't: Critical Race Theory and Qualitative Studies in Education*. Boulder, CO: Perseus Books.

Powell, Arthur, and Frankenstein, Marilyn (1997) *Ethnomathematics: Challenging eurocentrism in mathematics education*. Albany: State University of New York Press.

Remillard, Janine T. 1999. Curriculum materials in mathematics education reform: a framework for examining teachers' curriculum development. *Curriculum Inquiry* 29(3):315–342.

Skilton-Sylvester, Paul (1994). Elementary School Curricula and Urban Transformation. Harvard Educational Review, 64, 309–329.

Skovsmose, Ole (1994). *Toward a philosophy of critical mathematics education*. Dordrecht, Netherlands: D. Reidel.

Skovsmose, Ole (2005). *Traveling through education: uncertainty, mathematics, responsibility*. Rotterdam: Sense Publishers.

CHAPTER 27

METHODOLOGIES OF RESEARCH INTO GENDER AND OTHER SOCIAL DIFFERENCES WITHIN A MULTI-FACETED CONCEPTION OF SOCIAL JUSTICE

Jeff Evans
Middlesex University

Anna Tsatsaroni
University of Peloponnese

ABSTRACT

Our discussion of social justice begins by embracing both *distribution* and *recognition* aspects of social relations (Vincent, 2003). The first aspect considers the way that goods, knowledge, skills, rights, etc. are distributed among social groups; the second aspect focuses on aspects of society structuring social encounters and relations—modes of communication, 'treatment', respect. Our approach to social justice focuses not only on the dimension of gender, but

Critical Issues in Mathematics Education, pages 395–421

also on social class and ethnicity: we consider that concern with these latter forms is crucial for a full appreciation of the structural and the interactional aspects conveyed by the concept (Arnot, 2002; Gillborn & Mirza, 2000). In methodological terms we argue that researching the questions posed by a social justice agenda, will require both 'quantitative' and 'qualitative' methods—within any substantial research programme, and even within individual studies. Our emphasis on the theoretical aspects of formulating a research problem points to the importance of a third facet of social justice, to do with *representation* and power. We discuss several examples of fruitfully combining different aspects of social justice and multiple methods of research. In particular, we argue that considerations of social justice, and a related 'hybrid' methodology, will play a crucial role in our ongoing study of the (re)production of images of mathematics in popular culture, and in particular advertising.

INTRODUCTION

There is a range of concepts available for considering the fairness or otherwise of social arrangements in society, and their relationship to the education system and its effects on individuals and social groups. Here, we focus on *social justice*. Other work discussed below focuses on *equality*, or on *democracy*.

Indeed, *social justice* is a good concept to focus the discussion, because it is clearly related to the idea of equity or fairness, and because it is currently widely discussed in educational research (e.g., Vincent, 2003; Arnot, 2002; Burton, 2003; *Philosophy of Mathematics Education Journal*, 2007). It also appears to resonate with contemporary policy-makers' concerns, including the European Union's social agenda and the activities (or rhetoric) of many national governments around the world. The concept of social justice provides a way of grouping two distinct, but related, aspects: *distributive* justice and justice in terms of *recognition*.

This paper has several aims:

- To discuss a dual understanding of social justice in terms of distribution and recognition, in recent work in studies of educational policy and practice.
- To discuss several illustrations, from the sociology of education, and from adult mathematics education, which support our position of emphasising social class and ethnicity as key dimensions of social justice, as well as gender.
- To argue, drawing on these illustrations, that researching the questions posed by a social justice agenda requires both 'quantitative' and 'qualitative' methodologies.

- To show that addressing questions of social justice allows a broader investigation of issues in the teaching and learning of adult mathematics (formal and informal), drawing for illustration on our analysis of images of mathematics in advertising.

A DUAL UNDERSTANDING OF SOCIAL JUSTICE

The first, *distributive* facet of social justice focuses on the question of whether the individual (or social group) has (or has access to) material goods and services, and symbolic resources, such as knowledge or skills. Also important in our societies are the opportunities an individual (or a social group) has to gain access to, to participate in, or to succeed in, particular spheres of social activity, e.g., further and higher education, or certain occupations or grades of work.

The overriding concern of policy-makers in western societies from the 1960s onwards was equality, or more precisely, 'equality of opportunity' across social groups—meaning, at that time, across social classes. This made a distributive view of social justice dominant in social science analysis. However, in recent years, in mathematics education and educational research more generally, the tendency has been for social class not to be talked about directly (Secada, 1992; Skeggs, 1997; Vincent, 2003a), and sometimes in the educational research literature to be discussed in terms of 'ability differences'. The distributive facet of social justice, however, is especially important in situations where educational opportunities may be rationed and distributed according to 'ability', or 'ability to benefit'. Distributive concerns might suggest researching questions like:

- How are different forms of knowledge (e.g., mathematical, scientific), or certain educational credentials / qualifications distributed among the different social categories of class, gender, ethnic or language groups?
- How does provision of resources, or performance, differ between different social categories in school or college mathematics?

An interesting example is given by David Gillborn and Heidi Mirza, in *Educational Inequality: Mapping race, class and gender* (2000). This study, largely concerned with distributive issues, appears to use mostly quantitative methodology—based on a moderately large sample of official ethnic monitoring returns from local educational authorities, and a representative national youth survey. However, these researchers also draw on (their own and others') ethnographic research to focus on the experience of students in school. In particular they show that black pupils receive harsher *treatment* in discipline terms than others; and that teachers have *lower expectations* of Black students and *less positive assumptions about their motivation / ability* (p. 17). This suggests that there is also a need to study social justice and inequalities in *social interactions* in schools and classrooms.

Thus a second aspect of social justice concerns *'recognition* of difference', stemming from a concern with status-related inequalities, relating to manners of communication, 'treatment', respect for difference, and, in particular, avoidance of misrepresentation, stereotyping, and disrespect.

Since the 1980s, there has been increased interest in a 'politics of recognition' (Cribb & Gewirtz, 2003), understandable as a response to a range of 'New Social Movements'—concerning gender, ethnicity, disability, sexual preference—leading to assertions of / demands for 'women's liberation', 'black is beautiful', 'gay pride', etc. These have had important implications for education and social policy, expressed as a concern with 'multiculturalism', 'social inclusion' (Tett, 2003). Recognition concerns might prompt educational research questions like:

- How are different groups of students treated by teachers in educational institutions?
- What expectations and assumptions do teachers have about different groups in education? To what extent are these sometimes 'stereotypical'?

Examples of such concerns include, besides Gillborn & Mirza (2000), research programmes and individual studies in the sociology of education on issues of difference and identity, such as: Arnot (2002) on gender (see Section 3); Gillborn (1995) on ethnicity; Barton (1996) on disability.

We can use the discussion so far to suggest a number of provisional directions for further consideration. First, we might distinguish the bases of educational inequality / oppression for different groups. For some, inequality may be rooted *primarily* in an economic context, requiring redistribution: for example, social class, 'ability' differences, disability. For others, oppression may be generated *largely* in a cultural context, by lack of recognition/ mis-recognition, e.g., ethnicity, sexual orientation, religion, disability. In this connection, gender inequalities are central, since they can be seen as 'multiply generated' (Lynch & Lodge, 2002, p. 3), that is related both to redistributive and recognition issues. Nevertheless, there is a complex interrelation between these two aspects of social justice.

One recurrent criticism has been that social and sociological analyses of social justice or inequality have been too narrow, and have entailed, over the last forty years (in English-speaking countries at least), replacing concerns about social class, with those relating to gender, then with concerns about ethnicity, then disability and so on; furthermore that this has led to a proliferation of sub-fields, of research 'experts', and so on, thereby impoverishing the understanding of this most fundamental topic in educational research. However, we find it more appropriate to entertain the idea that, with each new form of inequality such as the currently held concerns with sexual orientation or disability, we do not simply have a replacement, but an *enriching* of the idea of social justice.

Therefore, as we shall show in the remainder of this paper, there are important reasons for us to embrace the notion of social justice in studying current educational arrangements, processes and outcomes. First, compared to the earlier conception of social and educational inequality, the notion of social justice points to a multi-faceted conception that considers questions about the structural determinants of social action, but does not limit the investigation to these. Second, it takes account of diverse forms of social difference, such as gender, social class, ethnicity, sexual orientation; while leaving open to consideration any other instance of social difference encountered, as to whether it might constitute the basis of a form of (in)justice. And, third, it expresses a commitment by those that use the concept to make efforts to combat the inequalities observed. This third reason, Cribb and Gewirtz (2003) refer to as the collapsing of the distinction between evaluation and action. They argue that a focus on evaluation alone carries the risk of representing social reproduction as inevitable. Furthermore, it is only by analysing examples of policies and practices aimed at promoting social justice that we can examine how tensions can be overcome in reality. This third reason for embracing the concept of social justice thus recommends that researchers focus on and scrutinise policies and practices prevailing in public life.

AN ENRICHED IDEA OF SOCIAL JUSTICE: AN EXAMPLE

Arnot (2002) provides an illustration of how the idea of social justice may be developed, in the course of her critique of social (class) reproduction theory (Bowles and Gintis, 1976). This is the idea that the social structure is perpetuated through ensuring broadly that—in Willis's (1977) now classic phrase—'working class kids get working class jobs', and similarly for middle class kids. Arnot's account allows us to see the importance of multiple aspects of social justice, based on multiple dimensions of social difference.

1. Arnot and others have criticised a notion of social class defined only in economic (or occupational) terms, and argued for inclusion of 'dynamic aspects of identity', and the role of patterns of consumption (spending), community and family values (Arnot, 2002, p. 205).

2. Thus the family, with its influence on gender relations, becomes crucial in processes of social class formation, through its role in forming identity, authority relations, perceptions of / resistance to schooling, mediation of differences in material circumstances. In this sense, she agrees with Bernstein (1977) that the 'long shadow of the family' also influences the 'expressive [affective] order of school'.

3. Youth cultures now soften and cross social class boundaries, and their class cultures are formed by young people acting during schooling as *mediators* between family class cultures and the cultures of employment; that is, they are not formed only by the latter.

4. Social class inequality may be affected by the differing opportunities of middle class professional and working class young women. The former now take responsibility for perpetuating their own class position (rather than aiming to do so via marriage), while also making different demands on family life (e.g., sharing of childcare), and different contributions to building up the family's capital (cultural and economic); working class women, in contrast, despite little if any improvement in the real opportunities available to them, paradoxically seem to appear to manage the contradictions by using the discourse of individualisation and choice (Arnot, 2002, pp. 212–14).

5. Black women have emphasised education and examination qualifications as a way of crossing class and breaking out of traditional gender and race classifications, and gaining an advantage in securing scarce local jobs (see also Mirza, 2005).

6. Nevertheless, in the national study of overall school attainment at age 16+, discussed in the previous section, Gillborn & Mirza (2000) found that social class differences had remained over the previous ten years, and, when social class differences were controlled for, there remained significant ethnic group differences. Gender differences were small overall.

The research discussed by Arnot illustrates the importance of theoretical development of social justice to encompass distributive and recognition (identity) aspects (illustrated here by point 1 above). Her overall account also shows how class, gender and ethnic differences inter-relate to have effects in subtle ways. And the findings she draws on are based at least partly on ethnographic (qualitative) work (points 2 to 5.) and partly on survey (quantitative) work (illustrated here by point 6).

Drawing on Arnot's account for illustration, we argue that researching the questions posed by a social justice agenda, and developing theorisations to begin to address them, will require both 'quantitative' and 'qualitative' methods—within any substantial research programme. Below we give further illustrations from our own ongoing work.

QUANTITATIVE AND QUALITATIVE METHODS

There is not space here to discuss in depth the difference(s) between 'Quantitative' and 'Qualitative' methodologies, as that distinction is currently deployed in social and educational research (see for example, Bryman, 1992). Here we simply refer to some of the distinctions most evidently in use by a substantial proportion of practising social and educational researchers; see Table 27.1.

The simplest way to try to distinguish the two types of method is to look at the form of the data, and the methods of data analysis used: numerical data, analysed statistically, indicates quantitative methods, and textual data, analysed 'semiotically' indicates qualitative methods. These distinctions are highly overt or 'visible'. Less visible, but still often acknowledged, is the degree of structuring by the researcher of the verbal social interaction with the participants during the data production and coding stages. These are the sorts of distinctions used by many to categorise differences between their work and that of others, and these are the basis of the working characterisation of the differences that we will use in this paper.[1]

In fact very often researchers, in justifying their choices of method, refer the issue back to questions of epistemology and ontology in order to argue that the two methods study completely different worlds: one to do with an objective world 'out-there' and independent of the observer, and the other a world of individuals as intentional beings making sense of their interactions with others. However, our approach is to focus on an adequate

TABLE 27.1   Possible Bases of a Quantitative/Qualitative Distinction

	Quantitative	Qualitative	'Visibility' of distinction
Form of data	numerical	textual	very high
Methods of analysis of the data	statistical	'semiotic'	very high
Degree of structuring (by the researcher) of interaction in the field	substantial (in standardised, and coded, tests, interviews or questionnaires)	relatively little (in semi-structured interviews and participant observation)	moderately high

theorisation of the problem at hand, in this case that of social justice so as to guide, using such theorisations, the methodological decisions to be made, at each crucial point in the research.

TYPES OF METHOD IN PRACTICE: A CONSIDERATION OF FURTHER EXAMPLES

To explore the use of, and the fruitful combination of, 'quantitative' and 'qualitative' methods in practice, and using our developing notion of social justice, we draw on an earlier study by the first author, Evans's (2000) study of adult numeracy and emotion; and a joint ongoing study of images of mathematics in popular culture; see Evans, Tsatsaroni & Staub (2007).[2] In both studies issues of inequality have been crucial in conceptualising the research object, but the use of additional theoretical resources in the second study has allowed us to put the issue of learning and social justice at the centre of our problematic.

Adults' Mathematical Thinking and Emotions

Evans (2000) studied adults' mathematical thinking and performance in a higher education institution. He produced survey and interview data on three cohorts of students studying mathematics, as part of their main study of a social science subject.

Evans studied *distributional* issues in mathematics education—by asking students to complete a questionnaire including some performance items, and by analysing the results through building a statistical model. He found that the size of gender differences in the performance items was initially quite substantial. But they decreased (and were no longer statistically significant for school leavers in the 18–20 age range), after controlling for a range of competing explanations such as social class, age, level of qualification in school maths, and affective variables, especially confidence and mathematics anxiety.

Besides these quantitative methods, the research also used semi-structured interviews (life history/clinical) to observe student problem solving and affective reactions. This allowed the problematising of categories and scales from the questionnaire, and more flexible and rich description of several concepts:

- *performance*, in a context-specific—more precisely, a practice-specific—way, as in how to understand a student's calculation of a restaurant tip as '37.2p', when the UK's smallest unit of currency is 1p (Evans, 2000, pp. 162ff)

- *social class*, elaborated to include not only an economic, work-based aspect, but also cultural and identity-based aspects (for example, pp. 230–231; see also the discussion of Arnot's elaborations above)
- *anxiety expressed*, distinguished from *anxiety 'exhibited'*, using insights from psychoanalysis (pp. 171–172).

These elaborated categorisations were used to examine the main relationships in a richer way (though for a smaller sample), in what Evans called 'qualitative cross-sectional' (or might have called 'quantitative semiotic') analyses. These 'hybrid' analyses combine an interpretive reading of textual material with a basic statistical analysis; see Table 27.2, where the categorisation of an interviewee as having *'expressed* anxiety' or as 'likely to have *exhibited* anxiety, but not to have expressed it' depends on a careful reading of the entire interview transcript, drawing on psychoanalytic insights about the way that anxiety can be 'exhibited', even if it is not expressed (Evans, 2000, pp. 144–145).

When the results of the categorisation on the latter anxiety variable are cross-tabulated with gender, the resulting 'qualitative cross-sectional' analysis in principle allows a comparison across gender groups (Table 27.2).

Thus, 100% of the women expressed anxiety during the interview, as compared with 75% of the men. This suggests a 25% difference in the percentage of students expressing anxiety in the interview setting. However, virtually all students (22 of 25) took the opportunity to express anxiety during the interview, so any gender differences must be seen as small and not very reliable (because of the small sample size).

Another way that Evans (2000) brought together quantitative and qualitative methods was in indicating which students might be 'over-achievers' or 'under-achievers'. This was done by calculating the *residual*, or the difference between the student's observed performance score and their performance score that would be expected, given their gender, social class, age, level of mathematics anxiety, etc., and using statistical modelling (see Ev-

TABLE 27.2 Expressing and Exhibiting Anxiety in Semi-Structured Interviews: Cross-Tabulation of Numbers with Gender (*n* = 25)

Coding	Males	Females	Total
Anxiety *expressed*	9	13	22
Likely to have *exhibited* anxiety, but not to have expressed it	3	—	3
Total	12	13	25

Source: Evans (2000), Table 9.7

ans, 2000, note 8, chapter 10, pp. 270–271, and also Boaler, 1997, pp. 138–139). Such 'over-achieving' and 'under-achieving' individuals could then be selected to be invited for semi-structured interviews, on the grounds that they might possibly be informative 'critical cases' for the analysis. Thus both quantitative and qualitative methods are here used for distributive analyses of social difference/social justice in the context of education.

The semi-structured life history interviews provided material on subjectivity/identity, and therefore, allowed for the *recognition* of students from various social groups. In these interviews, the student related earlier experiences where they were 'recognised', or not, as a learner of mathematics. The interviews gave students space to express anxiety and other feelings about mathematics to the interviewer, as illustrated above. The interview also allowed several students to describe, but also to celebrate, their 'recovery' of confidence and competence in mathematics during the first year of their college course (Evans, 2000, p. 239), an issue which relates to the agenda of social justice research.

These two aspects of the study together show the use both of quantitative and qualitative methods to investigate distributional and recognition aspects of social justice in mathematics education. In particular, the 'qualitative cross-sectional' method (Table 27.2) can be seen as a *hybrid*, having *both* quantitative and qualitative features. However, it is also important here to stress that revisiting the study with a more clearly formulated concept of social justice, rather than a narrower focus on structural inequalities, has helped to draw out more clearly the methodological features of the study and their implications.

Images of Mathematics in Popular Culture

Evans, Tsatsaroni & Staub (2007) investigate the images of mathematics/mathematicians portrayed in a small sample of advertisements in the UK national daily press. The main aim of the part of our overall project focusing on advertisements is to describe the images portrayed, the discourses on mathematics drawn on, and various characteristics of the advertisements, e.g., their differing 'appeals' (Leiss, Kline & Jhally, 1990). Here we extend our earlier analysis to examine two related issues:

- the extent to which dominant discourses on gender can be identified
- whether these are likely to reinforce or to challenge long-term discourses and established gender stereotypes.

Several key methodological issues are discussed in Evans et al. (2007), including the definition of a 'mathematical advertisement', how the fieldwork was organised, and the sampling design. UK newspapers are divided into 'quality' papers, 'mid-market', and 'tabloids', on the basis of their traditional styles of presentation and level of reporting and commentary; we chose three 'qualities' (*Times, Daily Telegraph, Financial Times*), one 'mid-market' (*Daily Mail*), and two 'tabloids' (*Sun, Daily Mirror*) on the basis of their readership, including breakdowns as to gender and class. Then, randomly selected two-week periods of back issues (from 1994 to 2003) of the selected papers were searched for appropriate advertisements in the National Newspaper Library.

Thus the sampling over the ten-year period was rather 'light',[3] because of resource constraints and the pilot nature of the study. Yet, despite the fact that over 500 editions of daily newspapers were examined, only nine advertisements judged to contain an 'image of mathematics (or mathematicians)' were produced by this systematic sampling process. They were supplemented by a further four gathered in an 'opportunity' sample over the same time period (1994–2004), to make up the sample of 13 advertisements to be studied here. Evans et al. (2007) analysed this corpus, with a focus on the images of mathematics and mathematicians that were portrayed. Here we re-analyse these advertisements, with a focus on the images of gender (and other social differences), and on issues of social justice.

Because of the very small sample available in the pilot study, it is not really possible to do a 'quantitative' analysis with this set of data. However, we are currently replicating the study with a larger sample, where the sorts of analyses that we point to here can be used with much greater effect. Thus, the sort of 'quantitative' analyses that we present here are intended to be suggestive only. First we might do a simple categorisation of the characteristics of the advertisements, beginning with the product being advertised (Table 27.3).

In both our systematic random and the overall sample, 'mathematical' (or 'scientific') portrayals appear more frequently in advertisements for cars and for business services. Evans et al. (2007) observed that, when we com-

TABLE 27.3   Product Category of the Advertisements

Product category	Number in overall sample (number in random sample)
Automobiles	4 (3)
Business Services	3 (3)
Food	2 (2) [1 campaign]
Consumer Telephone Services	1 (—)
Bank (job advertisement)	1 (1)
Rail Transport	1 (—)
Men's Cosmetics	1 (—)
Total	13 (9)

bine this with the information that a substantial percentage of buyers (for many models) of cars are male, as are a high proportion of the senior managers who are commissioning the purchase of business services, we should expect that appeals in advertising, including appeals to mathematics, in the UK will be 'gendered' in particular ways (cf. Williamson, 1978). Our aim here is to study this gendering. We might begin by studying the *context* of the different advertisements, that is the newspaper in which the advertisement appeared and the characteristics of the readership (Table 27.4).

Here we can see some variation among the newspapers as to the percentage of their readership that is male. On the basis of an assumed masculine gendering of mathematics, we might hypothesise a *positive* correlation of the percentage of male readers for a newspaper with the number of advertisements found that contained an image of mathematics.[4] However, of the newspapers that showed more than one mathematical advertisement, the *Guardian* and the *Times* had over 55% male readership, whereas the *Daily Mail* has a majority of female readers. And of those that had no mathematical advertisements, the *Financial Times* had the highest percentage of male readers, while the *Sun* and the *Daily Mirror* had 'average' levels of male readership.[5] Thus, on the evidence so far, always keeping in mind the small numbers, we might tentatively conclude that the incidence of mathematical advertisements does not seem to relate to the proportion of male (or female) readers.

TABLE 27.4   Advertisements Found in Editions of Daily Newspapers in Overall Sample

Newspaper	Number of advertisements found	% of readers who are male	% of readers who are middle class ("ABC1")
Times	4 (Concert Communication Services, BMW, Abbey National Bank, Sun Microsystems)	56.8	89.2
Guardian	3[a] (Mercury Telephone Services, Peugeot, South West Trains)	58.6	90.1
Financial Times	0	77.1	94.0
Daily Telegraph	1 (Thales IT & Services)	52.9	87.4
Daily Mail	4 (Quorn Foods (2), Jaguar, Daihatsu)	47.7	64.7
Sun	0	58.8	36.8
Daily Mirror	0	51.8	39.7
All Papers	**12**[b]		

Source: calculated from results of National Readership Survey (2005)
[a] *The Guardian* advertisements came from the 'opportunity sample'.
[b] One advertisement (Givenchy) was found by our study only on the company website, and therefore is excluded from this table.

However, if we take into account the social class mix of the newspapers, we can start formulating one or two working hypotheses for the next phase of research.

First, mathematical advertisements seem to appear only in papers where there is a high percentage of middle class readers (the 'qualities'), or at least a more 'middling' level of middle class readership (the mid-market papers): we might go on to formulate an explanation in terms of levels of education and some (presumed) familiarity with and knowledge of mathematics. This would explain the absence of advertisements in the *Sun* and the *Daily Mirror*—but not that in the *Financial Times*! We might conjecture that the latter anomaly might be due to the readership being composed of financially highly specialised—and numerate—groups of (largely) middle class males, but the investigation of this idea would require further information or data.

Second, we might conjecture that there might be some difference between the types of 'mathematical' advertisements appearing in the *Guardian* and the *Times*, as compared with those in the *Daily Mail*. What our (limited) data show is that 'mathematical' advertisements for cars appear in three of the newspapers (*Daily Mail, Times* and *Guardian*); whereas 'mathematical' advertisements for business systems appear only in the *Times* and the *Daily Telegraph*. That might suggest a provisional hypothesis—that advertisements for business systems are more highly gendered than those for cars.

It is important to try to assess whether there is any gendering of the individual advertisements, and, if so, in what ways. To do so, we can ask:

- whether the advertisement includes any recognisable or stereotypical masculine or feminine character
- whether there is any appeal to masculine or feminine discourses
- whether the advertisement addresses the reader as likely to be male/ in a masculine role, or alternatively female/in a feminine role.

We shall classify each advertisement as gendered, ambiguous, or apparently gender-neutral (Table 27.5).

TABLE 27.5 Degree of Gendering in Advertisements

	Clearly Gendered	Ambiguous	Gender-neutral	Total
Number of advertisements	4 (Mercury, Peugeot, Givenchy, Thales)	3 (Jaguar, R4Quorn R5Quorn)	6 (Abbey National job advertisement, Concert, Daihatsu, BMW, Sun Microsystems, South West Trains)	13

Within the 'clearly gendered' category, the Thales advertisement (in the *Daily Telegraph*, in 2000) contains a picture of (a sculpted head of) the eponymous Greek 'Mathematician. Philosopher. Astronomer. Merchant', and the Mercury (see Evans, 2003) and Peugeot advertisements (in the Guardian, in 1994 *circa* and 1999, respectively) contain stereotypical male characters. The Givenchy advertisement (2002 *circa*) juxtaposes discourses on mathematics with those on masculinity.

The two advertisements for Quorn (a vegetable-based meat substitute) are categorised within the 'ambiguous' category, because, as a pair, they were clearly conceived to produce a 'gender balance'. Thus, the R4 Quorn advertisement shows a 'stick-figure' woman dressed in a cooking apron in an advertisement advising readers what to eat on Wednesdays; however, at the same time she is pointing to a graph, purporting to indicate the 'time spent smiling on Wednesdays.' The R5 Quorn (published four days later in 2003, also in the *Daily Mail*) is similar, except that the person in the cooking apron is now a man! However, a closer examination suggests that the man's apron is smaller, also revealing a shirt and tie, and he is pointing to an apparently more complex graph.[6] These somewhat subtle differences may nevertheless be read to suggest that the man has a less 'hands-on' relationship with cooking than the woman, and that he may be able to handle basic statistical graphs more easily.

The Jaguar advertisement (Evans et al., 2007) has been classified as 'ambiguous', since the imagery generally, and the 'soaring' Jaguar in particular, may function to suggest a striving and competitive masculinity.

We categorised under half of the advertisements (6 of 13) as 'gender-neutral'. First, the Abbey National bank's job advertisement (*Times*, 2001): despite any gendered stereotypes of statisticians that may exist, our reading of this advertisement for a financial statistician is that it appears to be addressed to a (young) reader of non-specified gender; it also emphasises the company's intention 'to meet our disability symbol commitments'. The two car advertisements, for BMW and Daihatsu, make no reference to men or women, nor to discourses on gender. The same is true for the two advertisements for business systems, for Concert and Sun Microsystems (though we hypothesised above that advertisements for business systems might be more gendered than for those for cars).

In the South West Trains advertisement (*Guardian*, 2004) the narrative expresses concern for people who might be 'worried about your safety' and announces that £17 million has been spent on devices and staff to assure passenger safety. The reader is not addressed as gendered, nor as a member of any other vulnerable group of rail travellers.

Following these 'quantitative' analyses of the data—involving categorisation, frequencies, and the suggestions as to how we might use associations/cross-tabulations or correlations—we examine three of the advertisements categorised as 'clearly gendered' with more attention to their possible meanings in terms of the gender discourses underlying the text. The Peugeot and Givenchy advertisements have also been discussed in Evans et al. (2007), but it was the mathematical qualities of the images, rather than gender and social justice issues, that were foregrounded in that discussion. Let us first consider the Thales advertisement (Figure 27.1, published in the *Daily Telegraph*, Dec. 2000).

This advertisement is dominated by Thales's sculpted face and his name, which has also become that of the renamed company. We are assured that he was a real person, by the added detail of his dates of birth and death: he certainly belongs to the Classical Era. Next follows his list of his achievements

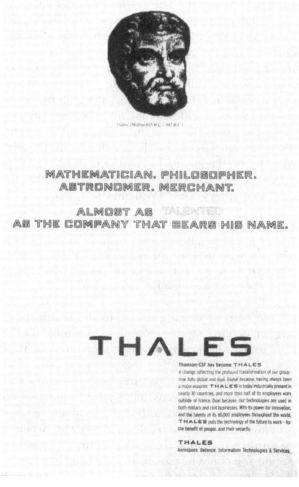

Figure 27.1 Thales advertisement (*Daily Telegraph*, 8 December 2000).

in a 'mini-CV': 'Mathematician. Philosopher. Astronomer. Merchant', each separated by full stops to give added emphasis. What is clearly considered his key attribute, 'Talented', is emphasised with a different typeface.

However, Thales, for all his talents, is displaced by the (superior) talents of the company in the line which asserts that he was 'Almost as Talented as the Company that Bears His Name'. This 'humorous' line is followed, by some space, perhaps to let its double-edged meaning sink in. The smaller text at the bottom then emphasises the transformation of the company to a 'global' and 'dual' one—concerned with 'both military and civil businesses'—guaranteed by the 'talents of its 65,000 employees throughout the world'.

The humorous strapline may function to allay possible 'worries' created elsewhere in the advertisement. The reader may feel anxious in the face of the long list of talents possessed by the ancient Thales, or concerned that one of the services provided by the modern Thales relates to the military and 'security'. Thus the appeal of the advertisement is a mixture—a 'rational' announcement of the change of name of the company, and its substantial capacities, and 'relief' from these sorts of anxieties. As for the ancient Thales, there is no doubt as to his masculinity, his talent or his achievements—and these qualities are meant to be associated with the company in this advertisement. His powerful and unambiguous masculinity is clearly established by the 'Greek god' image that heads the advertisement.

We next consider the Peugeot advertisement (Figure 27.2, published in the *Guardian*, August 1999). Evans et al. (2007) argue that this advertisement functions by creating a 'lack' (Williamson, 1978).

> The advert thereby establishes a 'need' for in-car air conditioning, by sketching a worrying fantasy of a vindictive public road service, staffed by aggressive, nasty workers, who want to make the reader's life a misery, even (or especially) on a holiday weekend. The relief is provided by the advertiser's cars, which have [...] 'personal air-conditioning'. (p. 46)

We can see the basis of this argument, in the domination of the picture by the large masculine figure—complete with huge traffic cone—and the 'calculating' tone of his words: "If I've got my sums right, these should cause a nice long tailback..."

The man with the cone is an apparently powerful and anxiety-inducing figure, but, on reflection, the advertisement subverts him in several ways. He is the scapegoat for the disruption, although rationally we know he isn't to blame: it is his employers and the road engineers who decide on where to put the cones. Further, the visual treatment is rather comical—he is almost caressing the cone and could be partnering the cone in a dance, but on the other hand, he is almost dwarfed by its size. Although the words attributed to him suggest that he thinks he can get one over on us (the readers), the

Figure 27.2 Peugeot advertisement (*Guardian*, August 1999).

resolution is that we get one over on him because we know that he is not really in control of the cones. We may indeed be tempted to laugh at him, because of the comic visual aspects (Bayley, personal communication, 2006).

Evans et al. (2007) suggest that the advertisement may point to class antagonism—of the road worker towards middle class car drivers. However, Bayley's comment in the preceding paragraph suggests that the other aspect of class antagonism is built into the very narrative of the advertisement, in that the middle class reader/car driver has the 'last laugh'. Thus, a consideration of the figure's masculinity is incomplete without an appreciation of his social class position. This confirms the arguments above that the gender aspects of social difference must be understood in connection with other aspects of identity.

π: **BEYOND INFINITY**
Deep in the nature of man is the will to go
further than any man has ever been before.
The quest is symbolised by the Greek letter π,
which evokes infinity. Men are still in pursuit of
the end of its innumerable string of decimals...
A perfume which is synonymous with the pioneering spirit, π celebrates internal force and
an adventurous imagination: energy and sensuality, unruffled calm and strength.

Figure 27.3   Givenchy advertisement (Givenchy, 2002).

The advertisement in our corpus with the most overt links with gender is
the Givenchy 'Pi' advertisement—unsurprisingly, since it is for a men's perfume, and perfume is one commodity that plays a large role in expressing,
and creating, gender difference (Figure 27.3).

This version of the advertisement was picked up on the company's website in 2002, part of a campaign that has attracted much attention from serious newspapers (e.g., MIT's campus newspaper) and from bloggers on the
internet; since that time, the product has been re-launched with different
colours and packaging, but with the same name.

Evans et al. (2007) note the 'sensuality' and sexiness of the pictures and
the text, which aim to reinforce the association of positive (masculine)
qualities with the perfume, for example by the references in the text to
'man...', 'men...'. The mathematical object, π, we are told, 'evokes infinity'; this is associated with the product's 'pioneering spirit' and 'adventurous imagination'. As for mathematicians, 'men' are claimed to be 'still in
pursuit of the end of its innumerable string of decimals': this allows the
text to assert the product's qualities of 'internal force', 'unruffled calm',
strength and energy.

Thus this advertisement not only picks up on gender stereotypes in the
wider society, but it also *reinforces* and *extends* the association of masculinity
with mathematics (see also Mendick, 2006), and with all sorts of other supposedly 'desirable' qualities.

We have thus sketched how the analysis of gender and (reproduction of) social differences using the corpus of advertisements can benefit from both 'quantitative' and 'qualitative' methods. In terms of social justice, this analysis appears to relate mainly to the *recognition* aspects, since what seems to be most at stake are the images of gender, and other social differences. However, we might ask how the portrayal of mathematics, and of the social divisions in relation to mathematics, affects the attraction to, or avoidance of, mathematics for individuals and diverse social groups—and hence the unequal *distribution* of types of mathematical knowledge, and satisfactory performance in school and college mathematics. Our discussion above gives an indication of ways to do further research into relations between different types of newspapers, notably those with different readerships, mathematics advertisements, and gender and other social differences—based on a larger sample that would allow one to extend quantitatively the picture shown in Tables 27.3, 27.4, and 27.5.

More specifically, greater attention has to be paid to the apparent absence of advertisements in which mathematics figures in the popular press. For it might be that mathematics as a body of knowledge is being 'silenced' in this popular cultural domain, rather than being considered as a resource for public discussion for the average person. This in turn would have clear implications for the learning of mathematics and would raise the question of who can be recognised as a learner of mathematics. Positioning the subject outside the field of mathematical knowledge in discourses of the public domain, in this case advertising, constitutes an exclusion that might reverberate in some way in the positioning within formal educational discourses. And, for example, we can see here that drawing on such discourses, as the current pedagogy and policies recommend and encourage teachers to do, is not something that automatically benefits all students. At the same time, this would raise questions as to whether companies should be allowed to use the cultural heritage of mathematics in ways that are likely to clash with the work of teachers aiming to make comprehensible the ideas and methods of mathematicians or other scientists.

EXTENDING THE ANALYSIS OF SOCIAL JUSTICE TO FURTHER ASPECTS

So far we have developed a dual aspect of social justice or equality. Writers within the field of sociology of education have considered, in addition to 'distribution' and 'recognition', a third facet of social justice. Lynch & Lodge (2002) focusing on 'equality' in schools, use the term 'representation' to refer to (equal or unequal) relations of power in decision making processes. Cribb and Gewirtz (2003) in discussing different facets of social justice from a sociological perspective on policy, call this third facet 'associational justice'. It 'calls for a consideration of the social and political context within which schooling takes place and of the possibilities of building higher levels of participation within these contexts' (p. 26). Bernstein (2000), proposing a model for examining and evaluating democracy in schools, suggests that 'there is likely to be an unequal distribution of images, knowledges, possibilities and resources which will affect the rights of participation, inclusion and individual enhancement of groups of students' (p. xxii). In particular 'participation' which operates at the level of the political, is defined by Bernstein as 'the right to participate in the construction, maintenance and transformation of order' (p. xx).

The references above suggest that there is a lively and renewed interest within the various strands of theory and research in the discipline of sociology of education on social justice. Indeed, this discipline has also a long tradition in raising and elaborating on social justice concerns. As an indication, it is worth reproducing here four sets of questions which, according to Roger Dale, have been formative to the sociology of education in the course of its development. These questions are:

- Who gets taught what, how, by whom, and under what circumstances, conditions, contexts and resources?
- How, by whom and through what structures, institutions and processes are these matters defined, governed, organised and managed?
- What is the relationship of education as a social institution to other social institutions of the state, economy and civil society?[7]
- In whose interests are these things determined and what are their social and individual consequences? (Dale, 2001, p. 27)

It can be seen that, while the first point relates immediately to distributional issues, the last three raise issues concerning politics and power. These are not completely covered by the distributional and recognition aspects of social justice, so to be adequately conceptualised they require a third dimension of analysis.

Thus the third aspect of social justice or equality, in its most general formulation, concerns power as the 'representation of interests':

- power relations at macro-levels of state and related institutions
- micro-level power relations between teachers and students (or 'age-related status')
- Meso-level power relations.

This last refers to the way pedagogical discourses (including quasi-pedagogical ones such as media 'texts' and advertisements) are constructed, and the way institutional practices are organised so as to include or to exclude individuals (and social groups) from legitimately and effectively participating within the terms of the discourse.

Overall, 'representation' is less developed as a concern within educational research than redistribution and recognition. And, within this facet of social justice, the macro- and micro- levels are better researched than the meso level referred to above.

In our own research project on mathematics in advertising we are seeking to approach the representation facet of social justice by asking questions about the construction/recontextualisation (Bernstein, 2000) of discourse and the space it allows for individuals to participate and actively engage with it. In particular we explore how substantial mathematical topics such as 'equations' are recontextualised in advertising practices; attending in particular to the way the design of such texts position the 'pedagogical' subject.

An example comes from our new sample of advertisements, from the Jordans '3 in one' breakfast cereal advertisement (April, 2007). The Jordan's '3 in one' breakfast cereal advertisement contains large, colourful pictures of three 'natural ingredients', and a picture of the cereal box. The main part of the layout is arranged in the form of an equation, which 'adds' (the pictures of) the three ingredients to 'equal' the cereal box (Figure 27.4).

The three ingredients are not named in the pictures, but the claimed effect (e.g., 'Taste', for the apples and raisins) is shown above the picture (and ingredient and effect are linked in the text). The natural quality of the 'wholesome' ingredients is emphasised in the largest prints of the verbal text, with 'nothing artificial added' in the main text, and with a 'Back to nature' slogan, placed next to the brand's name.

The main equation covering the space of the advertisement looks simplistic and childish, perhaps evoking the experience with arithmetic as remembered from early learning classrooms. In the picture of the cereal

Figure 27.4   Advertisement for Jordan's '3 in one' (*Observer*, 1 April 2007).

box, the equation is repeated, this time equated to a bowl full of cereal, the content of which is summed up in a curious (mixed genre) statement made up by the combination of a number (3) circling the words 'in one'. The pattern created by the repetition of the equation, each time with a little more information added on, gives the advertisement a rhythm which totally directs its reading: everything adds up (with a little help from other signifiers in the verbal text, such as 'plus' and 'sums up') to a full breakfast, and everything leads you to Jordan's new '3 in One' cereal product.

In terms of its mathematical content, then, this equation pretty much dispenses with mathematical symbols altogether. The three little piles of ingredients cannot really be added mathematically, though measurable qualities of them (e.g., weight) might be. Thus they could not 'equal' the product in the box. The ingredients might be mixed together physically, but, in order to 'equal' the product in the box, (industrial or other) transformation of some of the ingredients would be required (the oats and other grains of course have already been transformed for the pictures). This shows something of the operation of the Referent System 'Nature/ Natural', which is ideologically related to the Referent Systems of 'Science' and 'Mathematics': Culture, through Science, transforms Nature—and we are left only with 'the Natural' (Williamson, 1978).

If we scrutinise the advertisements presented above with the systematic tools of analysis presented by Bernstein (1990), we can see how the Jordans '3 in one' advertisement has been designed. In terms of its *classification*, we note that boundaries between the mathematical ideas expressed in the idea of the equation and the everyday knowledges about eating healthy breakfasts are blurred; similarly, the language is a mixture of words and mathematical symbols; therefore, there is a weak classification of specialised and everyday language. Turning to *framing*, the degree of control that an implied author and a reader—as constructed in the text—have over the communicative process, we could argue that very little control is left to the addressee: the options left to the implied reader are reduced by repeated equations all leading to the box enclosing the product.

We are suggesting that mathematics topics as recontextualised in advertising practices may constitute crucial texts for exploring the third facet of social justice. In particular they invite us to examine the power and control relationships in (quasi-)pedagogical communication settings; therefore to analyse the potential for the individual's participation and engagement in spheres of knowledge; that is to say in a symbolic world through which social order is maintained, reproduced or challenged.

From a methodological point of view, the research we have carried out thus far on the advertisements is mainly qualitative, semiotic analysis of text. However there is no necessary reason why such work cannot be conducted using a quantitative methodology. Indeed the work of Koulaidis and colleagues in science education is a case in point (Dimopoulos, Koulaidis, Sklaveniti, 2003). They have studied both textbooks and unofficial pedagogical sites such as science reports in the press and artefacts in museums; and they were able to assess quantitatively the extent to which such texts construct students as active participants in science learning. Such findings in quantitative form clearly can contribute to a wide-ranging analysis of the meanings and effectivity of such texts, including the advertisements that we study.

Indeed, in the case of our mathematics advertisement project, it was in terms of quantification of number of advertisements, location, and kind of readership that absences were identified. And it is this observation that has provided certain methodological insights concerning further sampling strategies. Thus we argue here that, just as it is difficult to separate in practice each of the three analytical aspects of social justice, so it is difficult to separate strictly in practice between 'quantitative' from 'qualitative' work. As shown in this article, a crucial basis for deciding the methodological strategy of a research project is the theoretical conceptualisation of the problem.

CONCLUSIONS

In this paper we have argued that a perspective on social justice and equality is fundamental for many important research topics in educational research today. It is also central to many policy concerns. Our approach includes not only gender, but also social class and ethnicity, as crucial dimensions of social justice.

We have argued that research into social justice and equality requires the use *both* of quantitative and qualitative methods, as illustrated here. These may sometimes be used together in one particular study (e.g., Evans, 2000 and our 'mathematical advertisements' project, discussed above), or sometimes woven together in a developing research programme (see discussions above of Gillborn & Mirza (2000), or Arnot (2002)).

On a theoretical level, there is a complex inter-relation between the three facets of social justice/equality which should be understood as inter-penetrating. As Fraser (2000) shows, the different aspects are inseparable for many groups. And in this paper we have shown the importance of working with a multi-faceted notion of social justice in researching mathematics education[8]—a conception which can guide researchers as to which methodologies are appropriate to use.

Thus we want to emphasise again that theoretical analysis is crucial in underpinning empirical work of a qualitative or a quantitative kind, or their combination, and that sociological research programmes provide an essential basis for this research. Although there are different theoretical perspectives within the former, a theory that would combine some of the criteria underlined in the extract below is a necessary guide to research.

> Theory [...] offers a language for challenge, and modes of thought other than those articulated for us by dominant others. It provides a language of rigour and irony rather than contingency. The purpose of such theory is to de-familiarize present practices and categories, to make them seem less self-evident and necessary, and to open up spaces for the invention of new forms of experience... (Ball, 1998)

ACKNOWLEDGEMENTS

An earlier version of some of the ideas in this paper was presented in a plenary address to the XXI annual symposium of the Finnish Association of Mathematics and Science Education Research, University of Helsinki, in Oct. 2003; we appreciate the feedback of participants at that symposium. The ongoing work on this project is supported by a grant from the British Academy. We also acknowledge the work of those who produced the advertisements (for Thales, Peugeot, Givenchy's 'Pi', and Jordans '3 in one') discussed here.

This chapter was originally published in *Adults Learning Mathematics—An International Journal, Special Issue on The Future of Mathematics in Adult Education from Gender Perspectives, 3,* 1, 2008, 13–31.

REFERENCES

Arnot, M. (2002). *Reproducing gender? Essays on educational theory and feminist politics.* London: RoutledgeFalmer.

Ball, S. (1998). Educational studies, policy entrepreneurship and social theory. In R. Slee & G. Weiner with S. Tomlinson (Eds.), *School effectiveness for whom? Challenges to the school effectiveness and school improvement movement.* London: Falmer Press.

Barton, L. (1996) (Ed.). *Disability and society: Emerging issues and insights.* Harlow, Essex: Addison Wesley Longman.

Bayley, M. (2006). Personal communication, November.

Bernstein, B. (1977). *Class, codes and control, vol. 3.* London: Routledge & Kegan Paul.

Bernstein, B. (1990). *Class, codes and control, vol. 4: The Structuring of pedagogic discourse.* London: Routledge.

Bernstein, B. (2000). *Pedagogy, Symbolic control and identity: Theory, research, critique* (Rev. ed.), Rowman & Littlefield, New York. (Original ed. 1996, Taylor & Francis, London.).

Boaler J. (1997). *Experiencing school mathematics: Teaching styles, sex and setting.* Buckingham: Open University Press.

Bowles, S. and Gintis, H. (1976). *Schooling in Capitalist America: educational reform and the contradictions of economic life.* London: Routledge & Kegan Paul.

Bryman, A. (1992). *Quantity and quality in social research.* London: Routledge.

Burton, L. (2003) (Ed.). *Which way social justice in mathematics education?* Westport CT: Praeger.

Creswell J. W. (2003). Research design: Qualitative, quantitative, and mixed methods approaches (2nd ed.), Thousand Oaks CA: Sage.

Cribb, A. & Gewirtz, S. (2003). Towards a sociology of just practices: An analysis of plural conceptions of justice. In C. Vincent (Ed.) (2003). *Social justice, education and identity* (pp. 15–29). London: RoutledgeFalmer

Dale R. (2001). Shaping the sociology of education over half-a-century. In J. Demaine, *Sociology of Education Today* (pp. 5–29). London: Palgrave.

Dimopoulos, K., Koulaidis, V. & Sklaveniti, S. (2003). Towards an analysis of visual images in school science textbooks and press articles about science and technology, *Research in Science Education, 33,* 189–216.

Evans J. (2000). *Adults' mathematical thinking and performance: a study of numerate practices.* London: RoutledgeFalmer.

Evans J. (2003). Mathematics done by adults portrayed as a cultural object in advertising and in film. In J. Evans, et al. (Eds.) *Proceedings of 9th International Conference of Adults Learning Mathematics—a Research Forum (ALM-9), 17–20 July 2002, Uxbridge College, London,* ALM and Kings College London.

Evans, J. (2007). On methodologies of research into gender and other equity questions. *Philosophy of Mathematics Education Journal, Special Issue on Social Justice in Mathematics Education,* 2(21), (September).

Evans, J., Tsatsaroni, A. & Staub, N. (2007). Images of mathematics in popular culture / Adults' lives: a study of advertisements in the UK press. *Adults Learning Mathematics: an International Journal,* 29(2), 33–53. Retrieved December 20, 2007 from http://www.alm-online.org .

Fraser, N. (2000). Rethinking Recognition. *New Left Review, 233,* 107–120.

Gillborn, D. (1995). *Racism and antiracism in real schools: theory, policy, practice.* Buckingham: Open University Press.

Gillborn, D. & Mirza, H. (2000). *Educational inequality: Mapping race, class and gender.* London: OFSTED.

Givenchy (2002). 'π: Beyond Infinity'. Retrieved May 6, 2002 from http:// www. parfums.givenchy.com/givenchy/uk/pi.html .

Leiss, W., Kline, S., & Jhally, S. (1990). *Social communication in advertising: Persons, products and images of well-being* (2nd ed.). London: Routledge.

Lynch K. and Lodge A. (2002). *Equality and power in schools: Redistribution, recognition and representation.* London: RoutledgeFalmer.

Mendick, H. (2006). Masculinities in mathematics. Maidenhead: Open University Press / McGraw-Hill Education.

Mirza, H. (2005). Race, gender and Educational desire, Inaugural Professorial Lecture. London: Middlesex University.

National Readership Survey (2005). National Newspapers & Supplements. Retrieved December 20 2007, from http://www.nrs.co.uk/open_access/open_topline/newspapers/index.cfm.

Philosophy of Mathematics Education Journal (2007), *Special Issue on Social Justice in Mathematics Education, Part 1* and *Part 2*, No. 20 (June) & No.21 (September). Retrieved December 20 2007, from http://www.people.ex.ac.uk/PErnest/ .

Secada, W. (1992). Race, ethnicity, social class, language, and achievement in mathematics. In D. Grouws (Ed.), *Handbook of research on mathematics teaching and learning* (pp. 623–660). New York: Macmillan.

Skeggs, B. (1997). *Formations of class and gender.* London: Sage.

Tett, L. (2003). Education and community health: Identity, social justice and lifestyle issues in community. In Vincent, C. (Ed.) *Social justice, education and identity* (pp. 83–96). London: RoutledgeFalmer.

Vincent, C. (Ed.) (2003). *Social justice, education and identity.* London: RoutledgeFalmer.

Vincent, C. (2003a). Introduction. In Vincent, C. (Ed.) *Social justice, education and identity* (pp. 1–13). London: RoutledgeFalmer.

Williamson, J. (1978). *Decoding advertisements.* London: Marion Boyars.

Willis, P. (1977). *Learning to labour: How working class kids get working class jobs.* Farnborough: Saxon House.

NOTES

1. There is not space here to continue this discussion, by bringing in 'low visibility', but important, differences, such as the difference in aims of quantitative and qualitative approaches, or the (related) commitment to different types of explanation—namely, explanation by causes and explanation by reasons/intentions, respectively; for further on these issues, see Bryman (1992) and Creswell (2003).

2. For another illustration using Boaler (1997), who also uses multiple methods, see Evans (2007).

3. Only four weeks in each of five years were selected for examination for each newspaper.

4. Or more precisely, with the newspaper's 'success rate' (percentage of issues which included a 'mathematical advertisement').

5. We could produce a formal measure of the correlation, through producing a scatter plot, or calculating the correlation coefficient, but here we have a very small set of newspapers that were systematically sampled for the study – six, excluding the *Guardian*, since the latter was sampled only in an 'opportunistic' way in this phase of the research.

6. However, both graphs are clearly based on fabricated, indeed 'silly', data (see Evans et al., 2007).

7. Here we would add: relations of the educational institutions to the media and popular culture.

8. Following Lynch & Lodge (2002), we might also explore whether the emotional biography of an individual is a further facet or focus of the notion of social justice, and we should explore its implications for the current research agenda of mathematics education.

CHAPTER 28

THE PRIVILEGE OF PEDAGOGICAL CAPITAL

A Framework for Understanding Scholastic Success in Mathematics

Carol V. Livingston
The University of Alabama

INTRODUCTION

Overview

The theme that runs through this work is three-fold. First, there is a quality that some students possess that enables them to arrive at the academic table better positioned to take advantage of our educational offerings. This work seeks to forward for general vocabulary usage a name for that quality so that we as educational researchers can acquire it as a tool not only in the field of mathematics research, but analogously in all subject areas. The term being introduced is pedagogical capital. Secondly, as educational standards in mathematics become the rubric upon which the success or failure of teachers and schools are measured, it is important to consider

Critical Issues in Mathematics Education, pages 423–455
Copyright © 2009 by Information Age Publishing
All rights of reproduction in any form reserved.

whether these curriculum standards contain the seeds of social justice or hegemony. If mathematical standards convey an unconscious privilege to one group at the expense of another, then equity is at issue. And finally, as a new and emerging theoretical framework, the concept of education in this work uses Pierre Bourdieu's sociological idea of a firmly grounded, true mixed-methods approach of using both qualitative and quantitative data to highlight one detail in the overall picture of what is currently the portrait of mathematics education. Together these three points suggest an interesting cultural study concerning an issue of social justice that has to date been neglected in mathematics educational research.

Background

Several years ago, on a school field trip to the Huntsville Space and Rocket Center, I climbed a rock wall. One section had large, closely spaced hand- and footholds, while the other two sections had successively smaller and more sparsely positioned holds. The first section was much easier to climb and I had been advised by my students to climb that section and to avoid the other two sections. But being me, I had to try, and it took me several tries before I was able to successfully climb all three sections. Eventually, I achieved each pinnacle, rang the bell, and then descended from the rocky tower. That rock wall serves me as an analogy for schools and for the hand- and footholds students use as they maneuver toward scholastic success. For most members of society, school plays a central role educationally. It is the primary place, and for many the only place, where the logical practices and systematic supports—those hand- and footholds—that construct a person's education occur. The quality of the education built in each student is summed up collectively as scholastic success. Success gets quantified—and even qualified—along a spectrum of low to high, with some falling into a murky category called under-achievement due to life's vicissitudes. Unlike my multiple attempts at the three sections of rock walls, for the most part we do not get multiple attempts at nor do we get a choice of which section of wall to climb as we complete our education. What we are offered, however, is a scholastic wall to climb. Whether the hand- and footholds are personally and judiciously placed is a function of the Bourdieuian field and cultural capital present when and where one arrives at birth.

The concept of education in this work rests on Pierre Bourdieu's notion that scholastic achievement is directly related to access to and participation in logical practices and systematic supports that form various types of capital. That field of capital is related to the interplay of education with the individual. It can be called a resource for access much like those hand- and footholds allowed me access to the top of the tower. Access implies a threshold or point of entry. According to Bourdieu, we enter a field of play at birth. Participation implies not only crossing a threshold, perhaps many, but also becoming involved in a logical practice, nay a series of logical practices that eventually construct an education. Call it the luck of the draw, but for many students, the hand- and footholds involved in scholastic success in mathematics are not positioned advantageously. Or as Bourdieu would say, all households "do not have the economic and cultural means for prolonging their children's education beyond the minimum necessary for the reproduction of the labor-power" (Bourdieu, 1986, p. 245) found in the home at the time of the child's rearing. Bourdieu considered the "domestic transmission of cultural capital" (Bourdieu, 1986, p. 244) to be the best-hidden and possibly socially most important educational investment that can be made in a child. In this work, pedagogical capital will be advanced as a subtype of cultural capital and used as a framework for understanding some logical practices and systematic supports that lead to scholastic success, and in particular, scholastic success in mathematics.

Bourdieu summed up his theory of logical practice in two words. Those two words are "irresistible analogy" (Bourdieu, 1980, p. 200). The term analogy is based upon the Greek word *analogia*, which implies a comparison with specific regard to a relationship. In linguistics, the use of an analogy is a process by which words or phrases are created, or re-formed, according to existing patterns in the language such as when a child says foots for feet or says flower-works for fireworks. In this way, new words can be formed that may fall into general usage, in this case, pedagogical capital. In logic, an analogy is a form of reasoning in which one thing is inferred to be similar to another thing by virtue of an established similarity in other respects as in that any four-sided geometric figure is a quadrilateral, but a square is a special type of quadrilateral. In this way, new words also fall into general usage, and again in this case, pedagogical capital. But proportionately speaking, logical analogies tend to follow from an immediate past very similar to legal precedent, whereas linguistic analogies tend to create, like a new precedent perhaps unveiled just at that moment, an immediate and thereafter-visualized future. One analogy is a threshold from the past and the other is a threshold to the future. Together they form a bridge.

Because Bourdieu connected his sociological ideas to his empirical research, a fairly rare combination in philosophy, he consistently bridged a threshold from a grounded past to a theoretical future that is high in social utility. His theory of practice creates an irresistible analogy with regard to how the function of *habitus* recreates its own field of play—or in other words, the same hand- and footholds constantly reappear in successive generations—grounded as it is in the subtleties of grammar, ritual and other logical practices. In this work, critical discourse analysis will be the vehicle by which these grounded subtleties emerge. Influenced by human capital theory, Karl Marx and perhaps even Mao Zedong, Bourdieu asserts through strong empirical correlation that each individual finds himself located in a social space demarcated by the types of capital fluidly or statically in possession at any given time, rather than by class alone (Bourdieu, 1986). A theoretical underpinning of this work is Bourdieu's idea that in order to be used successfully as a source of power or to direct researchers attention to areas of unconscious privilege for one group over another, the subtypes of cultural capital are in need of being identified and legitimized.

Summary

To that end, this work will address and forward for general vocabulary usage one specific form of cultural capital, that is, pedagogical capital, as a subtype of cultural capital that might offer an unconscious privilege to those students who possess it. Using a combination of sociological ideas, empirical research and justifiable correlation, the notion of pedagogical capital will be used as a theoretical framework for understanding some logical practices and systematic supports that lead to scholastic success. For the purposes of this research, that scholastic success will be in the area of mathematics, but this same style of analogy should predictably emerge in other scholastic fields. It is hoped that by use of irresistible analogy—and I will be generous in using the analogy of climbing up the rock wall—it will be derived through qualitative analysis (critical discourse analysis) and empirical analysis (inferential statistics) that students who possess pedagogical capital will display evidence that they enter some fields of educational play, particularly in conjunction with the mathematics curriculum, with a higher probability of becoming scholastically successful due to the relation and interplay of Bourdieu's notion of *habitus*, the distribution of cultural capital within the family, and the standards and norms of the institution.

PROBLEM

Introduction

The scholastic success of economically disadvantaged students in mathematics continues to be one of the greatest challenges faced by mathematics teachers and mathematics policymakers. Studies have consistently shown that a student's social and cultural background routinely influence whether that student will perform well in mathematics (Lamb, 1998). While prior reform efforts have brought to the surface the extent of the problem in mathematics education with regard to race or gender issues, there has been little success in the area of social and cultural disadvantage. Johnson (2002) asserts that "changing content and performance standards without fundamentally transforming educators' practices, processes, and relationships cannot lead to success" (p. 11). Thus, newly derived efforts aimed at eliminating achievement gaps and cultivating a culture of equitable scholastic success in mathematics in our schools needs to begin to acquire the tenor of meaningful and thoughtful educational discourse surrounding areas of potential privilege that continue to perpetuate conditions of underachievement among economically or otherwise disadvantaged students.

Purpose

Mathematics as a field of play has a long history of displaying underachievement among its economically disadvantaged students. The purpose of this study is to unveil one area where factors and/or resources facilitate or impede scholastic success in mathematics. Additionally, this proposed study is designed to reveal relationship(s) among those factors and/or resources. In order to promote social justice and to provide all children, regardless of class, with an equal life chance, mathematics teachers and mathematics policymakers must confront and address some very difficult issues, perhaps hegemonic and ingrained, within the mathematics curriculum. These issues adversely affect the performance of economically disadvantaged students at a time when demographic trends may be pointing to the continued growth of those populations. Amid the current rhetoric surrounding the rationale that no child is to be left behind, the school culture and any structural elements of the curriculum that might perpetuate these inequities of educational opportunity for students who are economically disadvantaged can no longer go unnoticed, nor can they continue to be politely dismissed.

Scholastic Success in Mathematics

Although there is no standard definition of scholastic success in mathematics, it is fairly well documented that more school failures are caused by mathematics than by any other subject and it has been thus for decades (Wilson, 1961; Aiken, 1970; Lamb 1998). For the purposes of this research, the performance descriptors of the math general rubric for the Alabama Reading and Mathematics Test (ARMT) will be used as quantifiers for scholastic success in mathematics (Alabama Department of Education, 2005). These levels are:

Level I Does not meet academic content standards: Demonstrates little or no ability to use the mathematics skills required for Level II.

Level II Partially meets academic content standards: Demonstrates a limited knowledge of content material.

Level III Meets academic content standards: Demonstrates a fundamental knowledge of content material.

Level IV Exceeds academic content standards: Demonstrates a thorough knowledge of content material.

These four levels of achievement mirror the style of indicator used by the US Department of Education on state education reports, with those four levels being Below Basic (Level I), Basic (Level II), Proficient (Level III), and Advanced (Level IV) (2005). Since Levels III and IV of the ARMT general rubric are the scores awarded to students who meet the requirements of being either at or above grade level in mastery of prescribed content standards, with regard to this research, these two levels will be said to constitute students who are demonstrating scholastic success in mathematics.

Changing Demographics

Mirroring the population of the United States as a whole, the population of students in our public schools continues to be predominantly of European ancestry, but as time passes, that proportion is constantly declining. What population reports show is that beginning in the 1990's, the growth of all racial/ethnic groups increased with the exception of the white child. At the beginning of that decade, white school children accounted for 73.6% of the total school population, but by the end of the decade this demographic cohort had declined to 68% (Gordon, 1977). There was a corresponding increase in the traditionally minority populations. Black and Hispanic children, particularly, are two to three times more likely to be living in poverty than white children (Hobbs & Stoops, 2002). Who our disadvantaged and marginalized students are, and what needs that they present, affects which issues related to education are most important for research. The American Educational Research Association (AERA), as the nation's leading research organization concerned with the production of knowledge related to education, has committed itself "to disseminate and promote the use of research knowledge and stimulate interest in research on social justice issues related to education" in their social justice mission statement (AERA, 2006). Indubitably, any striking demographic shift in the US population will significantly alter the diversity and ethnic breakdown of the student population in the public schools. Social issues regarding those emerging critical mass populations as they are in the process of change should not be discounted.

For instance, the state of Texas has undergone, and continues to feel the effects of, a major demographic shift in its student population. This change will significantly alter the diversity and breakdown of the student population in its public schools. Murdock *et al.* (2000), in a study of a mostly Hispanic demographic shift in Texas, maintained that major demographic trends and analyses serve to underscore the need for continual educational reform that will effectively confront and address the root causes of the elements that continue to generate performance gaps between student groups in public schools. Murdock was able to project the Texas demographic population trends from 2000 to 2040 comparing Anglo, Black and Hispanic populations. It is important that educational research address not only the school of today, but also the school of the future if that seems at all predictable.

While not as startling as the Hispanic demographic shift in Texas, Alabama is undergoing a similar change in its student population. In a report on state education indicators with a focus on Title I, the US Department of Education (2002) compiled the following data for Alabama's 1,135 public schools, as shown in Table 28.1.

As shown by the data, Alabama is experiencing a similar shift in the demographic make-up of its student populations. Also, during the year 1999–2000, the percentage of students eligible to participate in the Free/Reduced Price Lunch Programs were reported in Table 28.2.

TABLE 28.1   Alabama State Education Indicators with a Focus on Title I

Race/ethnicity	1993–1994	1999–2000	% increase/decrease
Am. Indian/Alaskan Native	5,906	5,141	13.0% decrease
Asian/Pacific Islander	4,320	5,195	20.3% increase
Black	259,700	265,300	2.2% increase
Hispanic	2,781	7,994	187.5% increase
White	453,268	445,852	1.6% decrease

TABLE 28.2   Alabama Schools by Percent Eligible for the Free Lunch Program[a]

Percentage of students eligible	Number of schools at level
75–100%	257
50–74%	390
35–49%	320
0–34%	381

[a] 19 schools did not report

There were no data available for comparison with the 1993–1999 Free/ Reduced Price Lunch Program eligibility specifically for Alabama. However, for the nation as a whole, the percent increase from 1996 to 2005 changed from 34% to 42 % for 4th graders, and from 27% to 36% for 8th graders (US Department of Education, 2005). If the socioeconomic differences between white and minority populations continues to shift in the direction that trends are heading, Murdock *et al.* (2000) contends, "The changing demographics of the South could lead to populations that are increasingly impoverished and lacking the human capital necessary to compete effectively in a global economy" (p. 8).

Also complicit in the changing demographics of American public schools is the sexual activity of women who delay or forego marriage. According to the National Center for Health Statistics, almost 4 in 10 of the nearly 4.5 million babies born in the United States during the year 2005 were born to unwed mothers (Centers for Disease Control, 2006). This represents the highest rate of out-of-wedlock births on record as the percentage has risen a full 12 points since 2002, and will potentially be the trend population demographic mirrored in the public schools beginning five years from now, as those children begin their school careers. Women-headed households are among the poorest of the poor everywhere in the world. So rather than simply concentrating endless efforts on the symptoms surrounding the achievement gaps that, at best, produce minimal advancement for economically disadvantaged students, mathematics teachers and policymakers should strive to create reform that effectively addresses the root causes; the economic future of the South and it's quality of life will depend upon the scholastic success of these students.

Summary

There exist many explanations as to why economically disadvantaged students are out-performed by their peers and why reform efforts have failed in the past. Cuban maintains that "reforms return (again and again) because policymakers fail to diagnose problems and promote correct solutions...policymakers use poor historical analogies and pick the wrong lessons from the past...and policymakers cave into the politics of the problem rather than the problem itself" (Cuban, 1990, p. 153). In this forthcoming work, the term pedagogical capital is being advanced as a subtype of cultural capital and used as a framework for understanding some logical practices and systematic supports that lead to scholastic success, and in particular, scholastic success in mathematics. A theoretical underpinning of this work will be Bourdieu's idea that in order to be used successfully as a source of power, or to direct researchers' attention to areas of unconscious privilege for one group over another, the subtypes of cultural capital are in need of being identified and legitimized. Referring back to my analogy of climbing the rock wall, as school children attempt to climb their scholastic wall, it is important to consider whether the hand- and footholds they have available to them are judiciously placed.

Questions

The following questions will guide the forthcoming qualitative and empirical research:

1. Can the term pedagogical capital, as an unconscious privilege possessed by some students and as an ideology in its own right, which is being advanced for general vocabulary usage, offer a compelling qualitative interpretation for some scholastic success in mathematics?
2. Is there any empirical data that would show that the mathematics curriculum has areas of privilege for those with pedagogical capital over those without it; for instance, are there any structural elements in the *Alabama Course of Study: Mathematics* (2003) where children in possession of pedagogical capital thrive while their peers who possess less pedagogical capital struggle to or fail to demonstrate scholastic success in mathematics?
3. Would an unconscious privilege such as this be in keeping with the equity principle as outlined by the National Council of Teachers of Mathematics in their *Principles and Standards for School Mathematics* (2000) or might it point to an area lax in social justice?

THEORETICAL PERSPECTIVE

Introduction

I first coined the term pedagogical capital for myself on the morning after the highly dramatic United States presidential election of 2004. That morning, newly re-elected George W. Bush made a statement during a television interview, "I've just earned political capital, and now I'm gonna spend it." I had been toying with what to call a quality I sensed as a teacher that some students possessed, but for which we seemingly have no vocabulary in education. That quality is the idea that some students come to school more in possession of an intangible that allows them to experience scholastic success in all subjects including mathematics. Upon hearing his statement that morning, I snapped my fingers and said aloud, "That's what I'll call it . . . pedagogical capital." It, or that quality, was a general feeling that some of my students arrived at the academic table better positioned to benefit from the educational process than others. As I had informally interviewed other teachers, it was apparent to me that all teachers recognized this unnamed quality that some students possessed. For the students who have it, it acts as nearly a financial asset so linguistically it made sense to use terms from the provenances of both education and capitalism, thus pedagogical capital.

Edmund W. Gordon, EdD

An immediate search for the term showed that a member of the advisory committee for the Coleman Report (Coleman *et al.*, 1966) and one of the founders of the Headstart program, Edmund W. Gordon, had first advanced the term in the spring of 2001 (Gordon, 2001). As I read Gordon's extensive writings and did further research, it occurred to me that the term had not fallen into frequent usage, despite its potential for social utility. During a series of emails, Dr. Gordon agreed to allow me to interview him. So after resigning from my teaching position in the summer of 2005, I traveled to his home in Pomona, NY, to conduct a videotaped interview. Before I used the term, I wanted to make very certain that we were on the same theoretical page.

A quiet-spoken man, Gordon patiently answered the series of 14 questions I set before him. First among them was how Gordon had come to consider education as an element of human capital theory and when had he begun to use Bourdieu's theoretical framework of cultural and social capital reproduced by *habitus*. It was Gordon who suggested to me that perhaps Bourdieu had been influenced by a short essay written by Mao (1937) on practices. Human capital theory is currently arising in the field of education as economic interests infiltrate. No longer do we have souls to educate, but rather we are readying human resources for later economic pursuits. They are not students; they are current and potential future assets. What Bourdieu added to human capital theory (Becker, 1993, Bourdieu, 1986) was to pull the fence away from purely self-interested economic exchange and to re-position the fence to include capital in its dis-interested guises of cultural and social exchanges.

According to Gordon, he stumbled upon Bourdieu largely from his exploration of what he considered our nation's major program of affirmative development, the *Servicemembers' Readjustment Act of 1944*—commonly known as the *GI Bill of Rights* (US Department of Veterans Affairs, 2006). He considers it to be the largest affirmative action effort in the history of the United States (Gordon, 2004). The components of the bill ensured that veterans of World War II had ample opportunity to improve the state of their education, health and finances. Admitting he now conceptualizes it as much more deliberate than perhaps it had originally been implemented, he saw a nation "poised on an almost revolution in its economic and industrial development" (Livingston, 2005). The reward for military service was the excuse, but the *GI Bill of Rights* had the affect of changing the culture from one that had recently arisen out of the depression, where many forewent education and thus the nation was not very intellectually oriented at the time. Millions of previously uneducated, undeveloped men and women were given the opportunity to be positioned ahead of everyone else in the world due to the veterans' preference programs which were enacted. "Most of these kids had been farm boys" (Livingston, 2005) and they had not been equipped for the economic and industrial expansion that would occur in the latter half of the twentieth century. In other words, post World War II, the lucky ones had gone to war.

Bourdieu's work matched rather than influenced Gordon's thinking and got him to question, what are the varieties of things that position one to benefit from education? In 2001, Gordon referenced Bourdieu's (1986) idea that culture and social networks, along with reasonable health, other educated humans, and sufficient money are forms of capital that are invested in the development of successfully educated persons. Gordon contends that access to these varieties of capital are not distributed equally, and that it is access to them that provides a not too subtle supplementary education (Gordon, 1999, Gordon, Bridglall, & Meroe, 2005) that is not in the realm of the school. Given that little difference can be found in the material resources currently available to schools, there is an inferred association between the effectiveness of a school and the amount of supplementary education that exists outside of the school. The authors refer to this supplementary education as a hidden curriculum. And viewing supplementary education as a resource, there are not many people who would argue against Gordon's logic of using pedagogical capital as a term when you are talking about human resources to invest in one's development. Gordon suspects it is the association with more traditional concerns around capitalistic provenance that keeps terms like capital and human resources out of the vocabulary of education (Livingston, 2005). However, the most effective schools seem also to be those populated with kids who grew up in highly, richly resourced environments—or those with an appropriate level of pedagogical capital.

However, Gordon was quick to agree that scholastically successful students could also come from modestly resourced environments provided that the parent provided an appropriate level of pedagogical capital—supports for appropriate educational experiences that come from the home of the student (Gordon, 2001). One of the examples of this that occurred is James Comers' little book, *Maggie's American Dream* (1989). Maggie was his mother, a domestic worker, who bore and raised him and his four other siblings. James is a physician and all of his siblings have graduated with doctoral degrees, having thirteen degrees between them. "But Maggie brought a lot of pedagogical capital to the rearing of these kids" (Livingston, 2005). It seemed to Gordon that students who brought the wealth of these various capitals to complement the learning situation were bound to do better than students who were lacking them. Students who lack this out of school wealth, who lack pedagogical capital, suffered educationally just as though they were attending a poorly resourced school.

The Coleman Report

The Coleman Report, for which Gordon was a member of the advisory committee, created a controversy over families when it was released in 1966. In it, Coleman, *et al.*, asserted that variances in the quality of school alone did not adequately explain variations in levels of achievement. As a matter of fact, the authors advanced the idea that differences associated with the families accounted for most of the variance in school achievement (Coleman, *et al.*, 1966). There was an understandably negative response from the minority groups highlighted in that report at the time, but missing in that action was the notion that the resources available to the majority families were not present or supported by the lifestyles of the minority families during the days of the civil rights movement. Those were the days when many school districts were still maintaining separate and unequal schools. Considering that the research in the Coleman Report resulted from an explicit directive written into *Public Law 88-352* (Eighty-eighth Congress of the United States of America, 1964), commonly known as the *Civil Rights Act of 1964*—and that it was one of the first forays into research related to educational policy rather than merely educational financing—it is understandable that issues having their root in the home were not given much consideration because they were not subject to institutional policy directives. As a quantitative report, the survey tallied such things about the family home as were there encyclopedias in the home, were there more than five siblings in the household, and had the mother graduated from high school. But due to the fact that the Coleman Report clearly showed that there were real and pronounced financial and policy disparities in need of address, particularly in the south where 54% of Negro school age children resided and were being schooled (Coleman *et al.*, 1966), there began to be an over-identification of education with what occurred in the school and less identification with education as a cultural phenomenon occurring also within the family. Other researchers re-aggregated the data from the Coleman Report and were able to show that for minority students, the quality of school was far more important than family resources at that time (Petti-grew, 1967). This may have been because for many of the minority families, school was most likely to be the only place where the logical practices and systematic supports for academic learning took place; the sole place where pedagogical capital presented itself.

This controversy did not faze Gordon, who by 1990 had developed a new construct he called affirmative development as a complement to vari-ous entitlement programs be it the *GI Bill of Rights* or Headstart (Gordon, 2001). Gordon conceived that the veterans' preferences program included affirmative development as well as affirmative action in the way of school-ing, home loans, and other preferential treatments. The veterans were a protected group as much by public policy as by public patriotism. Having been influenced by the musings of W. E. B. DuBois, and even though skin color and other cultural identifiers continued to create problematic social divisions, by 2001, Gordon had become convinced that the unequal dis-tribution of capital resources, the gap between those that have and those that have-not would be for the 21st century what racial and other social divisions had been for the 20th century. He had begun to believe that with-out the capital to invest in human development, it is impossible to achieve meaningful participation in our currently advanced technological society (Gordon, 2001).

Gary S. Becker, PhD

The best-known use of the terms human capital and human develop-ment in economics was a 1964 book by Nobel laureate, Gary S. Becker, entitled *Human Capital: A Theoretical and Empirical Analysis with Special Ref-erence to Education.* According to Becker, human behavior adheres to the same fundamental principles as economic exchange. In his scenario, hu-man capital is seen as a means of production as though each employee is an individual factory capable of producing labor. This was different from the Marxist notion that workers merely sell their labor power for wages. On a macro level, human capital can be substituted and should be con-sidered as a separate variable in the production function. One can invest in human capital and raise the quality of it as when an employer offers education benefits, training or medical treatment because the quality of human capital owned is directly related to that investment. As a matter of fact, for the employer, income will depend partly on the rate of return on the human capital owned at any one time (Becker, 1993). Thus, human capital is an asset like stock, whose value can rise and fall depending on the state of the field of the employees at any given time in the present or in the future.

The concept of human capital can be infinitely elastic, including hard-to-measure aspects such as personal capital (for instance, efficacy and disposition) and social networks (via family or fraternity). This can and does vary from employee to employee and are to be considered at the micro level. However, Becker distinguished only between specific and general human capital and thus allowed the theory of human capital to work without explaining the difficult to measure. Specific human capital refers to skills or knowledge that is useful from a single employee as in knowing how to create blueprints. General human capital such as literacy or mathematical fluency is useful from all employees (Becker, 1993). However, it was the consequence of investing in human capital that was Becker's noteworthy contribution to the field of economics. The supply of human capital in a region can be used to show and explain regional differences that cannot otherwise be explained by way of the supply of economic capital alone. For instance, two sweater factories on the same size property, with the same machinery, climate, and raw materials can have as a result different qualities of sweater. The difference can only be accounted for in the quality of the worker manning the machines. Although Becker's analysis has at times been controversial, it has shown a high degree of utility in the way social environments are determined by the interactions of the individuals.

Pedagogical Capital: Definition and Implications for Math Education

Leaning heavily on the writings of Bourdieu and Becker, Gordon summed up human educational investment capital as:

1. *Cultural capital*: the collected knowledge, techniques and beliefs of a people.
2. *Financial capital*: income and wealth, and family, community and societal economic resources available for human resource development.
3. *Health capital*: physical developmental integrity, health and nutritional condition, etc.
4. *Human capital*: social competence, tacit knowledge and other education-derived abilities as personal or family assets.
5. *Institutional capital*: access to political, education and socializing institutions.
6. *Pedagogical capital*: supports for appropriate educational experiences in the home, school and community.
7. *Personal capital*: dispositions, attitudes, aspirations, efficacy, and sense of power.

8. *Polity capital:* societal membership, social concern, public commitment, and participation in the political economy.
9. *Social capital:* social networks and relationships, social norms, cultural styles, and values. (Becker, 1993, Bourdieu, 1986, Gordon, 2001, 2004)

The above list gleaned from Gordon's extensive writings and collaborations seems to create a parallel between cultural capital and the other forms of capital, and in his more recent writings he maintains this parallel. Furthermore, by early 2005, and prior to my interview with him, Gordon, in conjunction with Bridglall and Meroe, no longer included pedagogical capital in the categorical list of capital as the context and preconditions necessary for achievement. They did, however, argue that academic achievement gaps were "not a problem of schooling alone (p. 17). For the purposes of the forthcoming research, the term pedagogical capital will be reintroduced and refined as a true subtype of cultural capital in the style recommended by Bourdieu. Pedagogical capital will be defined specifically as supports for appropriate educational experiences that come from the home of the student. The implications of defining this heretofore unnamed quality for the field of mathematics education are manifold.

While controversies still abound in educational research, the trend has been toward pinpointing specific resources that really matter for children's mathematical achievement. While not specifically a mathematics education study, in *Home Advantage: Social Class and Parental Intervention in Elementary Education* (*Home Advantage*) by Annette Lareau (2000), the daily workings of social class and how social class affects a parents' ability to pass along advantages to their children are presented in vivid detail. Lareau compares and contrasts what Bourdieu's notion of cultural capital means in the current American setting as middle-class and working-class parents direct their time and energy toward their children's scholastic success. In this book, Lareau challenges "the position that social class is of only modest and indirect significance in shaping children's lives in schools" (Lareau, 2000, p. 2). She argues that social class, as a variable independent from ability, not only can but also does affect schooling. For the purposes of this research, it will be shown that this variable affects the child's learning of important mathematics.

According to Lareau, "in every society, parents with cultural capital will try to transmit their advantages to their children, but the way in which they do so is likely to vary with the organizational form of schooling and with the types of performance and knowledge that are mostly highly valued" (Lareau, 2000, p. xiv). She explored the ways that parents who are a part of the middle-class culture draw upon resources that are specific to them, particularly concerning their knowledge of the inner workings of schools, as they help their children. Middle-class parents set out to help their children—they are very aware that the child's future success depends upon how well they do in school. They do not set out to conspicuously display class privilege by perpetuating curricular advantages ingrained within mathematics.. However, in the myriad ways that they help their children by investing their cultural capital, they inadvertently increase inequality. One of her overriding observations is the contrast in how the spheres of home and school are viewed by the two classes. Working-class parents viewed home and school as separate spheres of influence, while their middle-class counterparts saw them as closely intertwined (Lareau, 2000). In other words, in something as simple as assisting in the development of fluency in multiplication pairs, middle class parents assist in creating an achievement gap that can be empirically documented. It is the purpose of the forthcoming research to document this phenomenon.

When these parent-wrought advantages are compared through a careful investigation of teachers' perceived standards and preconceived attitudes, it is the parent of the middle-class child who has systematically inculcated the child with the mastery of a robust variety of academic skills and attitudes which are the end product of many incremental steps (Lareau, 2000). Or as Bourdieu would surmise, the parental unit(s), working in conjunction with the family field, has the time from birth onward to direct the flow of the student's *habitus*. As will be explained, inculcation and assimilation take time—time that must be invested personally. Like the acquisition of a suntan or of a swimmer's stamina, it cannot be done second-hand. This rules out many of the effects of the school and community because delegation is thus impossible (Bourdieu, 1986, 1998). Therefore, the majority of the yield in the form of scholastic success for a given child depends on the cultural capital in possession of the family and on the amount of that capital which is invested by that family. The eventual maths education of the child ends up being a relation between the inertial qualities of the family's accumulated capitals and those activities that occur within and without the school.

Another excellent example of family and group practices is Lareau's *Unequal Childhoods: Class, Race and Family Life* (*Unequal Childhoods*) published in 2003. This book contains groundbreaking research into families who are rearing children by one of two strategies, one Lareau calls concerted cultivation and the other she calls the accomplishment of natural growth (Lareau, 2003). Again, leaning heavily on the cultural capital theory of Bourdieu, Lareau is able to capture the texture, describe the quality, and to give a peek at the quantity of the inequality experienced by some children. It would be an easy leap to say that in the current institutional scenario of schooling, children who are being reared using the strategy of concerted cultivation possess pedagogical capital in both Gordon's and my notion of the term and those who are being reared using the strategy of the accomplishment of natural growth do not possess the same qualitative or quantitative amount of pedagogical capital. In a similar separation as in *Home Advantage*, Lareau documents that working-class and poor parents viewed the world of the parent and the child as separate spheres of influence, while their middle-class counterparts saw them as closely intertwined.

In *Unequal Childhoods*, Lareau (2003) argues that key elements of family life cohere to form a cultural logic of child rearing. Again, the disjointedness of the two styles of child-rearing separates the middle-class child from their working-class or poor peer. Middle class parents tend to adopt the strategy she calls concerted cultivation and working-class and poor parents, by contrast, tend to undertake the accomplishment of natural growth. "For working-class and poor families, the cultural logic of child rearing at home is out of synch with the standards of the institution" (Lareau, 2003, p. 3). Once again, it is the continual inculcation of the child by the middle-class parent that brings about in that child a practical mastery of the ways of the school; concerted cultivation fosters an academically friendly *habitus* and a *docta ignorantia* (Bourdieu, 1980, p. 102) that seems to act as a lubricant in the machinery of mathematics education while the accomplishment of natural growth acts more as a wrench in the workings.

While both concerted cultivation and the accomplishment of natural growth offer certain advantages/disadvantages as far as family relations are concerned, it is important to note that they are accorded different social values by some highly important and influential and hegemonic institutions in the lives of the parent and the child. The middle-class strategy, concerted cultivation, appears to offer a greater promise of being transubstantiated into social, educational, and then later financial profits than does the working-class and poor strategy of the accomplishment of natural growth (Lareau, 2003). This is probably because the *habitus* developed within the middle-class child over the years is more closely aligned with the dominant set of cultural repertoires and behaviors that comprise the standards and policies of institutions like the school and the work environment. A close look at the social class differences as mirrored in the mathematical standards of these institutions can provide an opening for some vocabulary for understanding inequity—particularly if those institutional standards give some cultural practices a nod over others in that they pay off in settings outside of the home. The cultural indicators in the instances of these two pieces of research are mostly closely summarized by a lack of separation between the spheres of the home and the school, and the spheres of the parent and the child, which in turn mirrors the expectations of the teacher and school.

As a parting shot to my interview with him, Gordon countered my comment that the pervasiveness of pedagogical capital seemed to create an achievement gap in all subject areas including mathematics that was more difficult to document than those created by race, gender or socio-economic measures—measures which are easy to differentiate for measuring. He looked at me, smiled and said, "It may be the only achievement gap that really matters" (Livingston, 2005).

Pierre Bourdieu: The Forms of Capital

In perusing the foregoing definition of pedagogical capital and its implications for mathematics education, human capital obviously involves more than just money. Most of the above is intangible and must be learned through the educative process of logical practices and systematic supports (Bourdieu, 1980). This is why racial isolation, gender isolation, and other types of isolation had the effect of lowering achievement pre-1964 and continues for those socially and/or economically isolated today. Like a not too subtly hidden Markovian chain, these capitals are best utilized when the indulgence of "time gives it its form" (Bourdieu, 1980, p. 98), or in other words, these capitals have their substance in that the person adept at requisitioning them out does not bear the mark "of having been taught" (Bourdieu, 1980, p. 103) how to use them. There is a practical mastery that can be as simple as having the efficacy to call a well-placed administrator if a child is struggling that seems so effortless—like a learned ignorance of privilege ("*docta ignorantia*") (Bourdieu, 1980, p. 102).

In his seminal work, *The Forms of Capital*, Bourdieu (1986) outlines only three types of capital, economic, cultural and social. He makes the argument that the "social world is accumulated history" and that "capital is accumulated labor" (Bourdieu, 1986, p. 241). It is the process of accumulation that, borrowing a term from Newtonian physics, Bourdieu calls the *vis insita*, the power of the innate force of matter to resist change and continue in its present form. In other words, the accumulation becomes a native strength with inertia of its own. Accumulation differs from the distinction between inherited properties like color of skin or eyes and acquired properties like learned knowledge and "manages to combine the prestige of innate property with the merits of acquisition" (Bourdieu, 1986, p. 245). As a matter of fact, he also argues that the time necessary for accumulation is the *lex insita* (Bourdieu, 1986, p. 241), the principle or law underlying the inherent order of the social world. Things don't happen instantaneously as might with a game of roulette. Capital is the reason why not all scenarios are equally possible. It takes time to accumulate, it has a tendency to persist in its present state, and it is a force inscribed in the objectivity of things.

Bourdieu accounts for the structure and functioning of the social world by introducing capital in the non-economic forms of cultural capital and social capital (Bourdieu, 1986). Alluding to the historical invention of monetary capitalism, economics had academically reduced exchanges to those purely mercantile in nature, which are for profit and contain a high degree of self-interest. Left out of these exchanges were the dis-interested, non-monetary exchanges of the type that give structure to the distribution of types and subtypes of all capital in the social world. At any given moment, this distribution of both material and immaterial capital is able to paint a picture of the structure of the social world. If material capital in the form of money is said to exist, then implicitly defined and brought into existence is its opposite, immaterial capital in the forms of cultural capital and social capital. His leading argument for the existence of all three types of capital is the very real way that one type of capital can ultimately change form into a profit that appears as one of the other forms of capital. Bourdieu contends that capital and profits must be grappled with in all their types and subtypes in order to establish laws whereby they are convertible from one to another like Newton's laws of motion. Economic capital is most readily converted into money and can be institutionalized as property rights. Cultural capital, the underlying theme of this paper, can be institutionalized in the form of educational credentials. Social capital is made up of connections and can be institutionalized as a title of nobility. In summary, capital presents itself in one of three types: economic capital, cultural capital and social capital.

The notion of cultural capital initially presented itself to Bourdieu "as a theoretical hypothesis which made it possible to explain the unequal scholastic achievement of children originating from the different social classes" (Bourdieu, 1986, p. 243). This starting point gave him a theoretical break from the view that academic success is an effect of natural ability and also provided a break from human capital theory that heretofore had been purely economic. Bourdieu goes on to describe cultural capital as appearing as one of three forms, embodied, objectified and institutionalized. He considered the embodied form of cultural capital to be dispositions of the mind and body which are accumulated and that are long lasting—his notion of *habitus*. The objectified form is where cultural goods take the form of accumulated pictures, books, dictionaries, etc. Cultural capital in its institutionalized form is a special type of objectification, for instance when it takes the form of educational qualifications that confer a type of real property of its own on an individual. As a subtype of cultural capital, pedagogical capital is unique in that it embodies a disposition towards mathematics, it objectifies certain mathematical tasks deemed important, and certain educational qualifications in possession of the parent at hand confer a real advantage to a particular math student.

In his writing, Bourdieu gives credit to economists for directly questioning the relationship between the rate of profit and educational investment in human capital theory. However, he chides them for only taking into account "*monetary* investments and profits" (Bourdieu, 1986, p. 243) such as the cost of providing an education and the cash equivalent of time devoted to study. Not included or explained in human capital theory at the time were the ways different households, social classes, or other agents allocated their investment resources between the time needed for "economic investment" and the time needed for "cultural investment" (Bourdieu, 1986, p. 244). By not doing so, the economists had failed to systematically account for the way that the chance of profit differs as a function of time, or in other words, to account for that time which is needed for the total volume of the investment to play out as a component of the total field of assets available for input into that function.

They also neglected to relate the wide range of educational investment strategies to a total system of strategies that are able to re-produce those same strategies, and to comprise a logical theory of practice for that re-production. Ultimately, what was not included in human capital theory is what might be the "best hidden and socially most determinant educational investment, namely, the domestic transmission of cultural capital" (Bourdieu, 1986, p. 244). The empirical studies of the relationship between academic ability and academic investment diagramed in human capital theory (Becker, 1993) omitted the idea that ability is itself the product of a function between an interaction of time and cultural capital. Bourdieu called it a "typically functionalist definition of the function of education" (Bourdieu, 1986, p. 244) and claimed it ignored the idea that the actual scholastic yield from educational strategies will depend upon the cultural capital already invested by the household, class or other agent. Thus, embodied cultural capital is converted into an integral part of the person, or in other words, into a *habitus*.

Pierre Bourdieu: Habitus

To understand the concept of *habitus*, Bourdieu draws heavily on his fieldwork in Kabylia (Algeria). In 1974, after this fieldwork had been completed, and during a series of elections in France while some editorial cartoons were being published in the local paper, Bourdieu discovered that the relationship between the native informant and the researcher did not allow for the transmission of unconscious schemes of practice. The explanations that natives offered for their own practices concealed even from themselves the true nature of their practical mastery of the schema, that is, they had a way of understanding that did not include the principles of practice or reproduction of those practices (Bourdieu, 1977). To him, the natives were more possessed by their habits, than in possession of them, which he refers to as *habitus*. For instance, in gift exchange as in to denote the birth of a child or for a wedding, counter-gifting is a part of the overall scheme. What the natives failed to include in their explanation was the temporal element of this exchange, or in other words the time interval between the gift and counter-gift is what helps to define the whole truth.

Bourdieu explains that the social world is made up of three types of knowledge: phenomenological knowledge which is the explicit truth of the primary experience like the exchange itself, objectivist knowledge which structures practices and representation and makes the exchange experience possible, and lastly, scientific knowledge which explains the relation in which the exchange structures are actualized and tend to reproduce the same types of exchanges in the future. Bourdieu goes on to explain how the temporal element of the gift exchange answers inquiries into the mode of production and the practical mastery of the exchange schema (Bourdieu, 1977, 1980). Gifts countered too rapidly will curtail future exchanges or can be experienced as an insult in a similar way as with gifts countered too slowly. Models that abolish the interval between gift and counter-gift abolish the strategy also. For instance, in the scheme where a man is expected to avenge a murder or buy back property from a rival family and does not do so, after a certain interval, his personal capital is diminished on a day-by-day basis with the passing of that time. In the case of a when a bride is sought, the strategy schema holds that the exchange of negative answer gifts should come sooner than that of using the strategy of prolonging the anticipation of a positive answer gift. It is the theory of the practice of exchange, or the strategy used to decide when to exchange, therefore, which is the most specific aspect of the exchange pattern.

And it is this practical mastery of the theory of practice that is killed off by simple models of exchange and reciprocity when the temporal element is not adequately accounted for. A model rather brutally highlights what the virtuoso of the practice innately knows in order to produce the correct action at the correct time in each appropriate case. With rules or a model, reciprocity becomes a necessity, the product of exchange becomes a project, and things that have always just happened can no longer not happen. But in the real world of practical mastery, not every death is avenged nor is every gift countered—and nor must they be. There is a certain harmony of *habitus* that cannot be adequately re-produced in the model using mechanical laws for the cycle of reciprocity alone (Bourdieu, 1977, 1980). The passage from the highest probability that an action should happen to the actual completion of that sequence is both a quantitative and qualitative leap out of proportion to what the model reduces to a simple numerical gap that must be filled. There is a parable used by the Kabyle, "the moustache of the hare is not the moustache of the lion" (Bourdieu, 1977, p. 13).

This and similar parables, along with the temporal element inherent in exchanges transubstantiates a "disposition inculcated in the earliest years of life and constantly reinforced" (Bourdieu, 1977, p. 15) by a series of checks and corrections from the entire group disposition or group *habitus.*

In other words, the group as an aggregate embodies the dispositions of the individuals. Bourdieu was impressed that individual members of the Kabylia were unable to recite a litany of explicit axioms for the theory of practice, but they obviously had an inertial knowledge of the system that enabled them to re-produce it in its entirety in different locations and at different times. They knew what to do and when to do it even as they could not explain how they contained this knowledge or how they had come to acquire it. Thus the precepts of their customs had an unusual scientific underpinning in that they would "awaken, so to speak, the schemes of perception and appreciation deposited, in their incorporated state, in every member of the group, i.e., the dispositions of the *habitus*" (Bourdieu, 1977, p. 17). Rules and models are always a second-best to actual practice and are only in the group *habitus* insofar as they correct and make good the occasional misfiring of practice. *Habitus* is capable of generating practice without any apparent express regulation or any other institutionalized call to order. In a way that Bourdieu calls the fallacy of the rule, practitioners follow the rules without explicitly knowing them. The explanations natives may provide for their own practices include a learned ignorance (*docta ignorantia*) or a practical way of knowing the art of the game without any knowledge of its underlying principles.

Pierre Bourdieu: Structures, Habitus, and Practice

To understand what structures construct *habitus* and thus define the practice, one must examine the economy of the logic in use. One of the fundamental effects of the *habitus* is to produce a commonsense world of dispositions endowed with security due to a consensus of meaning. The word disposition seems particularly suited to express what is meant by *habitus*. It is broad enough to convey the meaning that the *habitus* is the product of an organizing action and it has a meaning that is close enough to that of words such as structure. What this means is that dispositions and systems of dispositions are not random (Bourdieu, 1977). For instance, a lack of money in a Marxist scheme would produce the habit of not buying books, i.e., "I have no money to buy books, and therefore I have no need to buy books." In one way, it creates a virtue out of a necessity, but the organizing structure was the lack of money. If the underlying structure of the lack of money continues, and nothing else changes in the field either, the disposition will continue to be produced ad infinitum each time the opportunity to buy a book presents itself. Unlike the existentialist Jean-Paul Sartre, who made each action sort of an unprecedented confrontation between the agent and the world (Gregory & Giancola, 2003), Bourdieu was unable to dismiss the accumulative effect of these durable dispositions.

In other words, if one regularly observes a close correlation between the objective disposition, that is, the chance of access, and the subjective disposition, that is, the motivation or need, this is not because the agent consciously adjusts their aspirations to a mathematical evaluation of their chances of successful acquisition each time, it is because there is a propensity to privilege their earlier experiences (Bourdieu, 1977, 1980). Imagine for instance the difference between two students, one who has learned to take no for an answer and one who has learned how to whine in order to get his way. Structural exchanges must have occurred in each student's past to give rise to each disposition, which helps to explain the current practice. In practice, the *habitus* is structural history turned into current nature, which through current interaction re-produces the actual production of the practice. Or as Bourdieu eloquently states, "it is yesterday's man who inevitably predominates us" (Bourdieu, 1977, p. 79). The present is little when we compare it to our long past, which in the course of we were formed and from which we are the result. Thus *habitus* is structurally laid down in each agent by his earliest experiences and is the *lex insita* or underlying principle for current practice.

Pierre Bourdieu: Capital, Habitus, and Field

Bourdieu's is a philosophy of science that one can call relational. Although a characteristic with what are considered the hard sciences, relations are seldom incorporated into the field of play of the softer social sciences because objective relations are difficult to show. In this paper, the term pedagogical capital is being advanced as a subtype of cultural capital and used as a framework for understanding some logical practices and systematic supports that lead to scholastic success in mathematics. As such, it will be important to understand the relation between the habitus, the field of play, and capital resources. It is hoped that can be accomplished via an irresistible analogy. Capital, *habitus*, and field are the fundamental concepts of Bourdieu's social theory of practice. Economic capital and cultural capital work in conjunction to create a structural two-dimensional field of play where agents are then able to interact. The first dimension, economic capital, is viewed as the most important because as a family finds itself with more of it, there is a higher likelihood that the family will also not be deprived of cultural capital (Bourdieu, 1998). His is also a philosophy of action designated at times as dispositional in order to note that the inertial potential inscribed on the agent and the inertial potential inscribed on the structure of the field that agents find themselves in is in fact a relation taking place in the temporal mode. Practices, such as playing golf or even consumption patterns, are not in and of themselves independent of the field.

In other words, the field that agents find themselves in will likely be a determining factor in their practices and dispositions also. If this is drawn up as a map, distance on paper is equivalent to social difference and class difference. Bourdieu uses this analogy to explain left and right leanings in politics (Bourdieu, 1998). However, it is the structure of the capital, along with the relative amounts of economic capital versus cultural capital and whether or not certain subtypes of capital are present and to what degree, that can rather predictably determine what type of dispositions that the field is likely to produce now and in the future.

Pierre Bourdieu: Families

It seems to Bourdieu that there is a new capital, which has taken its place alongside economic capital. He calls this cultural capital. The structure of the distribution of both economic capital and cultural capital—and the re-production of that structure—is "achieved in the relation between familial strategies and the specific logic of the school" (Bourdieu, 1998, p. 19). Bourdieu considers families to be corporate bodies animated by a system of re-productive strategies that produces a similar family tomorrow and a still similar family in the distant future. In short, family is simply a word because there is no universal definition of family. The family contains a set of persons who share properties that include dispositions, social spaces, and capital. Bourdieu calls the interplay of the three the family spirit (Bourdieu, 1998). It is the combinatory relation between the three in one family with the three of another family that gives rise to our notion of class and success, sort of like east, west, north, and south on a map. According to Bourdieu, the family is the primary site of social maintenance and re-production. In other words, for the word family to even be possible, it must have some structural function in the social world. So what is it and how does it do it?

Families invest in a variety of re-productive strategies, for instance, fertility strategies, matrimonial strategies, succession strategies, economic strategies and educational strategies. Focusing on the educational strategies, and for the purposes of this research, mathematical educational strategies, families tend to invest more in school education as their cultural capital becomes more important to them and as the relative weight of their cultural capital to their economic capital becomes greater—or in other words, if schooling is a determinant in social position due to some economic reward, prestige, or mobility, then the family is disposed to find education important. Because the institution of the school is able to initiate social borders analogous to the old social borders that divided the nobility from the gentry, and the gentry from the common people (Bourdieu, 1998), some families are disposed to cross these class borders via education. But it is the family *habitus*, combined with economic and cultural capital, which tends to direct each successive generation to re-produce the same types of *habitus* in the next generation. Bourdieu never arrived at a theory for what causes one family to have a disposition of mobility and for another family to lack that disposition of mobility.

But if one looks at families as occupying a social space with a field of power, then the symbolic work necessary to produce a unified field is carried out by all members of the family, and possibly in particular by the parental unit(s). The inculcation of the child begins at birth. The parental unit(s), working in conjunction with the family field, have the time from birth onward to direct the flow of the student's *habitus*. Inculcation and assimilation take time, time that must be invested personally. Like the acquisition of a suntan or of a swimmer's stamina, it cannot be done second-hand. This rules out many of the effects of the school and community because delegation is thus impossible (Bourdieu, 1986, 1998). Therefore, the majority of the yield in the form of scholastic success depends on the cultural capital in possession of the family and on the amounts of that capital which is invested by that family. Eventual educational quality ends up being a relation between the inertial qualities of the family's accumulated capitals and those activities that occur in the school. Since we know that capital is unevenly distributed, all agents in the field of educational play do not have the same means to prolong or otherwise enhance the child's education. The embodied cultural capital and the embodied dispositions available in the family at the time of the rearing of the child stand the highest chance of being re-produced.

To reiterate, if the cultural capital in the family is distributed in a mathematically friendly manner, that is if there is an appropriate amount of pedagogical capital in possession of the family, and if perhaps the family dispositions are mobile, then there is a higher probability that the family unit will produce a student for whom scholastic success in mathematics will be the norm. To be looked at in this research will be the introduction of and the development of fluency in the multiplication pairs in third and forth graders. The implication of this research using pedagogical capital in the field of mathematics education is two-fold. The qualitative part of the research will offer insight into the teaching style and classroom interaction of the introduction of an important mathematical topic. The quantitative research will offer a portrait of the development of fluency in the multiplication pairs upon which much of the math content of the middle years depends. Together they will address a mathematical standard that might convey an unconscious privilege to one social class at the expense of another, which as an ideology creates a ripple in the fabric of social justice.

Ideology

The word ideology was first coined by Antoine Louis Claude Destutt, comte de Tracy, between the years 1801–1815 as a science of ideas (Head, 1985). Every society has ideologies that form the basis of public opinion and common sense. Dominant ideologies can appear to be very neutral, or if presented in a unique fashion, they can pack quite a sting. Antonio Gramsci advanced the notion that when most people in a society begin to think alike about certain matters, or to even forget that there are alternatives, then we arrive at the concept of hegemony (Bates, 1975; Purvis & Hunt, 1993). Michel Foucault wrote generously on the notion of ideological neutrality, and about the ways that dominant groups strive for power by influencing a society's ideology by broadcasting their own ideology, and therefore causing the society to become closer to what they want it to be (Rabinow, 1984). Karl Marx proposed that a society's basic dominant ideology presented itself as a part of its overall economic superstructure. This base/superstructure model is determined by what is in the best interests of the ruling class (Churchich, 1994). Karl Mannheim and Jürgen Habermas furthered Marx's view by showing that ideology is used as an instrument of social re-production (Somers, 1995). And then, after an intensive philosophical re-reading of Marx, Louis Althusser and Étienne Balibar proposed the concept of the lacunar discourse, or gaps in a discourse, to suggest that some dominant ideologies are not specifically broadcast, but are rather suggested by what is not told (Brewster, 1970).

Ideology then, as a form of cultural capital, has been extensively developed by Bourdieu as a mechanism of social re-production, and much like his concept of *habitus*, members of a society are generally more possessed by their ideologies than in possession of them (Bourdieu, 1977; Bourdieu & Passeron, 1979; Bourdieu, 1980; Bourdieu, 1986; Bourdieu, 1998). In this work, pedagogical capital, as a yet unrecognized and perhaps unconscious privilege possessed by some, and as an ideology in its own right that might re-produce existing inequities due to the nature of math standards becoming hegemonic, is being advanced for general vocabulary usage in the hope that it might offer an alternative explanation for some success and failure in mathematics education.

Equity

This past two-and-a-half decades has been witness to perhaps some of the most sweeping changes in mathematics education—not only in the content of mathematics, but also in the way educators and academicians present their material and develop student expectations. Since 1989, the National

Council of Teachers of Mathematics (NCTM) has systematically developed a tripodal support system with its standards that seeks to create a balance between curriculum, teaching, and assessment. This approach is similar to the workings of a tripod table. According to the authors, adjusting one of the three legs can level the tabletop, or plane surface. Whether by design or coincidence, the authors of *Principles and Standards for School Mathematics* (*Principles and Standards*) (NCTM, 2000) have provided educators with a workable tripod having curriculum, teaching, and assessment for legs. According to the ideology, by adjusting the legs, every student can be offered a level field of support from which to learn mathematics. The standards document, as a trilogy, offers educators a tangible guide to adequately match the needs of all students with a meaningful mathematics education. And as with any equitable balancing act, finesse, and timing are critical.

Equity is an ideology. Disparities in achievement outcomes based upon race, sex, or national origin have long been key issues with regard to educational equity and have been the target of a large body of research. As the field of mathematics education strives to get its act together and to flourish amidst the public's demand for more definitive solutions, we must continually address diverse questions of equity. At different times and in different cultures, mathematics education has attempted to achieve a variety of different objectives. In its latest policy-guiding document, the NCTM has equity as its first guiding principle as a part of "making the vision of the *Principles and Standards*...a reality for all students" (NCTM, 2000, p. 12). Specifically mentioned in this equity principle are high expectations for all students (not just some), accommodating differences for students who may have special needs, and enabling and providing a significant allocation of human and material resources.

A principal of mine once made the comment that the parents of our students do not send their worst children to school and keep their best ones at home—they send their only children to school to be our students. If we believe in success for all groups, then as standards become the basis for our mathematics curricula, they must set forth a truly democratic vision to promote equity and not become neutrally hegemonic in themselves. When these standards serve those who traditionally struggle to learn mathematics instead of only those for whom the standards are second nature, or a part of their *habitus*, then we are on the road to promoting real equity.

Unconscious Privilege

In 1990, Dr. Peggy McIntosh published an excerpt from a working paper for which she entitled *White Privilege: Unpacking the Invisible Knapsack*. In the article, she made an association with institutional strength, economic power, and white dominance by creating a list of 26 privileges, the style of which has been unabashedly duplicated by many others who also wish to put the dimension of privilege into discussion. McIntosh's analysis was instrumental in prompting discourse via distinguishing areas of actual privilege from other areas where people's rights were being trampled on. For the purposes of this theoretical framework, McIntosh's concept of unconscious privilege, privilege that persons have been subtly enculturated to be unaware of, will be used to lessen the obfuscation between those enjoying an unfair advantage and those suffering an unfair disadvantage with regard to the mathematics curriculum.

Autobiography

I entered into the Bourdieuian field of scholastic play as a working-class, white child. My mother was a high school graduate, but my father had dropped out of high school and passed the General Educational Development Test (GED) after being drafted into the army in 1956. They married soon after his enlistment was over, on her 19th birthday. I was born before the first year of their marriage had passed. There is a history of mental illness in my mother's family—her mother spent the last 35 years of her life institutionalized—and my mother has less serious, but still noticeable problems, of which include rarely speaking for long periods of time, so there was little verbal interaction between parent and child when I was growing up. When I was very young, I rarely got a store-bought dress or play clothes. Mine came to me as hand-me-downs from women my mother worked with and perhaps from the Goodwill because my mother shopped there on a regular basis—or my mother would sew them. While we owned the house we lived in when I was in elementary school, and while our street had newer homes, it was surrounded by a declining neighborhood with unpaved streets. We moved into a more affluent community during the summer following my 5th-grade year. By that time, my father had graduated from college and was employed again.

Based upon my childhood activities, I can rather confidently say that I was reared using Lareau's strategy of the accomplishment of natural growth and not with the more desirable strategy of concerted cultivation. There was both a clear divide between my childish world and that of the adults in my life, and a divide between my home and school. I never attended a kindergarten, so my first day of first grade was the first time I had ever been inside of a school—or had even eaten in a cafeteria. I could "read" before I started first grade, but what I remember was that I could memorize the words to a children's book and recite them back on the correct pages, since children's books have very few words and they have pictures that act as cues. I remember learning to decipher words in first grade and becoming fascinated with reading. However, there were few children's books in my house unless we borrowed them from the local library, so most of the reading I was able to do occurred at school—and I loved reading. When I was in second grade, my father made a decision to stop working as a welder to attend college using his *GI Bill of Rights* benefits before they expired, so my mother worked as a bank teller to support the family of two adults and three growing children. Both my first and second grade teachers were older women who had pianos in their classrooms and they would play whilst we sang—I learned to read the words to songs that way. In third grade, I had a new, young teacher who drilled us on addition and subtraction facts using flashcards. I was enthralled by the flashcards. However, there were no flashcards in my household at the time, so I made my own on scraps of paper and played "school" while my dolls, Kissy, Mrs. Beasley, and an unnamed koala bear, sat behind cardboard box desks in the bedroom I shared with my sister. A man down the street had stopped being a salesman about this time, and one day I pragmatically salvaged quite an assortment of order pads that had carbon paper from his trash and brought this treasure home to play "school" with for many months afterward. In 4th-grade, I was introduced to the multiplication pairs, and stood amazed the next day as my fellow students could recite the answers—when I was still struggling to know them additively. I went home and drilled my dolls on multiplication pairs that night and for many weeks afterward.

It was then that I learned that some children had help at home that I did not have. Prior to this, I thought all kids had families just like mine and that school was a child's job. Since there was such a clear divide between adult and child, and school and home for me, I couldn't quite get over the notion that involving parents in schoolwork was somehow cheating. I struggled for the next few years to gain fluency in the multiplication pairs, and struggled also with the other math concepts that were being introduced to me during that period, for instance division and fractions. I could do the work, but it took me much longer than my seemingly gifted peers to master the tasks. I was still doing much of my multiplication additively—I did not have a fluency in the multiplication pairs—and I was not working in concert with a pedagogical guide in my home. Of course, that was in the days prior to the introduction of the low-cost hand calculator, so in order to produce correct answers, I had to master the concept of the mathematics by hand, which seemed to nurture an appreciation for the underlying patterns that the language of mathematics was being used to describe. I began to love math.

But outside of school, and like the poor and working-class children in Lareau's study on the accomplishment of natural growth, I spent most of my time just playing games with other children or playing alone. It was not until the end of 7th-grade, when my most gifted classmates were inducted into the Junior High Honor Society, and I was not, that I began to take any of my schooling seriously and begin to do school work outside of the school. By the end of 8th-grade, I had become an honor student also. Truly, I was completely unaware that some children had developed a scholarly *habitus* early in life. Despite the fact that my father had completed his college degree by the end of my elementary years, college for me was never discussed in my house even during high school. At the time, I was well on my way to re-producing the low-level labor-power that my mother was involved in.

By the time I finished high school, a rather traumatic divorce between my parents, and the discovery of the availability of grants and scholarships for college had changed my trajectory. Later on, as a parent, I seemed have taken a page from Herbert Kohl's 1994 book, *I Won't Learn from You*, and I was determined that my own children would not struggle in the institution of school as I had. Thus, I adopted the child-rearing strategy of concerted cultivation, without knowing that it was typical to my newly established middle-class status as a schoolteacher, and I began to systematically pass on my cultural capital to my children by providing them with a very high level of pedagogical capital.

Summary

This work, which is both exploratory and theoretical in nature, is an excerpt of a work in progress. Used are the ideas of Gordon, Becker, Bourdieu, and McIntosh on capital, class, culture, *habitus*, and unconscious privilege to make sense of the ways in which some children arrive at the academic table better prepared to demonstrate scholastic success in mathematics than other children. Speaking very generally, Bourdieu's central point is that cultural practices obey the same rules as economic practices and are an integral element in the struggle for power and dominance and thus affect issues of equity. The term pedagogical capital, independently created by both Gordon and myself, but presented as a subtype of cultural capital by me, will be legitimized and then used to help direct future research to areas of potential unconscious privilege in the mathematics curriculum for one group over another which can give rise to inequity in mathematics education for disadvantaged students.

The fact remains, like the differing sections of my rock wall, the hand- and footholds that a student finds as they attempt to climb their mathematical and scholastic wall can be vastly different from child to child. To have pedagogical capital is to have closely spaced supports—perhaps even more than necessary for the task. To lack pedagogical capital is to climb a wall with very few supports—perhaps with supports not adequate to the task. When standards serve those who struggle to learn instead of only those who will learn, then we are on the road to promoting equity. Today, as schools are being called upon to prepare all children for success, the need has never been greater to produce insights that can assist all educators to create classrooms that provide all students with opportunities, not just the lucky few.

The hope in this forthcoming research is to enlarge the repertoire of explanations for why some students continue to find scholastic success in mathematics elusive. Rather than adopting the position that past research, standards, and policy have never addressed the root causes of students' failure to thrive mathematically, I will end with a quote from Thomas Alva Edison, and suggest that through the forthcoming research we give careful consideration to the alternative explanation that the privilege of pedagogical capital, as an unconscious privilege and as a framework for understanding scholastic success in mathematics, can offer.

"I have not failed. I've just found 10,000 ways that do not work."

BIBLIOGRAPHY

Aiken, L. R. (1970). Attitudes toward mathematics. *Science and Mathematics Education.* 40(*4*) p. 551–596.

Alabama Department of Education. (2003). *Alabama course of study: Mathematics.* Montgomery, AL: Authors.

American Educational Research Association (AERA). (2006). *Social Justice Mission Statement.* Washington, DC: Authors. Retrieved August 8, 2006, from http://www.aera.net/aboutaera/?id=1541.

Bates, T. R. (1975). Gramsci and the theory of hegemony. *Journal of the history of ideas.* 36(*2*), p. 351–366.

Becker, G. S. (1964). *Human capital: A theoretical and empirical analysis with special reference to education.* New York: Columbia University Press.

Becker, G. S. (1993). *Human capital: A theoretical and empirical analysis with special reference to education.* Chicago, IL: University of Chicago Press.

Bourdieu, P. (1977). *Outline of a theory of practice.* Translated by Richard Nice. UK: Cambridge University Press.

Bourdieu, P. (1980). *The logic of practice.* Translated by Richard Nice. CA: Stanford University Press.

Bourdieu, P. (1986). The forms of capital. Translated by Richard Nice. In *Handbook of theory and research for the sociology of education,* edited by J. Richardson, 241–258. Westport, CT: Greenwood.

Bourdieu, P. (1998). *Practical reason: On the theory of action.* Translated by Randal Johnson. CA: Stanford University Press.

Bourdieu, P. & Passeron, J. (1979). *The inheritors: French students and their relation to culture.* Translated by Richard Nice. Chicago: University of Chicago Press.

Brewster, B. (translator) (1970). *Reading capital (Part III)*. Retrieved June 14, 2006, from http://www.marx2mao.com/Other/RC68iii.html.

Centers for Disease Control. (2006). Births: Preliminary data for 2005. Washington, DC: National Center for Health Statistics. Retrieved November 26, 2006, from http://www.cdc.gov/nchs/products/pubs/pubd/hestats/prelimbirths05/ prelimbirths05.htm.

Churchich, N. (1994). *Marxism and morality: A critical examination of Marxist ethics.* Cambridge, MA: James Clarke & Co.

Comer, J. P. (1989). *Maggie's American dream: The life and times of a black family.* New York: Penguin Books.

Coleman, J. S., Campbell, E. Q., Hobson, C. J., McPartland, J., Mood, A. M., Weinfeld, F. D., & York, L. R. (1966). *Equality of educational opportunity.* Washington, DC: U.S. Government Printing Office.

Cuban, L. (1990). Reforming again, again, and again. *Educational Researcher, 19(1)*, p. 3–13.

Eighty-eighth Congress of the United States of America. (1964). *Public Law 88-352.* Washington, DC: Authors.

Gordon, E. W. (1997). Task force on the role and future of minorities. *Educational Researcher, 26(3)* p. 44–52.

Gordon, E. W. ed. (1999). *Education and justice: A view from the back of the bus.* New York: Teachers College Press.

Gordon, E. W. (2001). Affirmative development of academic abilities. *Pedagogical Inquiry & Practice.* 2. Retrieved November 22, 2004, from http://iume. tc.columbia.edu/reports/ip2.pdf.

Gordon, E. W. (2004). Affirmative student development: Closing the achievement gap by developing human capital. *News from the Educational Testing Service Policy Information Center.* 12(2).

Gordon, E. W., Bridglall, B. L., & Meroe, A. S. (2005). *Supplementary education: The hidden curriculum of high academic achievement.* Lanham, MD: Rowman & Littlefield Publishing Group.

Gregory, W. T. & Giancola, D. (2003). *World Ethics.* Toronto: Thomson & Wadsworth.

Head, B. W. (1985). *Ideology and social science: Destutt de Tracy and French liberalism.* Hingham, MA: M. Nijhoff Publishers.

Hobbs, F. & Stoops, N. (2002). US Census Bureau: Census 2000 special reports, series CENSR-4, *Demographic trends in the 20th century.* Washington, DC: US Government Printing Office.

Johnson, R. S. (2002). *Using data to close the achievement gaps: How to measure equity in our schools.* Thousand Oaks, CA: Corwin Press.

Kohl, H. (1994). *"I won't learn from you" and other thoughts on creative maladjustment.* New York: The New Press.

Lamb, S. (1998). Completing school in Australia: Trends in the 1990s. *Australian Journal of Education, Vol 42.*

Lareau, A. (2000). *Home advantage: Social class and parental intervention in elementary education.* Lanham, MD: Rowman & Littlefield Publishers.

Lareau, A. (2003). *Unequal childhoods: Class, race and family life.* Berkeley, CA: University of California Press.

Livingston, C. V. (2005). *Transcribed personal interview with Edmund W. Gordon, EdD conducted by Carol V. Livingston on Tuesday, July 19, 2005.*

Mao, Z. (1937). *On practice: On the relation between knowledge and practice, between knowing and doing.* Essay retrieved September 20, 2005, from http://www.etext. org/Politics/MIM/wim/onpractice.html.

McIntosh, P. (1990). White privilege: Unpacking the invisible knapsack. *Independent School,* 49(2), p. 31–35.

Murdock, S.H., Hoque, M.N., Swenson, T., Pecotte, B., & White, S. (2000). *The rural south: Preparing for the challenges of the 21ˢᵗ century.* Southern Rural Development Center, 6, p. 1–10.

National Council of Teachers of Mathematics. (2000). *Principles and standards for school mathematics.* Reston, VA: Authors.

Pettigrew, T. F. (1967). *The consequences of racial isolation in the public schools: Another look.* Paper presented at the National Conference on Equal Educational Opportunity in America's Cities, November, Washington, DC. ED 015–975.

Purvis, T. & Hunt, A. (1993). Discourse, ideology, discourse, ideology, discourse, ideology.... *British Journal of Sociology.* 44(3), p. 473–499.

Rabinow, P. (ed.). (1984). *The Foucault reader.* New York: Pantheon Books.

Somers, M. R. (1995). What's political or cultural about political culture and the public sphere? Toward an historical sociology of concept formation. *Sociological Theory.* 13(2), p. 113–144.

US Department of Education. (2002). *State education indicators with a focus on Title I.* Retrieved August 8, 2006, from http://www.ed.gov/rschstat/eval/disadv/2002indicators/alabama/edlite-alabama.html.

US Department of Veterans Affairs. (2006). *Born of controversy: The GI Bill of Rights.*

Washington, DC: Authors. Retrieved November 11, 2006, from http://www.gibill. va.gov/GI_Bill_Info/history.htm.

Wilson, G. M. (1961). Why do pupils avoid math in high school? *Arithmetic Teacher,* 6: 168–171.

CHAPTER 29

PUPILS OF AFRICAN HERITAGE, MATHEMATICS EDUCATION, AND SOCIAL JUSTICE

Kwame E. Glevey
University of London

ABSTRACT

This article focuses on pupils of African heritage in learning mathematics in the midst of the debate on race and inequality in educational provision. It discusses the widespread notions about the challenging nature of mathematics as a subject to learn and the persistent underachievement of pupils of African heritage within the context of the classroom in England. It highlights some of the distinguishing factors at the root of the perceptions pupils of African heritage have of themselves and their implications for learning mathematics. The article argues that highlighting the contributions made by people of African heritage to mathematical knowledge may offer an opportunity for pupils of African heritage to engage with the subject. Furthermore, genuine desire and care for how they are supported is vital if their underachievement in mathematics learning is to be addressed.

Critical Issues in Mathematics Education, pages 457–473
Copyright © 2009 by Information Age Publishing
All rights of reproduction in any form reserved.

457

INTRODUCTION

The underachievement of pupils of African heritage in compulsory education in England has received wide coverage in recent years (Richardson, 2005; London Development Agency (LDA), 2004; Figueroa, 2001; Gillborn and Mirza, 2000; Gillborn and Youdell, 2000; Gillborn and Gipps, 1996). While these researches highlight the general lack of educational advancement of pupils of African heritage, the need to draw special attention to their accomplishment in learning mathematics is essential.

The importance of mathematics in everyday life is reflected in the high status of the subject in the school curriculum. In England for example, the significant role of mathematics in the life of the individual and that of the community and its economic advancement has been emphasised and sustained in various reports over many decades. For example, Cockcroft (1982) was unambiguous in stressing the need for individuals to learn mathematics as follows:

> It would be very difficult—perhaps impossible—to live a normal life in very many parts of the world in the twentieth century without making use of mathematics of some kind. This fact in itself could be thought to provide a sufficient reason for teaching mathematics, and in one sense this is undoubtedly true. (p. 1)

The significance of mathematics in the life of the individual as observed by Cockcroft continues to be highlighted in reports on the teaching and learning of mathematics. The most current report compiled by Smith (2004) supported the view that:

> The acquisition of at least basic mathematical skills—commonly referred to as "numeracy"—is vital to the life opportunities and achievements of individual citizens. Research shows that problems with basic skills have a continuing adverse effect on people's lives and that problems with numeracy lead to the greatest disadvantages for the individual in the labour market and in terms of general social exclusion. Individuals with limited basic mathematical skills are less likely to be employed, and if they are employed are less likely to have been promoted or to have received further training. (p. 13)

Evidently, the attainment of mathematical proficiency is generally understood to be particularly vital if all British pupils including those of African heritage are to acquire the necessary basic skills for their future participation in society as responsible citizens. In view of the importance of mathematics in the lives of individuals, the need for genuine search to support the progress of pupils of African heritage in learning mathematics is of vital importance.

In this article, I will refer to Black pupils (that is, pupils of African, African-European, African-Caribbean, African-American or African-other) whenever possible as pupils of African heritage. This is a deliberate strategy for two reasons firstly, to limit my references to merely visible characteristics as the standard form of representation and secondly, to emphasize their commonly shared heritage. The reference to visible characteristics as the most common way to identify persons of African heritage contributes to some of the difficulties in focusing on issues relating to them (London Development Agency (LDA), 2004; Tickly et al, 2004). While it is important to acknowledge the cultural diversity and changing identities among people of African heritage (Appiah, 2005; Tickly et al, 2004; Tizard and Phoenix, 1993; Anim-Addo,1995), widely respected scholars of African heritage (Diop, 1989, 1978; Asante,1990) have emphasized the existence of traditional African-centred worldviews that form the basis for some of the distinguishing cultural characteristics and beliefs unifying people of African heritage. Graham (2001) argues that some of these cultural characteristics have survived the physical uprooting of African people through enslavement to remain a necessary part of their ethos, regardless of their geographical location. I will leave the above issues to one side for now and turn to the general experiences of pupils of African heritage in compulsory education in England as a prelude to the main focus of the discussion.

SCHOOL EXPERIENCES OF PUPILS OF AFRICAN HERITAGE

The concern about the general academic underachievement of pupils of African heritage was clearly and forcefully expressed over three decades ago by Bernard Coard that:

> The Black child acquires two fundamental attitudes or beliefs as a result of his experiencing the British school system: a low self-image, and consequently low self-expectations in life. These are obtained through streaming, banding, bussing, ESN schools, racists news media, and a white middle-class curriculum; by totally ignoring the Black child's language, history, culture, identity. Through the choice of teaching materials, the society emphasizes who and what it thinks is important—and by implication, by omission, who and what it thinks is unimportant, infinitesimal, and irrelevant. (Coard, 1971, p. 31)

Although considerable changes have occurred since the 1970s much remains the same in terms of their academic underachievement. In the preface to Brian Richardson's (2005) *Tell it like it is: How our schools fail Black children*, Herman Ouseley commented that the problems observed by Coard decades earlier have fundamentally remained the same. He noted that:

> Bernard Coard's work has withstood the test of time because the problems facing the African-Caribbean parents and their children have fundamentally remained the same. Racism, race prejudice and social inequalities are crucial factors in the perpetuation of educational practices which cause the system to fail the African-Caribbean communities. (Ouseley, 2005, p. 13)

Similarly, research by the Department for Education and Skills (DfES, 2005) recently concluded that boys of African heritage are twice as likely to have been categorised at school as having behavioural, emotional or social difficulty as their White English counterparts.

A number of researches have consistently identified key issues such as pupils' relationship with their teachers underpinning what has been acknowledged as the academic underachievement of pupils of African heritage at all Key Stages (LDA, 2004). For example, Gillborn's (1990) two-year study examining teacher-pupil interactions highlighted the negative experiences of African-Caribbean pupils in the form of high teacher expectations for bad behaviour, creating the conditions for them to receive more punishment than their White English or Asian counterparts. Similar research carried out by Sewell (1997) on how boys of African heritage survive schooling observed that a significant number of them are specifically subjected to negative labelling and stereotyping, and as a result are experiencing racism and sexism to a high degree in schools compared to their counterparts of non-African background. Similarly, Wright et al (2000) have highlighted experiences of racism perceived by some pupils of African heritage from some of their teachers of non-African heritage.

The overall effect for many pupils of African heritage was that, in spite of the fact that most start their primary education viewing school as a positive and enjoyable place for learning and adventure, their focus on learning decreased the longer they stayed in the schooling system (LDA, 2004). Secondary school aged boys in particular felt that they do not belong in their mainstream school due to the perception that their teachers generally ignored them which often left them frustrated and angry. Such lack of participation in school activities is often at the core of the misunderstandings between pupils and teachers which, in many cases, lead to exclusions from school (Majors et al, 2001; Wright et al, 2000).

Focusing on Local Education Authorities' (LEA) and schools' responses to the requirements of the Race Relations (Amendment) Act 2000 (RRAA) concerning their exclusion practices, Parsons et al (2004) concluded that a number of schools were judged to be making progress in implementing comprehensively the RRAA in relation to minority ethnic exclusions. However, while the disproportionality in rates of permanent exclusion for pupils of African heritage has fallen considerably over a number of years they are still excluded at about three times the rate of pupils of non-African heritage with devastating consequences for their attainment and participation.

ATTAINMENT IN MATHEMATICS LEARNING

The failure of pupils of African heritage to share in the dramatic rise in attainment at the General Certificate of Secondary Education (GCSE) examinations which occurred in the 1990s for their White English peers prompted an investigation (Gillborn and Mirza, 2000) which demonstrated how underachievement of pupils of African heritage becomes institutionalized through the tiering[1] system.

> Black pupils were significantly less likely to be placed in the higher tier, but more likely to be entered in the lowest tier. This situation was most pronounced in mathematics where a *majority* of Black pupils were entered for the Foundation Tier, where a higher grade pass (of C or above) is not available to candidates regardless of how well they perform in the exam. (Gillborn and Mirza, 2000, p. 17)

Since the publication of this report, comprehensive evidence on minority ethnic pupils in compulsory education is now available (DfES, 2005; DfES, 2006). The growing evidence establishes the lack of progress of pupils of African heritage, substantiating earlier findings (LDA, 2004; Gillborn and Mirza, 2000; Gillborn and Youdell, 2000).

Focusing specifically on mathematical attainment, pupils of African heritage consistently scored the lowest levels throughout the Key Stages[2] as indicated in Table 29.1.

The results highlight the attainment at Key Stage 1 for Black pupils was 84% compared to 91% for White pupils, 86% for Asian pupils and 96% for Chinese pupils. At Key Stage 2, their attainment fell to 60% compared to 73% for White pupils, 67% for Asian pupils and 88% for Chinese pupils. A Further fall to 54% in their attainment was recorded at Key Stage 3 as compared to 72% for White pupils, 66% for Asian pupils and 90% for Chinese pupils. At Key Stage 4 only 36% of Black pupils achieved five or more passes at the end of their General Certificate of School Examinations compared to 51% for White pupils, 53% for Asian pupils and 75% for Chinese pupils.

It is not clear to what extent pupils' ethnic classification affected the conclusions drawn by the research. For example pupils of mixed parentage were classified as: 1) White and Black Caribbean, 2) White and Black African, 3) White and Asian and 4) Any other mixed background. Whilst pupils with one Black parent were subdivided into two groups, there was no explanation for the non-representation of pupils with one Chinese parent if any, raising the question whether they are accounted for as any other mixed background or as unclassified. The potential issues associated with the assumptions underpinning the classification of minority ethnic pupils in the research require further scrutiny which is beyond the immediate concern of this article.

TABLE 29.1 Percentage of Pupils Achieving the Expected Level at Each Key Stage by Ethnic Group (DfES, 2005)

Ethnic Group	Key Stage 1: % Expected Level			Key Stage 2: % Expected Level			Key Stage 3: % Expected Level			Key Stage 4: % Expected Level
	Reading	Writing	Maths	English	Maths	Science	English	Maths	Science	5+ A *-C GCSEs
White	**85**	**82**	**91**	**76**	**73**	**87**	**70**	**72**	**70**	**51**
White British	85	82	91	76	73	88	70	72	70	51
Irish	84	81	91	82	78	90	75	75	73	60
Traveller of Irish Heritage	28	28	52	23	19	36	49	49	45	42
Gypsy/Roma	42	38	60	30	27	48	33	35	35	23
Any other White background	80	78	89	74	72	84	66	70	65	52
Mixed	**85**	**82**	**91**	**77**	**72**	**87**	**69**	**69**	**67**	**49**
White and Black Caribbean	83	79	90	73	67	85	62	62	60	40
White and Black African	86	83	90	77	72	85	69	68	68	48
White and Asian	88	85	93	81	78	89	78	78	76	65
Any other mixed background	85	82	91	79	75	88	71	71	68	52
Asian	**80**	**78**	**86**	**69**	**67**	**79**	**66**	**66**	**59**	**53**
Indian	88	86	92	79	77	87	77	79	72	65
Pakistani	76	73	83	61	58	72	57	55	47	42
Bangladeshi	75	73	83	68	63	77	58	57	48	46
Any other Asian background	82	80	89	73	74	82	70	78	69	59
Black	**78**	**74**	**84**	**68**	**60**	**77**	**56**	**54**	**51**	**36**
Black Caribbean	79	74	84	68	59	78	56	53	51	33
Black African	77	73	83	67	62	75	56	55	50	41
Any other Black background	79	75	86	71	62	79	58	55	54	34
Chinese	**90**	**88**	**96**	**82**	**88**	**90**	**80**	**90**	**82**	**75**
Any other ethnic group	74	71	85	63	67	75	59	64	58	46
Unclassified	76	73	85	69	66	83	63	67	68	47
All pupils	**84**	**81**	**90**	**75**	**72**	**86**	**69**	**71**	**68**	**51**

Nevertheless, the results of the research underline the seriousness of the underachievement of pupils of African heritage in learning mathematics. A similar result from America (Tate, 1997) also concluded that the mathematics achievement gap is slowly closing between white students and students of colour; however African-American students continue to perform at significantly lower levels than White students. Ginsberg et al (1997) studying preschool mathematical knowledge of children from several countries across different cultures found that most children including African-Americans demonstrate reasonable competence in informal mathematical thinking. However, while most children from other ethnic backgrounds go on to succeed at school African-American children in particular do not, even though their pre-school performances are similar to other successful groups. Research has shown (Osborne, 2001) that for members of groups for whom there are negative group stereotypes concerning the intellectual ability of the group, intense aversion is generated by schooling and the school environment long before the manifestation of the achievement gap.

The lack of success and progress of pupils of African heritage may be understood as the result of various factors interacting in very complex ways. For example, pupils' views about schooling, their rapport with their mathematics teachers, delivery of the subject matter, peer pressures, poverty, lack of family support all present serious potential sources for underachievement. While all these factors cannot be explored in this article, attention will be drawn to some of them and their impact on teaching and learning the subject. The general level of accomplishment of pupils of African heritage in learning mathematics raises difficult questions that demand honest answers. Is it simply the case that pupils of African heritage lack serious mathematical aptitude, or is the subject taught in ways that encourage them to fail?

If it is the case that pupils of African heritage do indeed lack significant mathematical aptitude, then, one possible way to explain this situation would be in terms of innate deficiency in intellectual aptitude of people of African heritage, as presented by Herrnstein and Murray (1994) in their book, *The bell curve: intelligence and class structure in American life.* This view however, is forcefully contested. Cheikh Anta Diop (1991), a nuclear physicist, argued in his book *Civilization or Barbarism: An Authentic Anthropology* that the abundant contribution of people of African heritage to our present understanding of mathematics cannot be ignored. Diop illustrated how particular notions of mathematical reasoning as expressed in geometry and arithmetic for example, essentially have their origins in the ancient cultures of Africa. Diop (1989, 1974) does this by employing rigorous scientific methodology to explore the commonly shared ancestry and cultural practices of the ancient Pharaonic Egyptians and present-day Africans. Using Ancient Egyptian hieroglyphic and hieratic texts, Diop (1991) provided numerous mathematical problems and their solutions to point out the high level of sophistication of Egyptian mathematics detailing its pervasive and enduring influence on Greek thought.

According to Gillings (1972) it is not uncommon in histories of mathematics to read that Egyptian mathematics (particularly their multiplication) was clumsy and awkward due mainly to their very poor arithmetical notation. In spite of such criticisms, Gillings noted the deep influence of Egyptian mathematical techniques throughout the Coptic and Greek periods and over a thousand years later in the Byzantine period. For Gillings the profound influence of Egyptian mathematics is obvious:

> How far have we progressed in multiplication since the times of the ancient Egyptians, or even Greek and Roman times? What are our grounds for being so critical of Egyptian multiplication, in which it was only necessary to use the twice-times tables? In English-speaking countries, at least, as late as the sixteenth century, it was not part of the school curriculum to learn any multiplication tables at all. (Gillings, 1971, pp. 16–17)

Bernal (1987) provides a response to Gillings's questions by noting that, by the 1680s there was widespread opinion that Africans were merely sub-humans with negligible intellectual qualities. Bernal argues that this perception of Africans led to the vilification of Ancient Egyptians:

> If it had been scientifically 'proved' that Blacks were biologically incapable of civilization, how could one explain Ancient Egypt—which was inconveniently placed on the African continent? There were two, or rather, three solutions. The first was to deny that the Ancient Egyptians were black; the second was to deny that the Ancient Egyptians had created a 'true' civilization; the third was to make doubly sure by denying both. The last has been preferred by most 19th- and 20th-century historians. (Bernal, 1987, p. 241)

If we accept Diop's (1991, 1989, 1987,1978,1974) thesis concerning the African origin of Ancient Egyptian civilisation, which he invites us to scrutinise, then this raises questions concerning the African contribution to mathematical knowledge and how we currently understand the origins of European mathematics. Genuine consideration of how best these issues can be addressed in the mathematics curriculum may well provide a basis, not only for positively supporting pupils of African heritage to view mathematics as part of their own heritage and engage with it, but may also lead to deeper appreciation of the subject for all learners. This brings us to the second part of the question posed earlier concerning the teaching of the subject in ways that disadvantage pupils of African heritage.

In our schools at present the LDA (2004) reported how a significant number of pupils of African heritage go through largely negative school experience as a result of the general antagonistic element in their managment. The care and attention they experienced, the quality of communication and levels of conflict were all less positive. Furthermore, the lack of high teacher expectations is a major factor in the underachievement of pupils of African heritage at school, resulting in less than adequate opportunities for them to engage with studying mathematics at a higher level:

> In particular, there is a striking association between ethnic origin and pupils chances of entry to the Higher tier. White pupils are four times more likely to be entered in the higher tier than their Black peers, meaning that African Caribbeans are almost completely absent from Higher tier mathematics...
> (Gillborn and Youdell, 2000, p. 120)

Consequently, the combination of low teacher expectations and the lack of opportunities for pupils of African heritage to fully engage with learning mathematics create ideal conditions for their underachievement in the subject. Recent evidence (DfES, 2006) gathered from the African-Caribbean achievement project involving 30 schools indicated that in spite of the success of the project in raising some awareness of African-Caribbean issues in schools, some teachers remain reluctant to fully commit to focusing on the needs of African-Caribbean pupils. Given this state of affairs, providing the right incentive as part of the entitlement for pupils to successfully learn at school is a challenge that must be confronted, if all British pupils regardless of their heritage are to be fully supported.

THE CHALLENGE FOR SCHOOLS AND TEACHERS

We can assume that the crucial role of schools and mathematics teachers in supporting and guiding their pupils to learn is beyond dispute. The issue, however, is how they support all their pupils to progress in learning mathematics. Teachers are not isolated from the society within which they live and work and as such share in the values of their profession as well as the wider community. Gates comments that:

> As a teacher of mathematics, one holds certain values and one articulates values through the forms of classroom organisation and the nature of interactions we have. In other words we hold implicit and explicit values and these explicit values might convey a set of implicit values or 'worth' which may even be contrary to those which we hold. Some values may be deeply embedded in the acceptable practices of a school. Such values would impinge also on the values we held to be important in and through the study of mathematics—and we give out these messages all the time. (Gates, 2002, p. 213)

One of the fundamental messages that teachers give out all the time is embedded in how they approach the teaching and communicating of the subject. Mathematics is often characterised as a subject that one can or cannot do (Gates, 2002). This view of the subject is sustained and reinforced by the authoritarian and divisive way in which pupils are often grouped in their lessons according to their mathematical abilities, thus making implicit assumptions about the quality of their reasoning. Walkerdine puts it this way:

> Success at Mathematics is taken to be an indication of success at reasoning. Mathematics is seen as *development* of the reasoned and logical mind. (Walkerdine, 1989, p. 25)

Considering the fact that pupils of African heritage are particularly underachieving in mathematics learning, does this not make it difficult for some teachers to maintain low expectations and negative outcomes for pupils of African heritage?

As already highlighted (Warren, 2005; LDA, 2004) most pupils of African heritage feel excluded or not belonging to their mainstream school. In mathematics learning, these pupils have very little to identify with and often what exist is misrepresented (Shan and Bailey, 1991). School textbooks show negligible positive role models for pupils of African heritage in spite of the abundance of such resources. These concerns led Claudia Zaslavsky (1979) to highlight the contribution of Africans to the science of mathematics in order to encourage African-American pupils to regard mathematics as part of their own cultural heritage. Similarly, the richness of African mathematics is at the core of the work of Paulus Gerdes (2006, 1999), in uncovering mathematical ideas embedded in African cultural practices. The development of learning materials that genuinely incorporate the historical role of Africans in our present understanding of mathematics is likely to contribute to the quality of the mathematics curriculum by providing good opportunities for all pupils to value the mathematics that we have inherited from other cultures.

Schools are beginning to search for ways to address the need to support pupils of African heritage to achieve their potential. In England, the National Curriculum provides the guideline within which schools function and as such it stresses the need for promoting pupils' spiritual, moral, social and cultural development through mathematics. As an example, the National Curriculum states that mathematics provides opportunities for cultural development:

> Through helping pupils appreciate that mathematical thought contributes to the development of our culture and is becoming increasingly central to our highly technological future. (DfES, 1999, p. 8)

Kassem (2001) however, concludes that with regards to ethnic minorities, the National Curriculum does not clarify whose culture should be promoted in view of the multicultural character of contemporary England. Kassem argues that although the National Curriculum deserves credit for acknowledging the contribution of many cultures to the development of mathematics still it fails to give a socially or historically critical stipulation towards its delivery. In other words, the multicultural aspect of mathematics is treated as an appendage to the main mathematics curriculum, and consequently, it falls short in offering real opportunities for pupils from ethnic minority backgrounds to advance.

Added to the lack of great concern for the provision of appropriate historical and cultural setting in the delivery of school mathematics, the impact of public examinations and league tables on schools (Harlen and Deakin Crick, 2002) is most likely to influence schools and teachers to adopt strategies that ultimately must conform to the delivery of mathematics in less than adequate historical and cultural terms. For Ernest (2001), the content of the National Curriculum provides insufficient opportunities to account for the diverse social and cultural experiences of pupils. Ernest argues that the structure of the National Curriculum is that of a single fixed hierarchy within which all pupils in state schools are expected to work their way through the curriculum materials irrespective of age, aptitude, interest and need. As a result it only serves to reproduce inequalities in social opportunities. The difficulties facing schools in supporting all their pupils to flourish is immense, and how they succeed in providing the needed support will depend in part on their appreciation of the ideological positions and tensions within which they function.

THE FUTURE AHEAD OF US

The poor performance of pupils of African heritage in attaining reasonable mathematical proficiency in spite of their competence in informal mathematical thinking before entering the educational system may be rooted in the severely disadvantaged background (Schama, 2006; Du Bois, 1989; Washington, 1986; Fanon, 1986; Davidson, 1980; Kay, 1967) from which they have emerged. Today remnants of such disadvantages encapsulated in current issues concerning race (and social class) continue to persist in education. (Mirza, 2005; Gillborn, 2002).

How racism, that is, the notion that human beings can be divided into various races based on physical and nonphysical attributes, is manifested in contemporary social relations, has more to do with how we distance ourselves from others, what Dei (1996) refers to as 'between self and other' or 'between us and them'. When such distancing is embedded in how an institution or organisation functions, a system is created that perpetuates racism throughout the institution or organisation. Macpherson summed up the concept of such a system as:

> The collective failure of an organisation to provide an appropriate and professional service to people because of their colour, culture, or ethnic origin. It can be seen or detected in process, attitudes and behaviour which amount to discrimination through unwitting prejudice, ignorance, thoughtlessness and racist stereotyping which disadvantage minority ethnic group people. (Macpherson, 1999, p. 23)

Concern for such a system operating in the provision of education in England has long been emphasized (Richardson, 2005; Gillborn, 2002; Rogers, 2001; Coard, 1971). Gillborn boldly raised the question regarding institutional racism as follows:

> If we can't discuss it, how can we defeat it? (Gillborn, 2005, p. 94)

For Gillborn, the issues about institutional racism in education must be uncovered and addressed so that the provision of quality education for all pupils, regardless of race, is justly achieved. To this end he maintains that a lot has been done to address some of the pressing issues within the educational system:

> But much has remained unchanged. We have more research than 35 years ago and we have more academically successful Black children—and yet we still endure a system that fails disproportionate numbers of Black children, excludes many from mainstream schooling altogether, and channels others into second-class courses deemed more appropriate by a teaching force that continues to be unrepresentative of the community it serves. The need for radical change is as pressing today as it ever was. (ibid, pp. 95–96)

In mathematics, the divisive three tier system of grading has now been replaced by a two tier model (Qualifications and Curriculum Authority (QCA), 2006) in order to create opportunities for more pupils to successfully achieve no less than grade C in the General Certificate of Secondary Education (GCSE) in the subject (see Figure 29.1). The tier chosen for pupils to follow will be based on what teachers expect pupils to achieve in their final GCSE assessments.

The first teaching for the new GCSE mathematics qualifications commenced in September 2006 and the first certification will be awarded in 2008.

Nevertheless, changes to the assessment model without changes to how teachers and schools support their pupils in the classroom is likely to yield no better results for pupils of African heritage. Creating an environment

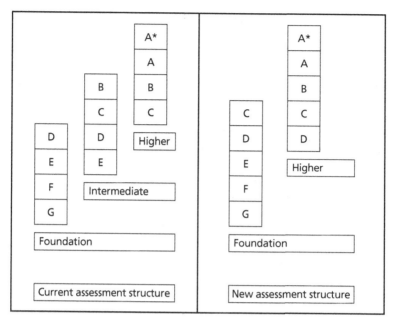

Figure 29.2 The new two-tier assessment model for GCSE mathematics (QCA, 2006).

that deeply values the contribution that people of African heritage have made to our understanding of mathematics can help to encourage pupils of African heritage to share in the rewards associated with learning the subject. In order to create helpful environments, important factors such as caring (Noddings, 2005) and compassion (Ladson-Billings, 1997) must form an integral part of the approach for supporting pupils of African heritage to succeed in learning mathematics at school. In providing such support for pupils the quality of care that is exercised is of vital importance. In particular, the care that parents show towards their children's progress at school; the care with which teachers support and guide all their pupils' learning; the caring ethos that schools development to support their teachers, pupils and their parents, and above all, the caring vision with which local and national government direct schools.

CONCLUSION

In drawing attention to the progress of pupils of African heritage in learning mathematics, I discussed the general school experiences of pupils of African heritage and then I focused on their attainment in mathematics. I explored some of the challenges that their underachievement present for schools and teachers.

The well-being and progress of every child in learning mathematics at school as indicated by Smith (2004) does matter especially in our present digital age. Some groups are progressing as expected whilst others are failing to do so. The persistent underachievement of pupils of African heritage is a challenge that must be confronted and defeated. It is not for want of research that this cannot be done; for much research has been gathered over the past four decades (LDA, 2004). While legislations are useful in persuading teachers to treat all pupils with dignity, the crucial importance of genuine care and compassion cannot be overlooked if real progress is to be made in supporting all pupils.

The challenge posed by the lack of academic success for pupils of African heritage and the ensuing approaches that are adopted in finding lasting solutions will have far reaching implications for the liberal democratic values of justice and equality that we uphold.

REFERENCES

Anim-Addo, J. (1995) *Longest Journey: A History of Black Lewisham.* London: Deptford Forum.

Appiah, K. A. (2005) *The Ethics of Identity.* Princeton: Princeton University Press.

Asante, M. K. (1990) Kemet, Afrocentricity and Knowledge. New Jersey: Africa World Press.

Bernal, M. (1987) *Black Athena: the Afroasiatic roots of classical civilization.* Vol. 1, *The fabrication of Ancient Greece 1785–1985.* London: Free Association Books.

Coard, B. (1971) *How the West Indian Child is made educationally subnormal in the British School System.* London: New Beacon Books.

Cockcroft, W.H. (1982) *Mathematics counts: report of the Committee of Inquiry into the Teaching of Mathematics in Schools.* London: HMSO.

Davidson, B. (1980) *The African Slave Trade.* Boston: Little, Brown.

Dei, G.J.S. (1996) *Anti-Racism Education.* Halifax: Fernwood Publishing.

Department for Education and Skills, (1999) *The National Curriculum for England.* London: QCA.

Department for Education and Skills, (2005) *Ethnicity and Education: The Evidence on Minority Ethnic Pupils. Research Topic Paper: RTP01-05.*

Department for Education and Skills, (2006) *Ethnicity and Education: The Evidence on Minority Ethnic Pupils. Research Topic Paper: 2006 Edition*

Diop, C.A (1987) *Precolonial Black Africa.* New York: Lawrence Hill.

Diop, C.A. (1974) *The African Origin of Civilization: Myth or Reality.* Chicago: Lawrence Hill.

Diop, C.A. (1978) Black Africa: The Economic and Cultural Basis for a Federated State. Connecticut: Lawrence Hill.

Diop, C.A. (1989) *The Cultural Unity of Black Africa.* London: Karnak House.

Diop, C.A. (1991) *Civilization or Barbarism: An Authentic Anthropology*. New York: Lawrence Hill.

Du Bois, W.E.B. (1989) *The Souls of Black Folk*. New York: Penguin.

Ernest,P. (2001) Critical mathematics education, in: Gates, P. (ed) *Issues in mathematics teaching*, pp. 277–293. London: RoutledgeFalmer.

Fanon, F. (1986) *Black Skin, White Masks*. London: Pluto Press.

Figueroa, P. (2001) 'Multicultural education in the United Kingdom: Historical development and current status'. In Banks, J.A. & Banks, C.A.M. (eds) *Handbook of Research on Multicultural Education*, pp 778–800. San Francisco: Jossey-Bass.

Gates, P. (2002) 'Issues of equity in mathematics education: defining the problem, seeking solutions'. In Haggarty, L. (ed) *Teaching mathematics in secondary schools: a reader*, pp. 211–228. London: RoutledgeFalmer.

Gerdes, P. (1999) *Geometry from Africa, Mathematical and Educational Exploration*. Washington DC: The Mathematical Association of America.

Gerdes, P. (2006) *Sona Geometry from Angola. Mathematics of an African tradition*. Milan: Polimetrica.

Gillborn, D. (1990) *Race, Ethnicity and Education—Teaching and Learning in multiethnic schools.*(London: Unwin Hyman.

Gillborn, D. (2002) *Education and Institutional Racism*. London: Institute of Education.

Gillborn, D. (2005) It takes a nation of millions (and a particular kind of education system) to hold us back. In Richardson, B. (ed) *Tell it like it is: How our schools are failing Black Children*, pp. 88–96. London: Bookmarks.

Gillborn, D. and Gipps, C. (1996) *Recent Research on the Achievements of Ethnic Minority Pupils*. London: HMSO.

Gillborn, D. and Mizra, H.S (2000) *Educational Inequality: Mapping Race, Class and gender: a synthesis of research evidence*. London: OFSTED.

Gillborn, D. and Youdell, D. (2000) *Rationing education: policy, practice, reform and equity*. Buckingham: Open University Press.

Gillings, R.J. (1972) *Mathematics in the time of the Pharaohs*. New York: Dover.

Ginsberg, P.H.; Choy, Y.E.; Lopez, L.Z.; Netley, R.; Choa-Yuan, C. (1997) Happy birthday to you: Early mathematical thinking of Asian, South American, and US children. In Nunes,T. and Bryant, P. (eds) *Learning and teaching mathematics: an international perspective*, pp. 163–207. Hove: Psychology Press.

Graham, M. (2001) The 'miseducation' of black children in the British educational system—towards an African-centred orientation of knowledge. In Majors, R. (ed) *Educating our Black Children: New directions and radical approaches*, pp. 61–78. London: RoutledgeFalmer.

Harlen W, Deakin Crick R. (2002) A systematic review of the impact of summative assessment and tests on students' motivation for learning (EPPI-Centre Review, version 1.1). In *Research Evidence in Education Library. Issue 1*. London, EPPI-Centre, Social Science Research Unit. London: Institute of Education.

Herrnstein, R.J. & Murray, C. (1994) *The bell curve : intelligence and class structure in American life*. New York: The Free Press.

Kassem,D. (2001) Ethnicity and mathematics education in: Gates, P. (ed) *Issues in mathematics teachin*, pp. 64–76. London: RoutledgeFalmer.

Kay, G.F. (1967) *The Shameful Trade*. London: Muller.

Ladson-Billings, G. (1997) It Doesn't Add up: African American Students' Mathematics Achievement. In *Journal for Research in Mathematics Education, Equity, Mathematics Reform, and Research: Crossing Boundaries in Search of Understanding,* 2 (6), 697–708.

London Development Agency, (2004) *The educational experiences and achievements of Black boys in London schools 2000–2003.* London: lda.

Macpherson, W. (1999) *The Stephen Lawrence Inquiry: Report of an inquiry by Sir William Macpherson of Cluny.* London: Stationary Office.

Majors, R.; Gillborn, D. and Sewell, T. (2001) The exclusions of Black children: implication for a racialised perspective. In Majors, R. (ed) *Educating our Black Children: New directions and radical approaches,* pp. 105–109 London: Routledge-Falmer.

Mirza, H.S. (2005) 'The more things change the more they stay the same': assessing Black underachievement 35 years on. In Richardson,B. (ed)(2005) *Tell it like it is: How our schools are failing Black Children,* pp. 111–118. London: Bookmarks.

Noddings, N. (2005) *The Challenge to care in schools: an alternative approach to education.* New York: Teachers College Press.

Osborne, J.W. (2001) Academic disidentification: unravelling underachievement among Black boys, in: Majors, R. (ed) *Educating our Black Children: New directions and radical approaches,* pp. 45–58. London: RoutledgeFalmer.

Parsons,C.; Godfrey, R.; Annan, G.; Cornwall, J.; Dussart, M.; Hepburn,S.; Howlett, K.and Wennerstrom, V. (2004) *Minority Ethnic Exclusions and the Race Relations (Amendment) Act 2000.* Department for Education and Skills: Ref No: RR616.

Qualifications and Curriculum Authority, (2006) *Changes to GCSE mathematics: A two-tier model (QCA/06/2645).* London:QCA.

Race Relations (Amendment) Act 2000, ISBN 0 10 543400 0

Richardson, B. (Ed) (2005) *Tell it like it is: How our schools are failing Black Children.* London: Bookmarks.

Schama, S. (2006) *Rough Crossings: Britain the slaves and the America revolution.* London: BBC Books.

Sewell,T. (1997) *Black Masculinities and Schooling.* London: Trentham.

Shan, S and Bailey, P. (1991) *Multiple factors: classroom mathematics for equality and justice.* Stoke-on-Trent: Trentham.

Smith, A. (2004) *Making Mathematics Count: Inquiry into Post-14 Mathematics Education (MMC).* Nottingham: DfES.

Tate IV, W.F. (1997) *Race, SES, Gender, and Language Proficiency Trends in Mathematics Achievement: An Update.* Madison: National Institute for Science Education, University of Wisconsin-Madison.

Tickley, L.; Caballero, C.; Haynes, J.; Hill, J. (2004) *Understanding the Educational Needs of Mixed Heritage Pupils.* Nottingham: DfES.

Tizard, B and Phoenix, A. (1993) *Black, white or mixed race?* London: Routledge.

Walkerdine, V. and The Girls and Mathematics Unit (1989) *Counting Girls Out.* London: Virago.

Warren, S. (2005) Resilience and refusal: African-Caribbean young men's agency, school exclusions, and school-based mentoring programmes, in: *Race, Ethnicity and Education,* 8 (3), 243–259.

Washington, B.T. (1986) *Up from Slavery*. New York: Penguin.

Wright, C.; Weeks, D.; McGlaughlin, A. (2000) *'Race', class and gender in exclusions from school*. London: Falmer Press.

Zaslavsky, C. (1979) *Africa Counts: Number and Pattern in African Culture*. New York: Lawrence Hill.

NOTES

1. Tiering refers to the different examination papers that pupils are entered to sit in their final General Certificate of School Examinations (GCSE) instead of a single common examination paper. See Figure 29.1.

2. Compulsory schooling in England is divided into 4 Key Stages with national assessments at the end of each Key Stage. Key Stage 1 covers pupils aged 5 to 7, Key Stage 2 covers pupils aged 7 to 11, Key Stage 3 covers pupils aged 11 to 14 and Key Stage 4 covers pupils aged 14 to 16 at which point they sit their final examinations (GCSE). At the end of each Key Stage pupils are expected to have reached a given level of achievement for their age.

CHAPTER 30

RESEARCHING, AND LEARNING MATHEMATICS AT THE MARGIN

From "Shelter" to School

Renuka Vithal
University of KwaZulu-Natal

INTRODUCTION

In this paper I draw attention to the mathematics education of that group of learners who are usually on the margins of society and also on the margins of mathematics education research, theory and practice. Specifically, these are children who for various reasons have left home, eke out a living on the streets of a city—referred to as "street children," and are often placed in "shelters" and "homes" (Chetty, 1997). I refer to research conducted related to providing mathematics education for such learners, to address firstly, the question of how such children come to engage particular experiences of mathematics education and secondly, the challenges and consequences of doing such research. The story told here is that of Nellie and Wiseman as researched and documented by Sheena Rughubar (2003),

Critical Issues in Mathematics Education, pages 475–484
Copyright © 2009 by Information Age Publishing
All rights of reproduction in any form reserved.

but I also reflect on broader issues of doing research and its processes and relations that involves working outside mainstream schooling and on working with research students in such settings.

This opportunity to work with children who are very much at the margins of society and of schools began with a project that the Faculty of Education participated in at the then University of Durban Westville.[1] It was a collaborative venture with the Dept of Health and Welfare and involved the City Council but the task for the faculty was that of setting up a school at what was a residential shelter for children who were in various ways recruited, rescued, sometimes arrested from the streets and brought here—called the Thuthukani Harm Reduction Centre (Amin 2001). This is where Sheena began her study closely investigating a small group of learners as they were taught mathematics. It was here that she saw Wiseman, who surprised her with his interest and capacity to do mathematics despite the conditions of his life and the quality of the learning environment.

The intention of the school at Thuthukani was to provide educational support to the learners so that they could be integrated into the mainstream public school system. Sheena intended therefore to follow a group of mathematics learners from the shelter school into a mathematics classroom the following year. But the next year none of the children she had interviewed and observed in the shelter could be found. She searched for several months and almost dropped out of the study. It was then decided to reframe the study and look for another learner from a shelter who had been placed in a public school. After much difficulty Sheena located and met Nellie and the study was rescued. The research focus shifted to investigating how mathematics is taught to and learnt by such learners in two different teaching and learning environments—a "shelter" and a "mainstream" mathematics classroom.

METHODOLOGY AND LEARNING SETTINGS

Both Thuthukani and Sanville Secondary School are characterised by considerable distractions and disturbances for learners. In the case of Sanville the school is close to an airport and both learners and teachers were observed having their interactions interrupted by the level of noise with various consequences such as loss of concentration and disengagement by the children. Thuthukani is in the middle of the city, located in a building in a state of disrepair. There is much traffic noise and intrusions of city life into the workings of schools especially since learners are allowed to leave and return at will. On some days the school does not function because meals are not provided to learners. Any school in its normal run of events copes with events and other disruptions which in various ways coincide and conflict with those faced by a learner within a classroom setting, sometimes significantly reducing actual teaching and learning time. This intensifies discontinuities in learning that learners like Wiseman and Nellie bring into the learning environment.

Mrs. James' classroom at Sanville is overcrowded with forty learners and barely any space to walk between desks that are arranged in traditional rows. Nellie sits in the middle of the class. There is strong discipline and structure that regulates behaviour and attendance in the school, which is also maintained in the classroom by Mrs James who is an experienced and qualified teacher. Nellie sits in the middle of the classroom, is generally quiet and is often observed not participating in the lesson. Thuthukani runs its school in a large hall, with little groups of learners working in classrooms without walls. Although the desks are clustered in traditional rows, learners have to concentrate hard to ensure that they can hear Mr Xulu and also put up with the regular streams of visitors, donors and administrators who walk through the hall throughout the class period. The repeated interruptions require patience and perseverance on the part of learners and teachers. Mr Xulu, a novice, recently graduated and qualified mathematics teacher, who is teaching virtually on a voluntary basis at the school, while hoping to find employment in a public school, is flexible, light-hearted and takes the disturbances in his stride. Wiseman does not miss school, usually sits alone and pays attention during lessons.

Teachers and researchers are seldom prepared for facing the erratic events of classrooms and also the poor material conditions of places like shelters. Nevertheless, a broad range of data in both settings were generated including: video recordings of lessons; photographs; interviews with Nellie and Wiseman; informal discussions with teachers and learners; written reflections and their class documents such as exercises and tests; and a researcher's journal. These were first analysed in terms of categories that examined the learner and his/her mathematics learning and the environment as well as the mathematics teacher, teaching and content. The analysis being focussed on here that emerged from the study is mainly that related to learning and the learners Wiseman and Nelllie (rather than their teachers and teaching); and this is extended to reflect on the research itself.

DISRUPTIONS, DISTRACTIONS AND DISCONTINUITY

For many learners in shelters or homes, disruption in their schooling in general and mathematics in particular, is marked by moving to several schools and very erratic attendance. Nellie and Wiseman are both fifteen years old, black, and appear to have had some primary schooling. Although it has been difficult to establish Wiseman's primary schooling level, Nellie has had a disruptive primary schooling having attended three different schools. Nellie is quiet in the classroom, and by her own admission scared to speak but volunteers much about her life to Sheena including the abuse that she had suffered at the hands of her mother which caused her to leave home and impacted on her schooling. Wiseman, on the other hand, is articulate in class, even correcting errors in the mathematics that the teacher makes, and volunteering to work out problems on the board, but does not speak about his life.

These discontinuities in learners' mathematical lives are also present in the learning settings. Nellie has been placed in a grade 8 classroom where the teacher is observed leaving the mathematics classroom to attend to other school functions. Mrs James' large class does not leave any time for her to provide any additional support to Nellie to bridge the gaps left by the discontinuities of the classroom or of Nellie's own life. The work Mr Xulu is doing with Wiseman's class is at grade 7 but he often seems unprepared relying only on a textbook. He has to cope with erratic attendance of learners and constantly changing groups to work with and the distractions of the learning environment.

Establishing background data for learners is not a simple process when school life is linked to painful personal life experiences. For the researcher working in these settings on the margin, disruption and discontinuity in their data production strategies are reflected in the lives learners or indeed even in their (non)availability as well as in the settings in which learning is taking place. Situations of poverty produce uncertainty because acquiring the basic necessities such as food or shelter take precedence over schooling or attending mathematics class. Overlaid with emotional and other injuries, mathematics learning is engaged within and against this whole life experience. Yet learners and teachers continue to do the work of mathematics education as do researchers. The question is how do practices and theories of learning mathematics take account of the whole, often disrupted life of a learner as they interact with specific mathematical tasks—the focus of much mathematics education research.

MATHEMATICAL VERSUS EMOTIONAL AND PHYSICAL NEEDS OF LEARNERS

Despite the hardships endured, both Nellie and Wiseman continue to attend school and mathematics classes regularly. However, Nellie was often observed falling asleep in the mathematics class or not paying attention and is embarrassed by the teacher and other learners when caught. The teacher admits "I don't know much about her background . . . but she lives in a home for children in the area" and states that she treats all the learners the same. Nellie, however, like many of these learners is working through experiences of abuse, neglect and poor health while trying to cope with schooling. She explains to Sheena how she was hospitalised when she fell ill in school. It is not surprising that such learners often lack confidence, have poor self concepts and low self esteem (Booyse, 1991). In many respects Wiseman is different. Not only does the teacher affirm him and regard him as one of his best students who will definitely be placed into one of the public schools, Wiseman is proud and derives confidence from his mathematical ability and assists others in mathematics in the class participating in discussions. Nellie and Wiseman experience the mathematics classroom in quite different ways. Falling asleep and being silent are ways in which to escape or disappear from the classroom when being forced to be there by the rules and rituals of a mathematics class. But a mathematics classroom can also be a place to feel good about yourself. Hierarchies of needs established in psychological studies do not fully explain why and how learners in poverty and violence situations continue to learn and want to learn mathematics.

Exploring mathematical experiences of these learners force researchers and teachers to engage much broader needs. When learners disappear from class or are found engaging in illegal or other activities, working with such children also has an emotional impact on the researcher. As the extent of the suffering endured by these children becomes known the researcher's questioning of her own participation in the research or educational endeavour and deeper values and life experiences often surface. Depending on the research paradigm the researcher is working in, dealing with this could include the generation and analysis of the researcher's biography and engaging issues of the ethics and politics of research more directly and explicitly. This often includes reflections of their own relationships with their students, parents or other life experiences and acting on these in reciprocal relations within the research process or as an outcome of the research. The point here is that mathematical needs cannot be examined or addressed in isolation from emotional, physical and other needs. Sheena, a mathematics teacher herself, repeatedly reflects on how this research experience has made her notice and redirect her gaze in her mathematics classroom; and reshaped her own practices and understandings in teaching mathematics in the mainstream.

ALIENATED AND SHARED IDENTITIES

Street children as a group develop their own identity within a particular sub-culture from having to survive in the harsh street conditions. When they enter the shelter they share those experiences which get played out in the shelter school in the construction of the learning environment. To this extent the notion of "community of practice" (Lave and Wenger, 1991) may be relevant and useful for explaining how Wiseman participates in the shelter math classroom. Even though this community may fragment along other lines of community such as "gang alliances" or identities of age, geographic urban-rural home etc., for the period they are in the mathematics classroom, they are participating in a particular social world that collides and coincides with these different identities. The teacher may not know the full individual histories of his learners but he is aware of their fragility as a group.

In the mainstream school children from shelters are often singled out and face discrimination from both teachers and other learners (Vithal, 2003). Nellie is no exception. If the classroom is deemed a community of practice then Nellie is clearly located outside this particular community: "I live at the home and they don't... I feel different... the other children they do not understand... they will laugh at me... tease me." She is marked as different in this classroom not only by virtue of living in a "home for children," she is a black learner in a school that is predominantly "Coloured"— an apartheid invented racial categorisation that still dominates to refer to people of mixed origins. The equality perspective that the teacher entrenches by claiming to treat all children equally further ensures that Nellie's different personal circumstances are not taken account of in supporting her mathematics learning. So she continues to be "othered" also by her (lack of) competence in mathematics.

No doubt researcher identity is productive of particular data with particular research participants in particular settings. Sheena's relation with Nellie could not be reproduced say by a male researcher. Nevertheless, researchers who come with a particular gender, race and social class identity, have to overcome their own prejudices and experiences of "street children" and develop empathy and understanding. This may be achieved by developing close relationships with individual children over time; and gaining knowledge about the whole life of a child and the severity of their life conditions and experiences. This is necessary to provide a much wider data set within which to place any analysis of their engagement with mathematics teaching and learning.

INTENTIONALITY, INTEREST AND INVOLVEMENT

How much genuine interest, enjoyment and involvement any learner invests in the learning is linked not only to background but also to how they see their present learning connected to a future life scenario—their foreground. Learners come with different dispositions which shape their "intentions-in-learning" both with reference to their backgrounds and their foregrounds (Olro & Skovsmose, 2001). While backgrounds have been overemphasized in explaining mathematics performance and participation, foregrounds have not been adequately factored into studies of learning. One way of understanding Wiseman's interest and investment in learning mathematics may be by noting his hopes and dreams for the future "I want to go to Moment High School. I want to be a scientist. I like science and mathematics." In the shelter schools there was no compulsion to attend school though non-attendance was questioned. Wiseman came to all classes and paid careful attention, even becoming annoyed when detecting errors made by the teacher.

Nellie's poor performance in mathematics and negative experiences of her interaction with the teacher can be related to her poorer levels of interest and involvement. Despite her low performance, Nellie claims to like mathematics. Nellie's intentions may described as broken or destroyed. As Sheena observes, "Nellie does not refer to anything in the future but rather continues to reflect on the past" (p. 92). The construct of intentionality is useful for locating and linking explanation for learning (or not learning) to aspects both inside and outside mathematics and the mathematics classroom.

Whatever the methodological design, researchers who bring also particular intentions to these settings often get much more deeply involved beyond and outside their research projects. The significantly impoverished situation of the learners and their environment compared to the resources, both physical and intellectual, that any researcher brings means that they are often in a position to contribute to improving the situation. Researchers have the possibility to make a much wider social situation available to learners as possibilities for the future. In this they confront in direct ways the objectivity–subjectivity dilemmas of their positioning in the research.

TRANSITIONS, CURRICULUM AND RELEVANCE

In the shelter school in which a teacher worked with a small group of between 6 to 10 learners meant that the curriculum could be organised much more tightly around the needs, performance and interests of learners. But the imperative to place these learners back into mainstream often resulted in rather traditional curricula offerings. The notions of "transitions" (Abreu, Bishop and Presmeg, 2002) may be useful for exploring the bridge between the practices engaged in the shelter school and those of mainstream school. The tension that this transition opened is that since not all learners in the shelter school were likely to be placed into public school, a "mathematics for life" versus a "mathematics for school" became visible.

For Wiseman this may be described as including elements of a "mediational transition"—where the shelter school learners "interact in an intentionally educational activity designed to change perceptions and meanings before involvement" (p. 17) in school mathematics, to facilitate their participation and experience of school mathematics.

Nellie on the other hand may be described as experiencing a "lateral transition"—"moving between two related practices in a single direction"— having much in parallel with that of "immigrant students in mainstream schools" (p.17). She moved from a predominantly "African school experience" to a different institutional culture of a "Coloured school" and having to reconstruct her identity as a learner who lives in a "home for children." Unlike Wiseman she is not accommodated or included in this setting, being lost in a large class of over forty learners.

Mainstream research education and training seldom prepares researchers for the trials and turbulences of facing contexts like shelters and learners on the margin in their research. Much of the focus in research has been in what Skovsmose (2004) calls "a prototype mathematics classroom" which are well resourced with well-behaved teachers and learners all interested and engaged in the mathematics. Often well-designed strategies collapse in the face of resistances, lack of trust or the impoverishment of the setting. Learners refused to have photographs taken for fear of media exposure because of criminal involvement or they resist writing a journal because of poor language competence or fear of having confidences betrayed that could have serious consequences for them.

MARGIN, POWER AND VOICE

The notion of margin is used in this paper in a number of ways. Shelter schools are one kind of margin that exits in relation to mainstream schools. Within classrooms, shifting margins and centres exist. Nellie is excluded and lives on the periphery of the classroom both in terms of mathematics and pedagogy. As a group of children, both Nellie and Wiseman are on the edge of society belonging to what Castells (1998) refers to as the "Fourth World" or regarded as "disposable people" (Skovsmose, 2003). Despite the harsh conditions of life both inside and outside schools and classrooms, these learners still choose in some sense to attend mathematics lessons. How then does mathematics and its mediation participate in their experience of life both inside and outside the classroom? And how is it represented in mathematics education theory research and practice?

The status of mathematics secures interest and through this power, success in mathematics translates into improved self-concept and self esteem. This is because doing well in mathematics provides not only a gateway to a better life but also bestows prestige on the learner given how it is valued in schools and societies and by the learners themselves. Both Nellie and Wiseman state that they like mathematics and want to succeed in it. Notwithstanding the background each brings into the learning setting, inclusion into or exclusion from mathematics is to a large extent mediated by the teacher. Both teacher attitude and teacher knowledge (in its broad sense) is critical in how empowerment and disempowerment are enacted in a mathematics classroom. Mrs James does not show the caring that Mr Xulu does, while Mr Xulu lacks competence in the content that Mrs James demonstrates. Wiseman is excluded by limited access to mathematics and Nellie by a pedagogy that that does not recognise and account for her difference.

Doing research in such settings draws attention to the plight of these learners to a different audience of mathematics education researchers and practitioners. Issues of voice and who speaks for whom has been extensively debated, especially in gender studies and must always be raised when attempting to speak for those who are powerless and voiceless in society. Yet research and researchers can and must speak for such learners: firstly to address the enormous disadvantage and suffering of such learners but secondly because it redirects the researcher and practitioner's gaze in the mainstream to other margins by developing a different more empathetic gaze on those learners who fail to learn mathematics. It forces researchers to develop a more caring and creative research approaches with a sharper concern for the ethics and politics of the research setting and research relationships. The focus on the "outliers" in research sites and participants opens for possible refutations and development in the theories of teaching and learning mathematics, a practice well-established in mathematics yet surprisingly lacking in mathematics education research.

As diversity increases, inclusion and exclusion become more acute in mathematics classrooms, requiring teachers and researchers to broaden explanations for failure and success in mathematics learning. Psychological perspectives typically locate such explanations in the learner him/herself; and in mathematics education research tend to keep the focus narrowly on the mathematics and its learning. However broadly the notion of margin is understood, by focussing on the margin, new insights could be gained for the centre; and such insights allow mathematics educators and researchers to account in more authentic ways for the diversity in their classrooms and schools in more equitable ways. The failure to learn mathematics lies, perhaps more significantly, outside mathematics, its teaching and learning, than inside. This assertion points to an imperative to bring political, social, cultural, economic and other perspectives into a closer dialogue in mathematics education research, theory and practice with the more dominant psychological perspectives.

REFERENCES

Abrue, G. de; Bishop, A. J. and Presmeg, N. C. (2002) Mathematics learners in transition. In Abrue, G de; Bishop, A. J. and Presmeg, N. C. (eds.) *Transitions between Contexts of Mathematical Practices.* Dordrecht: Kluwer Academic Publishers.

Alro, H. and Skovsmose, O. (2002) *Dialogue and Learning in Mathematics Education: Intention, Reflection, Critique.* Dordrecht: Kluwer Academic Publishers

Amin, N. (2001) *An invitational education approach: student teacher interactions with at-risk youth.* Unpublished MEd. Dissertation, University of Durban-Westville.

Booyse, A. M. (1991) The environmentally deprived child. In Kaap, J. A. (ed.) *Children with problems: An orthopedagogical perspective.* Pretoria: van Schaik Publishers.

Castells, M. (1998) *The Information Age: Economy, Society and Culture. Volume III. End of Millennium.* Oxford: Blackwell.

Chetty, V. R. (1997). *Street Children in Durban: An Exploratory Investigation.* Human Sciences Research Council. Pretoria: HSRC Publishers.

Rughubar, S. (2003) *The mathematics education of youth-at-risk: Nellie and Wiseman.* Unpublished MEd. Dissertation, University of Durban-Westville.

Skovsmose, O. (2004) Research Practice and Responsibility. ICME 10 Survey

Skovsmose, O., (2003) Ghettorising and globalisation: A challenge for mathematics education. Publication No. 39, May. Denmark: Centre for Research in Learning Mathematics.

Vithal, R. (2003) Student Teachers and "Street Children": On becoming a teacher of mathematics. *Journal of Mathematics Teacher Education,* 6, 165–183.

NOTE

1   The University of Durban-Westville is now merged with the University of Natal and is called the University of Kwazulu-Natal.